Springer Complexity

Springer Complexity is an interdisciplinary program publishing the best research and academic-level teaching on both fundamental and applied aspects of complex systems—cutting across all traditional disciplines of the natural and life sciences, engineering, economics, medicine, neuroscience, social and computer science.

Complex Systems are systems that comprise many interacting parts with the ability to generate a new quality of macroscopic collective behavior the manifestations of which are the spontaneous formation of distinctive temporal, spatial or functional structures. Models of such systems can be successfully mapped onto quite diverse "real-life" situations like the climate, the coherent emission of light from lasers, chemical reaction-diffusion systems, biological cellular networks, the dynamics of stock markets and of the Internet, earthquake statistics and prediction, freeway traffic, the human brain, or the formation of opinions in social systems, to name just some of the popular applications.

Although their scope and methodologies overlap somewhat, one can distinguish the following main concepts and tools: self-organization, nonlinear dynamics, synergetics, turbulence, dynamical systems, catastrophes, instabilities, stochastic processes, chaos, graphs and networks, cellular automata, adaptive systems, genetic algorithms and computational intelligence.

The three major book publication platforms of the Springer Complexity program are the monograph series "Understanding Complex Systems" focusing on the various applications of complexity, the "Springer Series in Synergetics", which is devoted to the quantitative theoretical and methodological foundations, and the "Springer Briefs in Complexity" which are concise and topical working reports, case studies, surveys, essays and lecture notes of relevance to the field. In addition to the books in these two core series, the program also incorporates individual titles ranging from textbooks to major reference works.

Series Editors

Understanding Complex Systems

Founding Editor: S. Kelso

Future scientific and technological developments in many fields will necessarily depend upon coming to grips with complex systems. Such systems are complex in both their composition – typically many different kinds of components interacting simultaneously and nonlinearly with each other and their environments on multiple levels – and in the rich diversity of behavior of which they are capable.

The Springer Series in Understanding Complex Systems series (UCS) promotes new strategies and paradigms for understanding and realizing applications of complex systems research in a wide variety of fields and endeavors. UCS is explicitly transdisciplinary. It has three main goals: First, to elaborate the concepts, methods and tools of complex systems at all levels of description and in all scientific fields, especially newly emerging areas within the life, social, behavioral, economic, neuro- and cognitive sciences (and derivatives thereof); second, to encourage novel applications of these ideas in various fields of engineering and computation such as robotics, nano-technology and informatics; third, to provide a single forum within which commonalities and differences in the workings of complex systems may be discerned, hence leading to deeper insight and understanding.

UCS will publish monographs, lecture notes and selected edited contributions aimed at communicating new findings to a large multidisciplinary audience.

More information about this series at http://www.springer.com/series/5394

Otto E. Rössler · Christophe Letellier

Chaos

The World of Nonperiodic Oscillations

Springer

Otto E. Rössler
Faculty of Science
University of Tübingen
Tübingen, Germany

Christophe Letellier
Normandie University - CORIA
Saint-Etienne du Rouvray, France

ISSN 1860-0832 ISSN 1860-0840 (electronic)
Understanding Complex Systems
ISBN 978-3-030-44304-7 ISBN 978-3-030-44305-4 (eBook)
https://doi.org/10.1007/978-3-030-44305-4

This Springer imprint is published by the registered company Springer Nature Switzerland AG
The registered company address is: Gewerbestrasse 11, 6330 Cham, Switzerland

Credit: Hugues Aroux

Preface

To chaos, I came like the virgin to the child (to quote a proverb). In 1975, I had given a talk in Vienna on biological clocks when Art Winfree told me that I had given better talks before—I, therefore, should take a look at "chaos." I had no idea what he meant, but 5 weeks later a big folder arrived with all papers and preprints existing up to that day on the subject of chaos in the modern sense, with the accompanying message that I should make something out of it since he momentarily lacked the time to do it himself. This was so overwhelming a gift that I felt obliged to try and generate a chemical reaction—kinetic analog with corresponding nonnegative rate equations as told to do. This proved to be a task vastly overtaxing my abilities. But I was in the bind to have to deliver for a good friend. So despair turned into chutzpah, and I tried gluing together two flows on a letter-Z-shaped sheet, one level laid above the other, with jumps at the edges, and with the height difference causing a shift of the two-dimensional oscillator assumed valid on the one sheet. It couldn't work. And unexpectedly the outcome was even simpler than the Lorenz attractor, which I was meant to reproduce in an abstract reaction system. So it was all Art's fault, and my attempt not to leave unreciprocated the tremendous effort made by a close friend.

The next gift from heaven was the presence of a modern analog computer in the department in which I worked. I had been sent to a computer course offered by the *Electronic Association Inc.* (EAI)—the biggest analog computer company at the time—which in fact helped me lose my inhibitions also toward the Dornier machine that was subsequently bought by our department. So "playing with equations" had become an available realistic option. Since nonlinearities are a notorious problem with analog computers, interactively simplifying the theoretical ("singular-perturbation") equations written down first in the two-sheet model as mentioned, was a natural option as well as a necessity. This lead to the simple equations that subsequently got recorded in the "chaos" movie which my wife, who had grown enthusiastic about the new reality, produced jointly with a coworker, Thomas Wiehr, from her lab in the Medical Policlinic of the University of Tübingen. The "chaos" movie also has sound (the third variable recorded on the soundtrack). Two years later, "hyperchaos" got recorded in the same fashion once more, which likewise can be watched and listened-to on the "Chaos" movie of 1976/79 that is on YouTube.[1]

The "sound of chaos" as well as that of hyperchaos turned out to be very familiar from daily life. The three speeds with which it was recorded (available on the Dornier machine) showed that snoring and a hoarse voice and piercing noises—but also wind-driven oscillations in another part of the body—were all chaotic. When there is a tiny lump on one of the two lips of the larynx, for example, spoken language necessarily becomes "hoarse," that is, chaotic. Nature is replete with three- and four-variable nonlinear dynamical systems of dissipative type. Also the irregular puttering of a motorcycle or car in the idling mode belongs here, not to mention cardiac irregularities. An abstract of ours written jointly with Herbert D. Landahl, a friend and former coworker of Nicholas Rashevsky, about cardiac arrhythmia came out of this work as well, presented in Japan in 1979. Another implication was endocrine chaos, discovered simultaneously independently by Colin Sparrow in England.

Many decades later, the new fundamental science of Cryodynamics, sister of deterministic Thermodynamics, would trigger a bet by the whole scientific community, on its being invalid because a multibillion dollar experiment based on the opposite assumption proved to have potentially deleterious consequences for the whole world. From the point of view of chaos theory, such counterintuitive "re-injections" make perfect sense. Society is also a dissipative dynamical system itself given over to unpredictable bifurcations and risks. Imagine the historical fact of warfare being adopted in irregular intervals in history.

Misunderstandings can thus be classified as generating dangerous societal developments. The new science of deductive brain theory, implicit in the deductive biology invented in dialog with Konrad Lorenz in 1966, subsequently allowed for the discovery of the interactive "smile explosion." It can be understood as arising between two brain equations or, rather, their carriers. Gregory Bateson was the first person to understand and appreciate the mechanism in 1975, but he was also the last. It explains the creative misunderstanding of the suspicion of benevolence being at work on the other side, invented by the toddler. So the small child becomes a carrier of a manifold, once more. The spontaneous invention of personhood, by the suspicion of benevolence being present on the other side is nothing but another analog to the two sheets that creatively allow for chaos in a tangle of deterministic operations. Here, two mirror-competent autonomous optimizers with cognition invent the conjectured existence of a benevolent intention being present on the other side. It is not chaos, and brains are not manifolds, but the analogy is clear. This particular application of chaotic visualization would enable the formulation of the "personogenetic bifucation" in close analogy to the two manifiolds involved in chaos

[1] https://www.youtube.com/watch?v=Tmmdg2P1RIM.

generation. This "sociological" realization of the letter-Z-shaped chaos-genesis is still very much in its spirit. Only that the invention of the suspicion of benevolence being encountered generates the exact opposite of chaos—unlimited trust—much as between Art and Christophe and me. But I now do better stop.

Tübingen, Germany Otto E. Rössler
November 2019

When I started my Ph.D. thesis, I started quite early to read Otto's papers, not the two well-known ones, but those full of beautiful hand-drawn pictures as shown in Fig. 2.5, p. 27. I always enjoy when I can draw a picture explaining the concepts that I am manipulating in my researches. Otto's drawings helped me a lot to understand branched manifold and their relationships with first-return map to a Poincaré section. I then worked for more than 10 years almost exclusively on the topological characterization of chaotic attractors, in particular with Robert Gilmore from Drexel University (Philadelphia).

In the early 2000s, I was working with some biologists from Rouen (Camille Ripoll, Janine Guespin, and Michel Thellier). They introduced me to René Thomas (Brussels) with who I enjoyed many exchanges. Unfortunately, we were never successful to maturate sufficiently our results related to a relationship between René's feedback circuits and branched manifolds for publishing them. At that time, René was in contact with Otto and recommended me to him. Otto kindly accepted an invitation to give a talk in Rouen. During the lunch after the talk, we had an amazing discussion about almost everything. I enjoyed so much the easiness with which Otto was flowing in the world of ideas. We kept in touch since that time.

During one of my visits in Tübingen, Otto's wife, Reimara, mentioned the existence of a manuscript written in the early 1980s which was never published. Otto submitted it to Springer which rejected it! It would have been the second book about chaos theory, the first one was written by Igor Gumowski and Christian Mira and published (hélas en français) in 1980.[2] Otto never submitted its manuscript to another publisher... It took about one year for seeing this manuscript, the time that Reimara recovered it, and that I returned to Tübingen for another visit. And then... an amazing journey in Otto's mind, full of astonishing discoveries. So many things were still new and not expressed with so much clarity.

[2] I. Gumowski & C. Mira, *Dynamique chaotique: transformations ponctuelles, transition ordre-désordre*, Cepadues Editions, 1980.

Otto and Christophe enjoying in digging in Otto's archives. Tübingen, 2010.

I immediately proposed to edit the manuscript, retyping everything, (re)computing all the figures. My target was to leave the text as much as possible in its original form and just to provide high-resolution pictures. I could say that for 90% of the pages, there is change neither in the text nor in the simulations which were obtained in one shot from the information reported in the manuscript. In some cases, the pictures were only drawn by hand with parameter values and initial conditions reported: even for these ones, the computed pictures were matching with the drawn ones. Unfortunately, for a few pictures, I was unable to reproduce them. With Otto, we tried to dig in his archives, sometimes we found the missing information, and sometimes not. In these last cases, it turned to be complicated to construct examples as initially imagined by Otto and we had to make some modifications of the text. This is limited to the two sections devoted to the blue sky catastrophe, and to the two sections where chaos in the Bonhoeffer–van der Pol equations is discussed. We tried to remain as much as possible close to the initial approach. During the edition of this manuscript, I added some references (marked with a "*") and some footnotes, particularly when Otto anticipated some results which came a few years later.

I wish to thank Jürgen Kurths who supported me for completing this project and helped for having this book finally published with Springer. He invited us to add an appendix about some realizations of chaos in the real world. They mostly came from my own researches performed in collaboration with many collaborators. I started by adding Otto's own contribution in finding chaos in the Belousov–Zhabotinski reaction and the one provided by one of Otto's great friends, John Hudson (Charlottesville) with who I collaborated—before meeting Otto—for investigating experimental data from an electrodissolution. I wish to thank Thomas Klinger for having provided the data from a plasma experiment. With Jean Maquet (Rouen), Robert Gilmore and Luis A. Aguirre (Belo Horizonte), we found a chaotic model from the sunspot numbers. With Jean we got a metastable chaotic model from the records of the Hudson Bay Company about population of Lynxes. I ended this appendix with an application of the techniques developed within the paradigm of chaos to the heart variability. This was developed with Emeline Fresnel and Emad Yacoub (who were my Ph.D. students).

With Valérie Messager (Rouen), we developed a short autobiography with some psychological insights. We used Otto's astrological chart as guidelines.

With Otto, we revisited the hierarchy of chaos he proposed in the 1983 and the one proposed by Gerold Baier and Michael Klein from Tübingen (who edited a book for Otto's 50th birthday).

I do hope that you will enjoy this book as much as I did myself in editing it and exchanging it with Otto.

Normandie, France
November 2019

Christophe Letellier

Contents

Chapter 1
The Phenomenon of Chaos

1.1 Introduction

Simple systems can produce complicated behavior. The reason lies in the structure of space and time. Simple 'weaving rules' suffice to determine very complicated motions.

Language produces a name even for incomprehensible things. Since times old, the word 'chaos' has in different languages been used to describe either an undescribable complexity or an incomprehensible primordial state. The modern mathematical use of the same term turns out to be compatible with these origins.

In Greek mythology, chaos was the first primordial entity, arising right before gaia and eros (Hesiod, 700 B.C.). Chaos literally means something like yawning gap; gaia, the (female) earth and eros then jointly gave rise to darkness and night, brightness and day, as well as heaven, mountains and oceans.

In Greek philosophy, on the other hand, Anaxagoras (416 B.C.) proposed a quite different view of the early states, a view which eventually (in hellenistic times) changed the meaning of the word chaos. Anaxagoras saw the primordial state as being undifferentiated not because of its abysmal emptiness, but because of its containing everything in a state of perfect mixture. The emergence of simple entities from this perfect state was then the problem. Anaxagoras' *deus ex machina* was the *nous* (mind) who alone was fine enough to be immiscible and as such to be able to suddenly interfere, by starting at one point in space and time a roughly circular ('perichoretic', that is, remains around, recurrent) motion. The latter, through its cyclic unwinding, was then responsible for the separation, out of the mixture, of the celestial and material entities as we know them. 'Recurrence' thus was responsible for generating simple entities out of chaos. Today, we would rather say that recurrence generates chaos out of simple constituents, but this is only because we look at time the other way round.

In the first chapter of the Bible, the primordial earth was *tohuwabohu* (two words, coupled by *wa*—and—, which both mean empty), when God's (female) mind hovered

O. E. Rössler and C. Letellier, *Chaos*, Understanding Complex Systems,
https://doi.org/10.1007/978-3-030-44305-4_1

over the male waters shortly before the light appeared. The traditional translation of *tohuwabohu* since Hellenistic time is, of course, chaos.

In Chinese mythology, as related by Chuang-Tzu (390 B.C.), see [4], there was an old emperor whose name (Hun-Tun) means chaos (mixture). He had no sense organs. Two imperial friends of his, Shu and Hu, pitifully tried to provide him with the necessary seven organs (eyes, ears, nose, mouth), by operating on him seven days. When the last sense organ was provided, Hun-Tun died. The name of the two friends, combined into one word (shu-hu), means lightning, that is, a sharp fracture in the world which even today allows people to catch a glimpse of the primordial reality behind the world. According to another, related myth, Hun-Tun means a leather-bag filled with blood (in a mixed state), and the role of the sharp lightning (mind?) is played by a fiery arrow [4].

In Japan, *Kon-Ton* today means both chaos in the sense of perfect mixture and the name of the old emperor who had no sense organs. The word at the same time has the connotation of a complete (unseparated) understanding of the world, without being in need to open one's eyes.

The meaning of perfect mixture, both in the West and in the East, is highly compatible with the modern mathematical interpretation of the word chaos, as will be seen next.

1.2 History of the Phenomenon Nowadays Labelled 'Chaos'

The phenomenon to be discussed was again first envisioned in a celestial context—that of celestial mechanics. The interaction of three graviting masses, the so-called three body problem, had proved analytically unsolvable in the early nineteenth century. Poincaré [21] first realized that in the three-body problem, as treated by Hill [10], an over finite times nonrepetitive, highly complicated motion can exist. Poincaré also saw that a single such complicated solution to the equations (which in his nomenclature was characterized by a certain property; cf. Sect. 6.6), 'homoclinic' implies the presence of an infinite number of further solutions of the same type.

Poincaré's trick lay in looking not at the trajectories themselves but only at a cross section ('surface de section') through the whole bundle of orbits. This trick to this day considerably simplifies the understanding of chaotic motions (see next sections).

Poincaré considered conservative (frictionless) systems only. Hadamard[1] [7] somewhat later discovered the presence of complicated locally straight lines (geodesics) on saddle-like surfaces. This class of systems is again of conservative type

[1]Jacques Hadamard (1865–1963) was a French mathematician who proved an inequality on determinants, leading to the discovery of *Hadamard matrices* when equality holds. In 1896 he proved the prime number theorem, using complex function theory and published his work on geodesics in the differential geometry of surfaces and dynamical systems. He studied in 1898 geodesics on surfaces of negative curvature, thus introducing symbolic dynamics. He continued his career by working on the problems of mathematical physics, in particular partial differential equations, the calculus of variations and the foundations of functional analysis.

(see [1, 19]). Birkhoff[2] [2] then proposed to term 'recurrent' the involved, only 'approximately closed' motion—because the trajectories keep recurring toward an appropriate cross section (being in this respect analogous to periodic motions), but on the other hand do not necessarily trace out a simple point pattern in that cross section (as all periodic and quasi-periodic motions do). The analogy to mixing porridge in a bowl (by way of inducing a perichoresis, this time in forward time) comes to mind again.

Recurrent motions need not to be conservative. That is to say, the area cut out by an arbitrary bundle of trajectories need not to be the same when they recur toward the cross section (but may have shrunk, for example). This implication of Poincaré's approach was first seen by his pupils Fatou [5][3] and Julia [12].[4]

Historically, a dichotomy applies at this point. On the one hand, people studied recurrences in their own right, that is, they simply looked at fixed transformations of a certain region into itself (without necessarily keeping in mind that this region was originally supposed to be a cross section through a continuous motion). After Fatou and Julia, who studied transformations of a complex variable, simple real maps were considered by Hopf [11],[5] by the Neumann-Ulam school [28], by Smale [26, 27],

---

[2]George David Birkhoff (1884–1944) proved Poincaré's "last geometric theorem," a special case of the three-body problem. In 1923, Birkhoff proved that the Schwarzschild geometry is the unique spherically symmetric solution of the Einstein field equations. Birkhoff discovered in 1931 the ergodic theorem.

[3]Pierre Fatou (1878–1929) was a French mathematician working in the field of complex analytic dynamics. In particular, he investigated recursive processes like $z_{n+1} = z_n^2 + c$, where $z = x + iy$ is a complex number. Fatou focused his attention to the case where $z_0 = 0$.

[4]Gaston Julia (1893–1978) was a French mathematician who devised the formula for the Julia set. His works were popularized by French mathematician Benoît Mandelbrot; Julia sets and Mandelbrot fractals are thus closely related. Gaston Julia had the unusual habit of wearing a patch to cover the center of his face. The patch was intended to hide extensive disfiguration of his nose and the surrounding region, suffered during Julia's service in the French Army during World War I.

[5]Eberhard Hopf (1902–1983) was a German mathematician and astronomer, one of the founding fathers of the ergodic theory and a pioneer of bifurcation theory. He also made significant contributions to the subjects of partial differential equations and integral equations, fluid dynamics, and differential geometry. Otto met Hopf, speaker of honor at Okan Gurel's New York chaos conference, in New York (May 1977). Hopf had told the audience that he had lost his memory and that one should tell him if he said the same thing twice in short order. Due to that he therefore unfortunately could no longer do math. While returning at home, Otto traveled with him to Kennedy Airport. When waiting for an hour together before the gate, Otto bridged the time by, on the spur of the moment, drawing for him baker's transformation onto a tiny little piece of scrap paper: Two squares with an arrow between them and a vertical divided in the left one and a horizontal divide in the right one, and a round-about back arrow. Hopf was excited and told Otto that he remembered distinctly John von Neumann (1903–1957) who was Hungarian and then became an American. Von Neumann was a pure and applied mathematician and physicist. Among others, he contributed to mathematics (functional analysis, ergodic theory, geometry, topology, and numerical analysis), physics (quantum mechanics, hydrodynamics, and fluid dynamics), economics (game theory), computing (Von Neumann architecture, linear programming, self-replicating machines, stochastic computing), and statistics. He was an important member of the Manhattan Project and the Institute for Advanced Study in Princeton. So Hopf remembered von Neumann drawning this very picture on the blackboard in Princeton in 1949, telling the audience that "*this has been invented by Eberhard Hopf*".

Myrberg [20], Sharkovsky [25], and many others [6]. It was also in this context that the notion 'chaos' was first used [14, 17].

In an independent second line of studies, continuous systems (presumably possessing such discrete maps as cross sections) were considered. The first example of a non-conservative continuous system showing Poincaré's complicated behavior (later known as chaos) was the periodically forced van der Pol oscillator, studied by Cartwright and Littlewood [3]. These authors found that if they subjected one of the simplest nonlinear two-variable oscillators [15, 30] to a periodic perturbation, highly complicated responses were possible. While something like this had also been observed empirically by von Holst [31], who coined the term 'relative coordination' for what he saw in his physiological experiments,[6] Cartwright and Littlewood [3] showed explicitly that for certain parameter values there existed an infinite number of periodic solutions of differing periodicities ('subharmonics') as well as an uncountable number of nonperiodic solutions (that is, solutions with an uncountable periodicity). These two properties if jointly present in a system were later proposed to be sufficient in order to allow one to speak of 'chaos' [14].

As a side remark, it is worth mentioning that the Cartwright–Littlewood oscillator, as a periodically force (and hence non-autonomous) system with two state variables, can also be written as a conventional (autonomous) system of three variables (see Sect. 6.1).

Astonishingly, it took almost twenty years until another three-variable example, the Lorenz' equation was discovered [16]. About ten years later, this equation began to be appreciated suddenly for its complicated motions in simple systems. McLaughin

[6]Erich von Holst (1908–1962) was a German behavioral physiologist. Holst worked with the zoologist Konrad Lorenz (1903–1989) concerning the processes of endogenous generation of stimuli and of central coordination as a basis of behavioral physiology. In the 1950s, he founded the Max Planck Institute for Behavioral Physiology at Seewisen (Bavaria) that Otto Rössler visited in 1967 (the center was then directed by Lorenz). From his studies of fish that use rhythmic, synchronized fin motions while maintaining an immobile body, he developed two fundamental principles to describe the coordinative properties of "neural oscillators": (i) *Beharrungstendenz* which is a tendency of an oscillator to maintain a steady rhythm (this would include movements such as breathing, chewing and running, which Holst called states of absolute coordination) and (ii) *Magneteffekt* which describes as an effect that one oscillator exercises over another oscillator of a different frequency so that it appears "magnetically" to draw and couple it to its own frequency. The result of interaction and struggle between *Beharrungstendenz* and *Magneteffekt* creates an infinite number of variable couplings, and in essence forms a state of *relative coordination*. As in his paper published in 1939 [31], he saw chaos produced by periodic forcing of fish fins. The brain was disconnected to let the spinal cord control the motion of the tail fin while a front fin was periodically forced as shown in the figure below.

and Martin [18] stressed the importance of the Lorenz equation[7]; Haken [8] found another physical application of the Lorenz equation (in the laser); and May [17] was motivated by Lorenz's 1963 paper to search for 'chaos' in models of biological populations.

More recently, there is a growing appreciation that the continuous Cartwright–Littlewood equation and the continuous Lorenz equation are not isolated phenomena, but rather represent just two special cases in a very large class of nonlinear systems with analogous behavior (see [22] and references therein).

The common label 'chaos' used in much of the recent literature has so far served a unifying function. Whether or not there indeed is a single principle behind the behavior of all the different examples presently known—as will be argued for in the following—is a solution of which the readers of this monograph are asked to contribute.

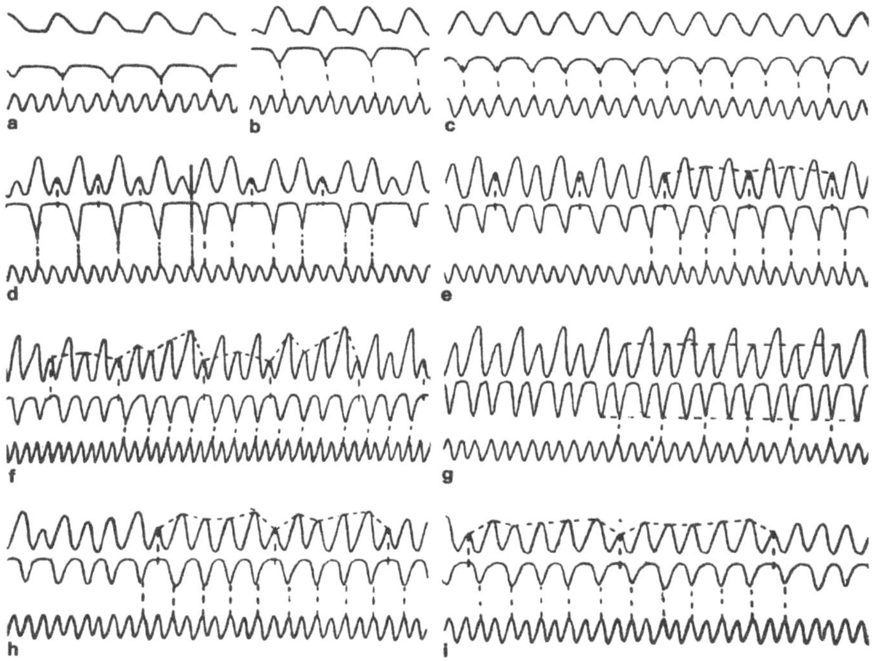

Surgus, pectoral fin (curve 1 and 2), tail fin (curve 3); all segments of curve taken from the same specimen. The nine specimens show ten differently shaped oscillations (in d there is a spontaneous transition from one periodic pattern into another). The different dashed lines are provided to facilitate the overview on the periodic patterns and to point out the relationship between the three rhythms.

[7]David Ruelle—who published with Floris Takens an important contribution concerning the turbulence (see Sect. 10.5)—considered that the Lorenz attractor was an example of the "strange attractor" they predicted. He thus popularized the Lorenz system as a benchmark system for studying "turbulent" behaviors [24].

1.3 The Re-injection Principle

'Chaos' is a natural thing. It occurs almost as readily in nonlinear three-variable systems as 'oscillation' occurs in two-variable systems (and 'self-regulation' toward a steady state does in single-variable systems). Indeed, chaos almost gives the impression of being part of a hierarchy which implies 'new behavior' whenever one more variable is admitted. If this were true, chaos could be considered the next higher form of dynamical behavior beyond oscillation: it would be a 'higher oscillation' in the same sense in which oscillation could be called a 'higher steady state'. And the type of behavior coming next (in four dimensions) should deserve the name 'higher chaos', and so forth. This hypothetical hierarchy indeed exists. Its properties can be demonstrated best with the aid of the principle of reinjection [23].

The human mind—although perhaps not the mind of an orang utan or a whale—is that of 'flatlander' (cf. [9]). It is therefore advisable to stick to two-dimensional explanations and pictures wherever possible. If need be, one should rather take two or more two-dimensional pieces and fit them together somehow than elevate oneself toward intricacies of a truly three-dimensional representation. The reinjection principle caters to this penchant.

The reinjection principle is even better understood in 'thread land', however. If we were thread-landers, with the constraint that motion is always unique (that is, one is not able to go backwards on the same thread), we nonetheless could still generate the main new behavior that is possible in two dimensions—by pasting together two threads. More specifically, we would define a 'threshold' on each thread at which a transition toward the other is defined. From the point of view of each thread, we would then have a 'reinjection' (with the other thread acting as an auxiliary device) back into the former thread. The consequence would be a (for a thread-lander) hard to grasp new kind of behavior: going round in a cycle.

Thus two zero-dimensional thresholds (allowing for reinjection between two one-dimensional threads) make 'typical two-dimensional recurrent behavior' (namely oscillation) possible and understandable from the point of view of one dimension, then analogously two one-dimensional transition thresholds (allowing for reinjection between two two-dimensional planes) should make possible and understandable 'typical three-dimensional recurrent behavior' (whatever this might be). And in the same vein, two two-dimensional transition thresholds between two three-dimensional flows should generate 'typical four-dimensional recurrent behavior' (whatever this is), and so forth. Such is the power of the reinjection principle.

The two-dimensional ('flatlanders') case is the one in which the runner re-defines himself onto another plane upon reaching a certain threshold, and then back! This is, obviously, the decisive case in the present context (for the whole hierarchy depends on its existence and simplicity). The two-dimensional situation is illustrated in Fig. 1.1. The picture has been drawn in such a way that the 'alternative plane' is shown elevated into a third dimension. But this is nothing a true flatlander will be able to notice. What he is bound to experience, however, is a strange complication of his formerly simple running path.

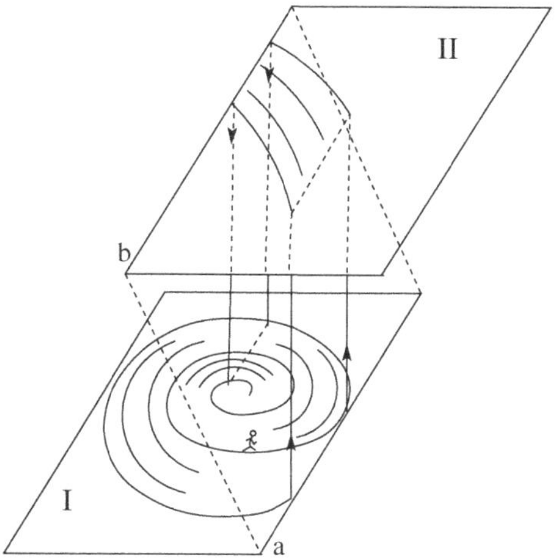

Fig. 1.1 A flatlander's running rink. There are two alternative two-dimensional rinks (I, II) and two 'redefinition' thresholds (a, b) at which an 'injection' into the other rink occurs. The resulting complicated running path illustrates the power of the 'reinjection principle'. The path exhibits 'spiral type chaos'

There are a number of ways open how to exploit Fig. 1.1 for chaos theory. One could, as a first example (i), focus on the one-dimensional cross section that is formed at the arrival on the upper plane (see the dashed line there), then focus on the similar one 'downstairs', and finally formulate the transition law that holds true for the combined map. Alternatively (ii), one might look at the single 'folded paper strip' that is formed if the two connecting strips (formed between the two planes) are added to the strips of paths up- and down-stairs. A combination of a Möbius strip and an ordinary strip, glued together side by side (with a spray of paths roughly parallel to the edges inscribed upon each could then be invoked as an explanation). Or one might, (iii), write down a couple of two-variable differential equations (for the trajectories upstairs and downstairs, respectively) and formulate the two thresholds as constraints. Finally (iv), one can look at the whole thing as an example of a single 'relaxation type' system of three differential equations (with two 'slow variables' and one 'fast switch'. Indeed, there are mathematical theories on (i) 'one-dimensional maps' (see, for example, [14]); on (ii) 'branched manifolds [32]; on (iii) 'constrained differential equations' [29]; and also on (iv) 'relaxation type dynamical systems' [13, 33]. They can each be used to better understand the running path of our flatlander.

However, rather than invoking any of these weapons at this point (where we have only one particular example), it is more efficient first to think about alternate possibilities for obtaining a non-trivial three-dimensional perichoresis/running around through 'overlaying with an overlap' two two-dimensional flow patterns. A pair of scissors, a glue stic, a pencil, and a few sheets of white stationary are all that one needs. Every time one is really intrigued by the game, a new variant should come up. And even if the obtained running scheme should be well-known already in some of its aspects, some others may not.

1.4 The Taffy-Pulling Machine

The construction principle of the last section appears to be easy to implement in terms of equations and realistic systems. This will be attempted in the next section. At this place, another everyday example is to be presented first because of its appeal to intuition. Its connection to the preceding class will turn out to be 'razor thin' (we will see that it can be considered as a model of what happens at the edge of one of the two former running planes).

Figure 1.2 shows a so-called taffy-pulling machine that is in use in sweets shops and candy factories since the last century. Two bars, each mounted excentrically on an axis, are being rotated in phase by a motor. Each bar has two little handles, as depicted, such that a piece of taffy (or rather, that sticky mass of caramel candy that is yet to be turned into taffy by way of mechanical mixing) can be laid over them as shown in Fig. 1.2a.

A number of the ensuing positions (which are cyclically repeated) are shown in Fig. 1.2b. The sticky mass is thereby elongated automatically, then folded over on two sides, then brought (in effect, glued) together, and so forth. Looking at the real machine while it operates is even more mind-boggling than is looking at Fig. 1.2b.

The function of such a machine is, of course, to mix. Mixing, however, implies two things: (i) bringing together points that in the original mass were widely separated; (ii) separating points that originally were close together. Since the domain is bounded, point (i) apparently cannot be realized without point (ii).

In principle it should be possible to describe the change of position of every point in the initial mass, from one round to the next. The volume of the taffy would hereby stay constant, so that the present system would qualify as a typical 'conservative' mixing system Fig. 1.2c illustrates this point.

Nonetheless, the main properties of the system depicted in Fig. 1.2 will, apparently, not be affected if the volume of the taffy is assumed to shrink (due to loss of fat, say) during the stretching process. The position if readily recognizable marker particles in the dough could still be followed under its iteration, and the two characteristic properties of mixing would still be observable. This 'modified version' may then be interpreted as a model of cross section through a 'non volume-conserving chaotic flow' of the very type of Fig. 1.1, as will be seen shortly.

Before leaving the subject of the taffy-puller, it seems appropriate to add a quantitative observation, as a diversion: How many rounds of the machine of Fig. 1.2a does it take until the (bold) left half and the (dashed) right half of the original bar of pre-taffy visible in the top of Fig. 1.2b are completely mixed (in the sense that a 'molecular sandwich' of taffy has formed in which one layer of 'left' molecules alternates with one of 'right' molecules)? Or equivalently, if there is one turn per second: How many years (or months or days or hours) are required?

The answer is of course, less than one minute. Assuming an initial thickness of 1 cm and supposing that the stretching and vertical thinning is by a factor of two on the average (while the width stays constant as it is the case), we after n rounds obtain

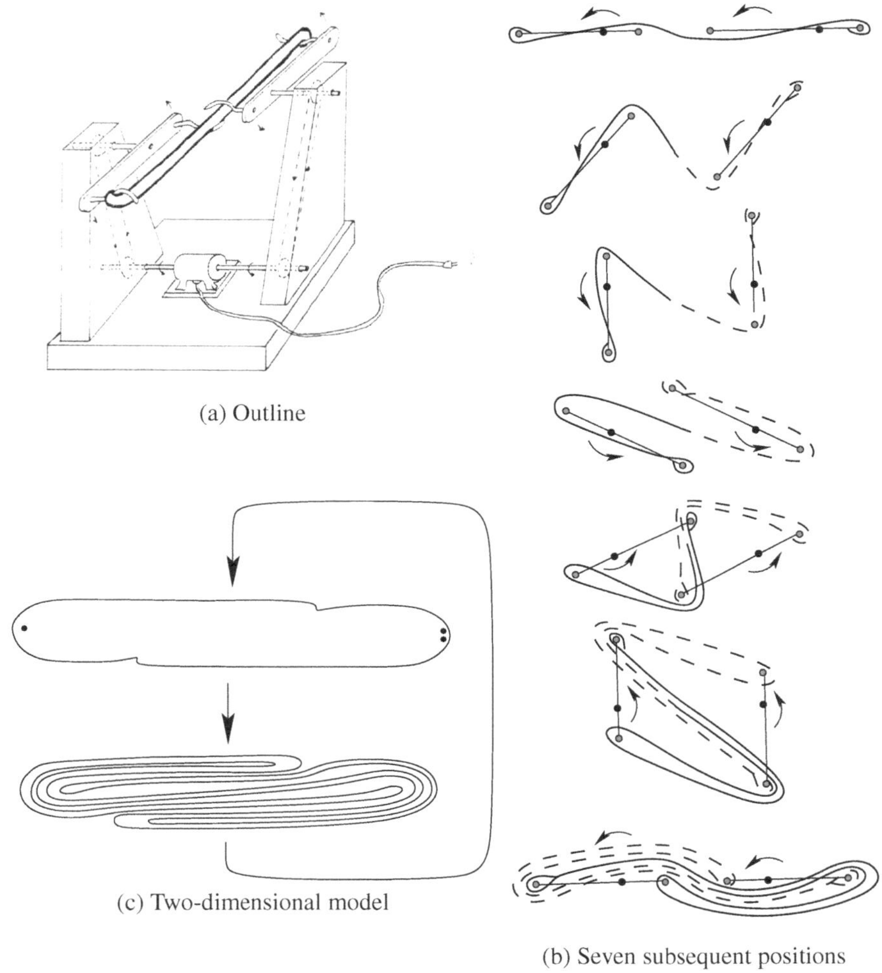

Fig. 1.2 Taffy-pulling machine as in use in some sweetshops. **c** Two-dimensional model (area-preserving homeomorphism). The dashing of the right-hand portion in **b**, and the marker dots and the gaps in **c**, are to facilitate understanding

stripes with an average width of 2^{-n} cm. Molecular width (about $10^{-8} \approx 2^{-24}$ cm) is obtained with $n = 24$, that is, after less than a half minute of watching.

We can conclude this section with a more general statement of a qualitative type. Reinjection along "merging" two-dimensional stable manifolds of oscillatory saddle points appears to be no less successful in producing chaotic flows in both abstract and natural systems, than is rejection based on one-dimensional thresholds in two-dimensional singular-perturbation subsystems (or at least one such component) of the spiral and screw examples of dissipative chaos. Both principles are equally natural, although the above "touching class" is—owing to the mutuality that is required—more complicated in terms of both the overall architecture of the flow ("doublewingedness") and the minimum number of nonlinear terms required in the underlying equations: at least two. A third general "compository recipe" for generating chaotic flows was the—again simpler—"breathing oscillatory principle".

It goes without saying that "distorsions" can be introduced in a wide variety of ways, without doing damage to the complexity and beauty of the dissipative chaotic flows created.

References

1. V.I. Arnold, A. Avez, *Ergodic Problems of Classical Mechanics* (Benjamin, New York, 1970); First Edition in French (Théorie Ergodique des systèmes dynamiques, Gauthier-Vilars, Paris, 1967)
2. G.D. Birkhoff, On the periodic motions of dynamical systems. Acta Math. **50**, 359–379 (1927)
3. M.L. Cartwright, J.E. Littlewood, On nonlinear differential equations of the second order: I. The equation $\ddot{y} - k(1 - y^2)\dot{y} + y = b\lambda k \cos(\lambda t + \alpha)$, k large. J. Lond. Math. Soc. **20**, 160–189 (1945)
4. A. Christie, *Chinese Mythology* (Paul Hamlyn, London, 1968), p. 44
5. P. Fatou, Mémoire sur les équations fonctionnelles. Bulletin de la Société Mathématique de France **47**, 161–271 (1919)
6. I. Gumowski, C. Mira, *Dynamique Chaotique* (Cepadues Editions, Toulouse, 1980)
7. J. Hadamard, Les surfaces à courbures opposées et leurs lignes géodésiques. Journal de Mathématiques **5**(4), 27–73 (1898)
8. H. Haken, Analogy between higher instabilities in fluids and lasers. Phys. Lett. A **53**(1), 77–78 (1975)
9. H. Helmholtz, *Handbuch der Physiologischen Optik* (1866); English translation: *Helmholtz's Treatise on Physiological Optics*, vol. 3, ed. by J.P.C. Southall (Optical Society, New York, 1926)
10. G.W. Hill, Researches in the lunar theory. Am. J. Math. **1**, 5–26 (1878)
11. E. Hopf, *Ergodentheorie (Ergodic Theory)* (Springer, Berlin, 1937), p. 42
12. G. Julia, Mémoire sur l'itération des fonctions rationnelles. Journal de Mathématiques Pures et Appliquées, vii **4**, 47–245 (1918)
13. J. LaSalle, Relaxation oscillations. Quaterly Appl. Math. **7**, 1–19 (1949)
14. T.-Y. Li, J.A. Yorke, Period-3 implies chaos. Am. Math. Mon. **82**, 985–992 (1975)
15. A. Liénard, Etudes des oscillations entretenues. Revue Générale de l'Electricité **23**, 901–912, 946–954 (1928)
16. E.N. Lorenz, Deterministic nonperiodic flow. J. Atmos. Sci. **20**, 130–141 (1963)
17. R.M. May, Biological populations with nonoverlapping generations: stable points, limit cycles, and chaos. Science **186**, 645–647 (1974)
18. J.B. McLaughlin, P.C. Martin, Transition to turbulence of a statically stressed fluid system. Phys. Rev. Lett. **33**, 1189–1192 (1974)
19. J. Moser, *Stable and Random Motions in Dynamical Systems* (Princeton University Press, Princeton, 1973)
20. P.J. Myrberg, Iteration of the real polynomials of second degree, III. Annales Academiae Scienctiarum Fennicae A, **336**(3), 1–18 (1963) (in German)
21. H. Poincaré, Sur le problème des trois corps et les équations de la dynamique. Acta Math. **13**, 1–271 (1890)
22. O.E. Rössler, Chaos, in *Structural Stability in Physics*, ed. by W. Güttinger, H. Eikemeier (Springer, Berlin, 1979), pp. 290–309

23. O.E. Rössler, Chaotic behavior in simple reaction systems. Zeitschrift für Naturforschung A **31**, 239–264 (1976)
24. * D. Ruelle, The Lorenz attractor and the problem of turbulence. Acta Phys. Austriaca (Suppl.) **16**, 221–239 (1976)
25. A.N. Sharkovsky, Coexistence of the cycles of a continuous mapping of the line onto itself. UKrain'kii Matematichnii Zhurnal **16**, 61–71 (1964)
26. S. Smale, Generalized Poincaré's conjecture in dimensions greater than four. Ann. Math. **74**, 391–406 (1961)
27. S. Smale, Differentiable dynamical system. Bull. Am. Math. Soc. **73**, 747–817 (1967)
28. P.R. Stein, S.M. Ulam, Non-linear transformation studies on electronic computers. Rozprawy Matematyczne **39**, 1–65 (1964)
29. F. Takens, Implicit differential equations: some open problems. Lect. Notes Math. **535**, 237–253 (1976)
30. B. van der Pol, Oscillation hysteresis in a triode generator with two degrees of freedom (in Dutch). Tijdschrift van het Nederlandsch Radiogenootschap **2**, 125 (1921); also Philosophical Magazine, vii, **43**, 177 (1922)
31. E. von Holst, Die relative Koordination als Phänomen und als Methode zentral nervöser Funktionsanalyse. Ergebnisse der Physiologie **42**, 228–306 (1939)
32. R.F. Williams, Expanding attractors. Publications Mathématiques de l'Institut des Hautes Etudes Scientifiques **43**, 169–203 (1974)
33. E.C. Zeeman, Differential equations for the heartbeat and nerve impulse, in *Towards a Theoretical Biology*, vol. 4, ed. by C.H. Waddington. (Edinburgh University Press, Edinburgh, 1972), pp. 8–67

Chapter 2
Simple Chaos

2.1 An Equation for Chaos

The design principle underlying Fig. 1.1 is easy to interpret in terms of realistic systems. The principle reads: 'combine an oscillator with a switch and you are likely to obtain'.

The principle is illustrated in Fig. 2.1. Here a limit cycle oscillator in two variables (x, y) is depicted underneath a single-variable switch (z). While the lower diagram is a state space diagram, showing the combined behavior of the two variables of the oscillator for a fixed set of external parameters, the upper diagram is a combined state space (z) and parameter space (P) diagram. It exhibits two switching thresholds [14] which jointly generate a hysteresis loop: the two switching events occur at different values of P so that depending on 'from where one comes' (hysteresis means 'after effect'), two different paths are followed. Although the notion of a 'parameter' strictly speaking implies fixed values, it is well-known that a sufficiently slow 'changing parameter', moving up and down, likewise will force the switch through a (only slightly softened) hysteresis loop. See, for example, [10].[1]

Since the switch can be driven by any sufficiently slowly changing variable (sufficiently slowly with respect to the internal dynamics of the switch), it is for example possible to use the variable x of the autonomous oscillator in Fig. 2.1b as the 'parameter' which controls the switch. Such devices were apparently first considered by Sëmen Chaikin (1901–1968) [2] who used them to transform a nearly sinusoidal oscillation (of the two-variable oscillator) into a square-wave type relaxation oscillation (on the level of the switch). Chaikin devised a simple electronic circuit in which the steady states of the switching variable z underwent a cusp-type bifurcation (first an $\int$-shaped, then an S-shaped curve of steady states; [27]) in dependence on an external parameter, so that the z-variable could show either a near-sinusoidal wave form or that of a relaxation oscillator (so-called square waves). Since these were the

[1]This is related to singular perturbation theory. See [11, 12].

O. E. Rössler and C. Letellier, *Chaos*, Understanding Complex Systems,
https://doi.org/10.1007/978-3-030-44305-4_2

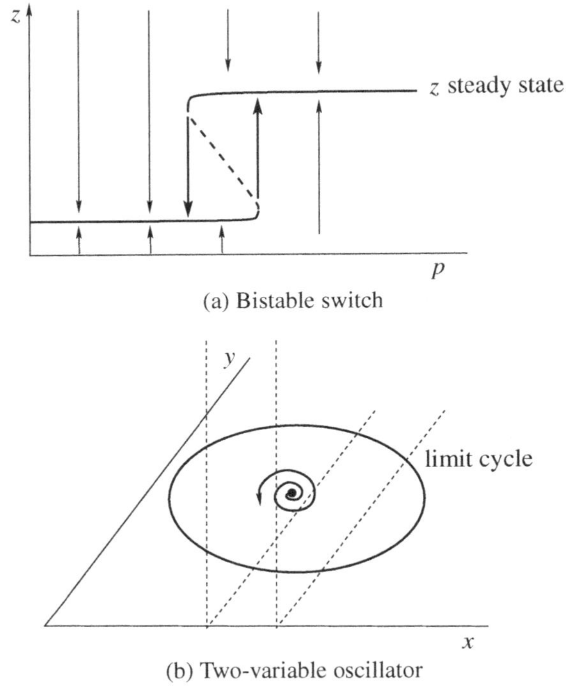

Fig. 2.1 (**a**) A bistable 'switch' showing hysteresis in dependence on a parameter P. (**b**) A two-variable oscillator of limit cycle type. If the motions in (**b**) are sufficiently slow compared to those in (**a**), the variable x may take the role of the parameter P. z steady state designates the curve of steady state points of z in dependence on P; dashed portion: unstable (repelling) steady states

major types of oscillation known at the time, the system was dugged a 'universal circuit' by Chaikin.[2] Yet there is even a third possibility in the same class of systems.

So far, the switch was assumed to be only passively driven by the oscillator. The flow 'upstairs' (with the switch 'on') therefore was no different from that 'downstairs', meaning that when looking down at the combined system 'from the top', one could see the same picture as if the variable z was absent. However, this is not the only possibility: introducing a 'feedback' from the switch toward the slow subsystem suffices to render the flow upstairs different from that downstairs. The upper flow may for example be shifted a little to the front, that is, in the direction of decreasing y. In that case we would have an immediate implementation of Fig. 1.1. Such a feedback effect, from the switch toward one of the variables of the driving oscillator, is indeed easily realizable [16].

The following equations

$$\begin{cases} \dot{x} = -y - 0.95z \\ \dot{y} = x + 0.15y \\ \varepsilon\dot{z} = (1 - z^2)(z - 1 + x) - \delta z \end{cases} \tag{2.1}$$

[2]The universal circuit is discussed in [1].

describe one particular such systems which is even somewhat simpler than was suggested by the ingredients of Fig. 2.1, for the shifted reinjection makes it unnecessary for the oscillatory subsystem (x, y) to self-limit its amplitude. Instead of the limit-cycle oscillator (with self-limited amplitude) depicted in Fig. 2.1, thus an unstable oscillator could be used in Eq. (2.1): without z, the linear first two variables simply generate a spiral of ever growing amplitude (in accordance with Fig. 1.1).[3]

The 'switching variable' z in Eq. (1.1) has the property that for smaller and smaller values of the parameter δ, the manifold of steady states underlying the hysteresis loop (obtained by simply putting the left-hand side of the third line of Eq. (2.1) equal to zero) becomes arbitrarily sharp-edged (letter-Z shaped). In the limit of δ approaching zero, z can only assume the stable values $+1$ and -1, respectively. (There are two additional solution branches besides the 'letter Z' if this particular switch is used; but these 7- and L-shaped, additional steady state curves lie outside the region of the 'letter Z' and are unattainable from this central regime, so that their presence does not matter in the present context. For a more complicated function on the right-hand side of the switch indeed lacks such additional branches, see [22].)

With the coupling constant 0.95 in the first line replaced by zero, z would be driven by x only passively (Chaikin's principle). With the z-term added, however, any change in z causes a displacement of the pseudo-steady state of the slow system (x, y) along the y-axis (since $y_{\text{st.st.}} = -0.95z$), as desired.

After these preliminaries, it is time to show that Eq. (2.1) indeed behaves as suggested. This is done most easily by putting it into a computer and solving it numerically. The result is shown in Fig. 2.2 [15]. It appears that the running path of the flatlander of Fig. 1.1 indeed does not close over large times.

2.2 Robustness

Equation (2.1) could also have been solved analytically (in the limit of both ε and δ approaching zero)—by pasting together solution branches of the two linear systems that are defined by the first two lines of Eq. (2.1) if z is replaced by 1 or -1, respectively. This (or something like it) will be easier to do with another example, however (Sect. 2.8). Right now, it is more significant to try and to understand two empirical facts that are brought up by the preceding simulation: (i) the parameter δ could be given a finite value (0.03 in Fig. 2.2) and (ii) so could ε (0.03).

The first point is not astonishing since 'linearity of the two sub-oscillators' (which disappears if $\delta > 0$) is obviously inessential, a slight positional shift of one of the two thresholds will change the transition law in a similar way as if a nonlinear compression of one of the two spiral flows had occurred. The second point (admissibility of

[3] Basically, this system has a structure which is similar to the so-called Rössler system [18] where the third equations has the general form $\dot{z} = a + g(x, z)$ where $g(x, z)$ is a nonlinear polynomial in x and z. Such a function is here more complicated than the Rössler system to evidence the S-shape curve of steady state.

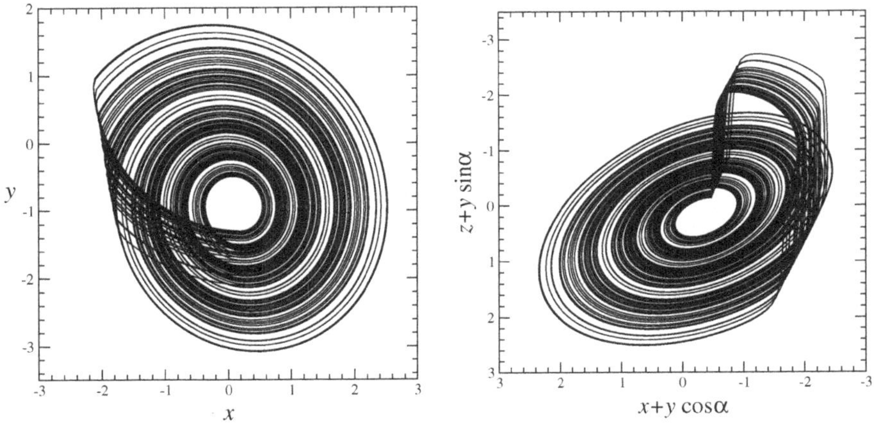

Fig. 2.2 Spiral chaos in Eq. (2.1). Compare with Figures 1 and 3. Two different projections of trajectorial flow in three-dimensional state space (x, y, z as functions of t) are presented. Original numerical simulation of Eq. (2.1) were performed using a standard Runge–Kutta–Merson integration routine (cf. [9]) implemented on an HP 9845A desktop computer with plotter 9872A. Parameters: $\varepsilon = \delta = 0.03$. Initial conditions: $x_0 = 1$, $y_0 = 0$, $z_0 = -1$. $\alpha = \frac{\pi}{3}$

$\varepsilon > 0$ as well) is surprising, however. For it is known that any finite value of ε immediately turns the piecewise two-dimensional, non-invertible 'paper flow' that existed for $\varepsilon = 0$ (and was described by the first two lines of Eq. (2.1) plus the algebraic constraint 'right-hand' side of the third line equals zero') into a fully three-dimensional flow. (The latter may be termed 'flint-stone flow' in view of the fact that everything has a finite thickness now; cf. [23]). Moreover, the genuinely three-dimensional new flow is automatically invertible: The mere fact of three right-hand sides all being smooth functions guarantees that exactly one trajectory (solution hairline) passes through almost every point in state space, and that once closely adjacent trajectories stay close together for all times almost everywhere. These are the two well-known 'uniqueness' and 'continuous dependence on initial conditions' theorems of the theory of ordinary differential equations (see, for example, [7]).

Thus, the seemingly innocuous transition between the limiting ($\varepsilon = 0$) and the non-limiting ($\varepsilon > 0$) case in Eq. (2.1) makes for an interesting example of non-uniform convergence (see Sect. 2.6). However, there is a second interesting problem being opened up by Eq. (2.1), at its present parameter values: the solutions of the limiting equation on the one hand and an arbitrarily close non-limiting equation on the other are not close everywhere. This can be seen by looking at the discontinuous 'gap' which in the running rink of Fig. 1.1 exists between the 'last non-switched' trajectory downstairs and the 'first switched' one: the latter runs a completely different course even though the two were at first (before approaching the edge "a") arbitrarily close. This gap is immediately filled by a continuous curtain of 'intermediary trajectories' as soon as ε becomes nonzero, however small. This again follows—in the absence of a saddle point near the edge (which could be built in, but makes the equation more complicated; see Sect. 2.8)—from the 'continuous dependence on initial conditions' mentioned above. Indeed a closer look at Fig. 2.2 shows that there is one trajectory (or rather, trajectorial segment) which passes right through this former 'no man's land' (or rather, its periphery): see the somewhat tilted right-most trajectory in the handle-shaped loop of Fig. 2.2.

The two convergence problems that are opened up by Eq. (2.1) nonetheless at the same time illustrate the close relationship between the two flows determined by Eq. (2.1) with $\varepsilon = 0$ and $\varepsilon > 0$, respectively. The reinjection principle therefore appears apt to generate not only interesting limiting flows (for which it was designed for) but also interesting non-limiting flows.

2.3 A Prototype Example for Spiral Chaos

The last-made observation (survival of reinjection chaos under 'de-idealization') deserves to be stretched to its empirical limits. This can be done by continuing simulating Eq. (2.1) while chipping away at its two extremal properties (the smallness of ε and the ideal Z-shape of the slow manifold).

In this interactive process, one finds not only that ε can be increased considerably, but also, that the nonlinearities of the former 'switch' can be markedly reduced, leaving eventually a single threshold whose underlying letter L-shaped (instead of letter Z-shaped) nullcline is generated by a single remaining quadratic term. The ensuing simplified equation [20] is

$$\begin{cases} \dot{x} = -y - z \\ \dot{y} = x + a\,y \\ \dot{z} = b + z(x - c)\,. \end{cases} \tag{2.2}$$

A numerical simulation of this equation with parameter values $a = 0.15$, $b = 0.2$ and $c = 10$ is displayed in Fig. 2.3.

The behavior of Eq. (2.2) as visible in Fig. 2.3 can in analogy to Fig. 2.2 still be explained in terms of the qualitative behavior of the system's components, although the explanation has to be couched in non-quantitative terms now. For example, it is still true that the variable z tends toward a near-zero stable state as long as the 'pseudo-parameter' x stays below its threshold (which is 10 now) as shown in Fig. 2.4. As soon as x surpasses this threshold, z starts growing autonomously until x spontaneously falls below the threshold again (so that z is forced to come down). Moreover, by virtue of the first line, any rise in z has again a shifting effect on the x-y motion, displacing its 'steady state' (supposed z were a constant), in the y-direction by $-z$. Hence any 'lifted' trajectory will, once more, touch down at a more 'internal' position than it would in the absence of the z-term in the first line. The flow of Fig. 2.3 is in accordance with these reasonings.

There is an additional qualitative observation possible: trajectories touching down from a greater height (that is, a larger value of z) will, in their approach toward the

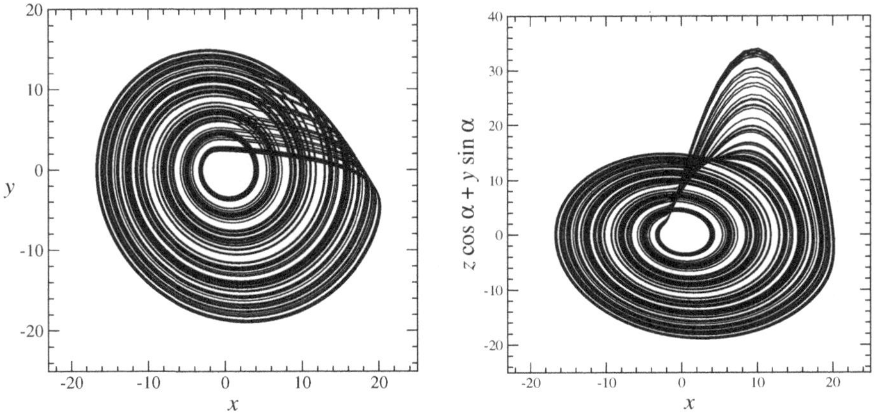

Fig. 2.3 Spiral type chaos in Eq. (2.2). Compare to Fig. 2.2. Parameter values: $a = 0.15$, $b = 0.2$ and $c = 10$. Initial conditions: $x_0 = 10$, $y_0 = 1$ and $z_0 = 0$. $\alpha = 4.75\pi$

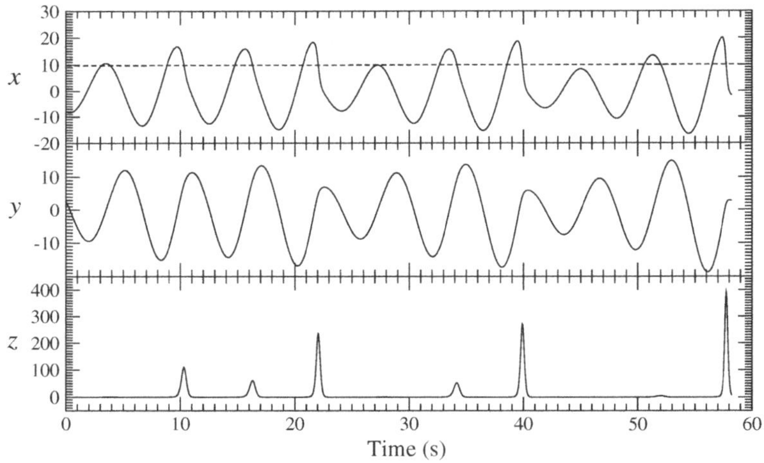

Fig. 2.4 Time series of the spiral type chaos in Eq. (2.2). Parameter values: $a = 0.15$, $b = 0.2$ and $c = 10$. Initial conditions: $x_0 = 10$, $y_0 = 1$ and $z_0 = 0$

attracting $z = 0$ plane, keep staying above solutions that started at the same x and y position but from a lower height. This means that it is in fact necessary to keep track of whole (finite-thickness) 'layer' of trajectories: as they approach the 'threshold' ($x = 10$), then rise, then come down again. In other words, the whole complicated behavior actually takes place within a 'cake' rather than within an infinitely thin, folded and glued-together 'paper sheet'. This fact is illustrated by Fig. 2.5.

The 'cake picture' (Fig. 2.5b) implies that there is not a one-dimensional cross section as through a paper flow (see the arrow marked P in Fig. 2.5a), but a two-

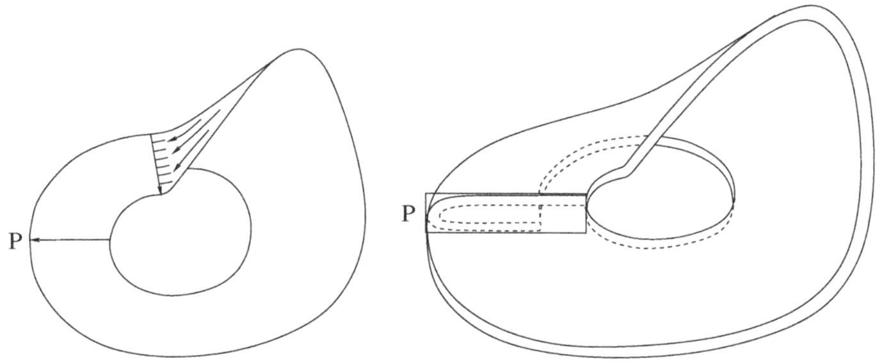

(a) Paper flow (b) 'Cake flow'

Fig. 2.5 The flow of Fig. 2.3 rationalized (**a**) as a 'paper flow', and (**b**) as a 'cake flow'. The corresponding cross sections P are one-dimensional and two-dimensional, respectively

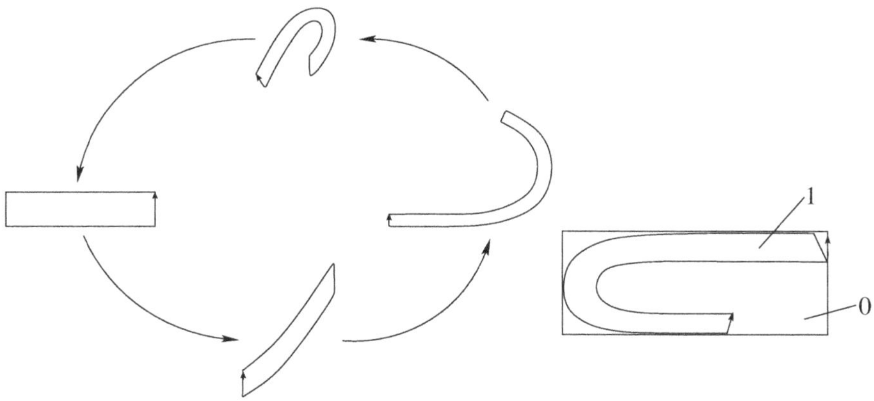

(a) Outline ('rotating taffy-puller') (b) Resulting 'walking-stick' map

Fig. 2.6 Deformations of a two-dimensional cross section through the flow of Fig. 2.3b as it is 'running around'. 0 = zeroth iterate (original domain), 1 = first iterate (image). The arrow on the right-hand side of the initial rectangle serves to facilitate identification

dimensional one. The distorsions to which it is subjected while travelling around (Fig. 2.5b) are re-drawn in more detail in Fig. 2.6a: after one round, the outline of an initial rectangle is found lying **inside** the original domain. The thus postulated two-dimensional map [16, 20] is shown once more in Fig. 2.6b.

2.4 A Second Main Equation

The map of Fig. 2.6b was thought to be generated by a smooth flow devoid of any discontinuities (saddle points) in the region of interest. This means that it must be a diffeomorphism, that is to say, a map that is uniquely invertible (bijective) at every point, whose local features are only smoothly distorted between one iteration and the next (so that 'corners' remain corners, for example), and whose inverse possesses the same two properties [4]. Non-area preserving diffeomorphism like that of Fig. 2.6b (yet of a more symmetrical—'horseshoe'—shape) have been a subject of intense mathematical investigations since Smale [24, 25] first envisaged their possible occurrence in smooth flows [3]. An explicit example of such a folded map has, nonetheless, been lacking until quite recently.

If one is not reluctant to use a computer as an instrument of verification, the following 'recipe' how to produce a map like that of Fig. 2.6b appears worth trying out: After labelling map's axes a and b, respectively, (i) make the 'next a' a folded (single-maximum) function of the last a, and (ii) assure that the 'next b' is non-inverted with respect to the last b along the ascending part of a's function and inverted along its descending part. These twin constraints condense into the following prototype equation [22]:

$$\begin{cases} a_{n+1} = \gamma a_n(1 - a_n) - b_n \\ b_{n+1} = (\delta b_n - \varepsilon)(1 - 2a_n)\,. \end{cases} \tag{2.3}$$

Here the last term in the first line provides for the necessary back-coupling (from b to a). The term in front of it generates a parabola (as a simple single-maximum function; point (i). In the second line, the second bracket changes sign with the slope of a's parabola while the first bracket produces both a compression ($\delta < 1$) and a displacement ($-\varepsilon$) of b; jointly (as a product) the two brackets implement point (ii).

Finding appropriate parameter values for γ, δ, and ε is not very difficult. The result of one such trial-error process is shown in Fig. 2.7.

The appropriate initial box (neither too big nor too small) was also choosen by using extremal values of a and b. Instead of the rectangle shown, also an ellipse could have been sought (or a folded sausage, or any shape approximating one of the higher iterates from the outside).

Obviously, all points of the initial domain (zeroth iterate) come to lie wholly within the first iterate, and thereafter within the second (which itself lies wholly within the first), and so on. This applies also to an initial 'raisin' as it is followed up (see Fig. 2.7c).

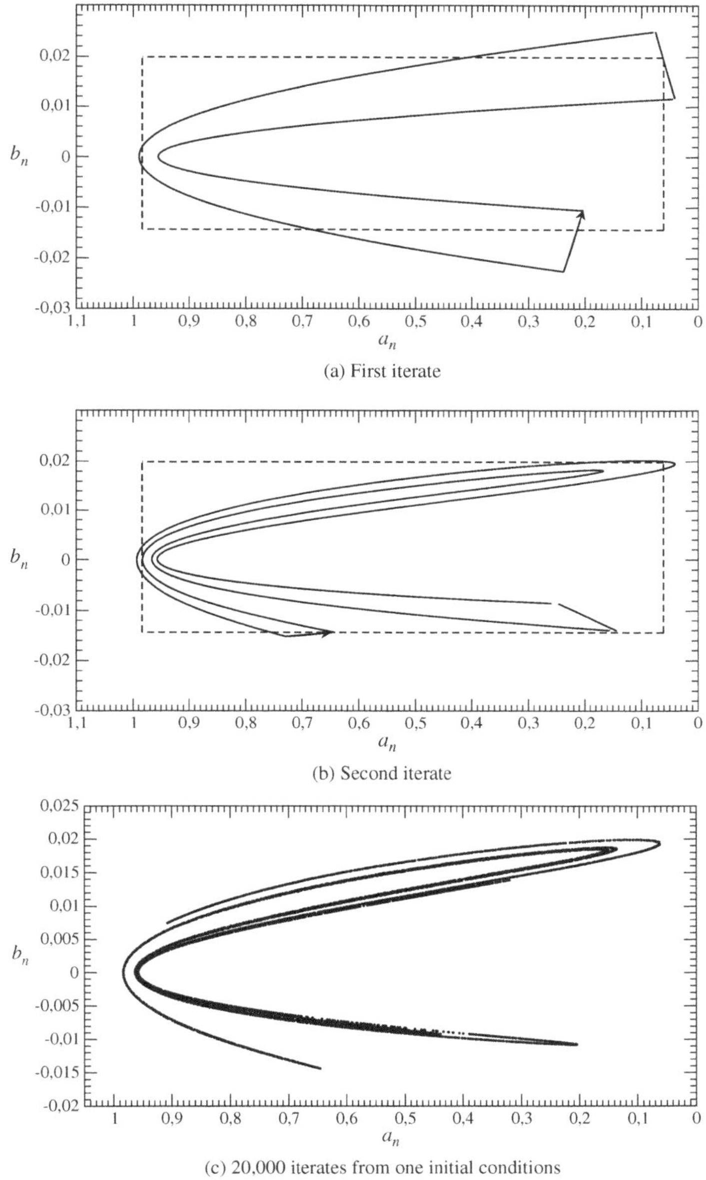

Fig. 2.7 An explicit walking-stick map (cf. Fig. 2.6b). Calculation of Eq. (2.3). Parameters: $\gamma = 3.9$, $\delta = 0.4$, $\varepsilon = 0.02$. First (**a**) and second (**b**) iterate of the initial box shown in dashed line. The initial box was defined by the extrema of each variable obtained when the 20 000 iterates (**c**) were obtained from initial conditions $a_0 = 0.98$ and $b_0 = -0.0294$

2.5 Two Special Cases

The walking-stick map of Eq. (2.3) admits two extremal cases that are known from the literature: the logistic map on the one hand and Hénon's equation on the other. The so-called logistic map [8] has the form

$$a_{n+1} = \gamma a_n(1 - a_n)\,. \tag{2.4}$$

It is obtained as a limiting case to Eq. (2.3), if in this equation the constant ε approaches zero while at the same time, δ is left unchanged. A numerical calculation is shown in Fig. 2.8.

The calculation was performed with $\varepsilon = 0.02 \cdot 10^{-50}$. Accordingly, the height of the whole map (width of the b-side of the box) is $0.04 \cdot 10^{-50}$ which is very small. If one compares Fig. 2.8 with Fig. 2.7, one sees that nonetheless not too much has been changed: it is still the picture of a taffy-folding map. The main difference is that one line of the original domain (namely, the line defined by $a = 0.5$) has been compressed to a point, which amounts to a local violation of invertibility.

The similarity is unexpected since the first line of Eq. (2.3) is a single-variable equation and hence by definition cannot describe a two-dimensional taffy-puller. However, the first line of Eq. (2.3) is indeed 'equivalent' to Eq. (2.3) (in an unilateral sense) in the limit of ε to zero. This is because Eq. (2.3) could as well have been written as

$$\begin{cases} a_{n+1} = \gamma a_n(1 - a_n) - \varepsilon \overline{b}_n \\ \overline{b}_{n+1} = (\delta \overline{b}_n - \varepsilon)(1 - 2a_n)\,, \end{cases} \tag{2.5}$$

with $\overline{b} = \frac{b}{\varepsilon}$, as can be verified by insertion. And Eq. (2.5) becomes ineffective in the limit. Conversely, nothing prevents one from adding the second line of Eq. (2.5) to the first line of Eq. (2.3) in the interest of better understanding a certain taffy-puller—Eq. (2.3)—up to one of its limiting cases; such an 'embedding' incidentally proves helpful also in other contexts (see Sect. 2.8).

If Fig. 2.8 is interpreted in terms of Eq. (2.5), the height of the boxes no longer equals $0.04 \cdot 10^{-50}$ but rather ε^{-1} times this values, namely 2. The same pictures (and the same height) apply if $\varepsilon \equiv 0$ (if the computer accuracy were unlimited, there would be a very small difference).

A second extremal case to Eq. (2.3) is obtained in the limit of δ approaching zero. While ε was responsible for the width of the image in horizontal directions (as expressed in the width of the knee), δ similarly governs the vertical direction. A numerical calculation of Eq. (2.3) with $\delta = 0$ is shown in Fig. 2.9.

One sees that of the vertical distances along the two sides of the initial box, nothing but horizontal distances remain after one iteration. This is because in Eq. (2.3) with $\delta = 0$, the 'next b' is the same for all points that have the same a-value (that is, lie on the same vertical line). As a consequence, it would no longer be correct to say that the first iterate, after having been thinned and stretched, was in any realistic sense 'bent over itself' before being put back into the box. The first iterate rather

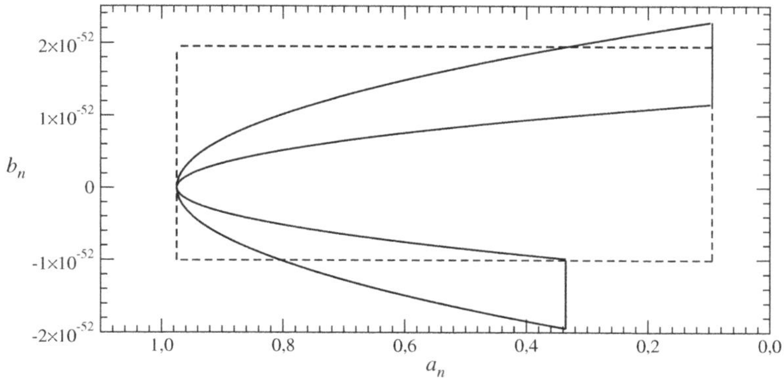

(a) Zeroth and first iterate

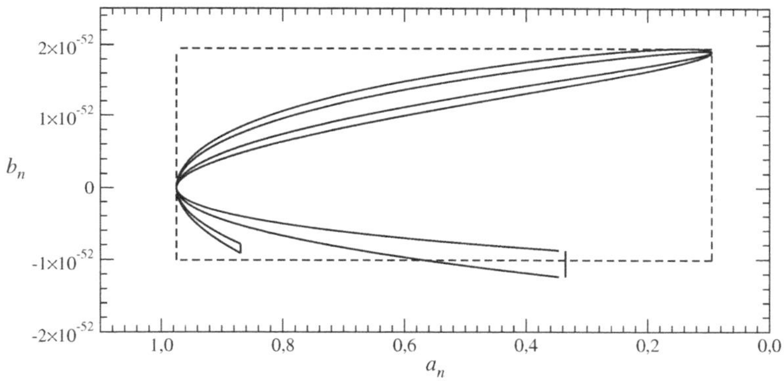

(b) Zeroth and second iterate

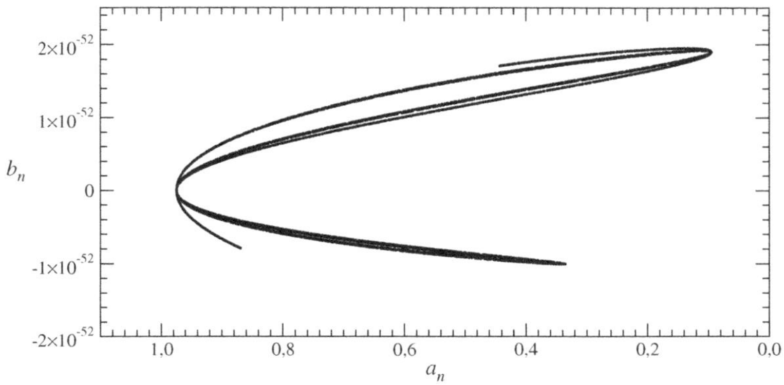

(c) 1200 iterates of the lower right corner point

Fig. 2.8 Calculation of Eq. (2.4) as ε approaches zero. Parameters are as in Fig. 2.7, except for ε which is $0.02 \cdot 10^{-50}$. Equation (2.5) yields the same pictures without requiring an enlargement of the b-axis (see text)

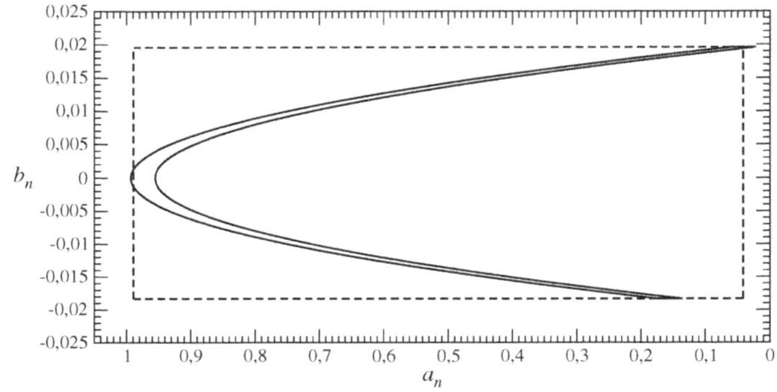

(a) First iterate

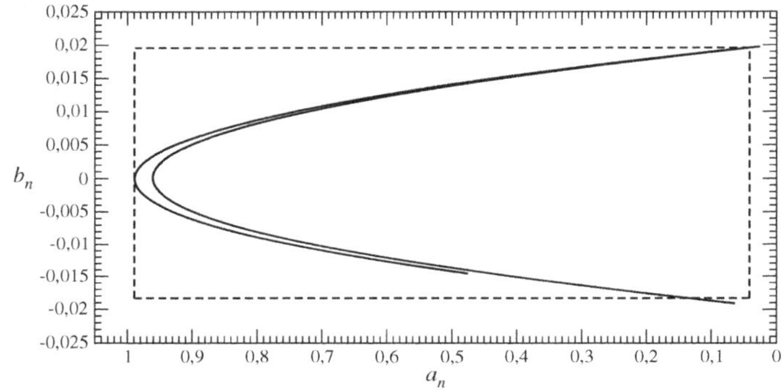

(b) Second iterate

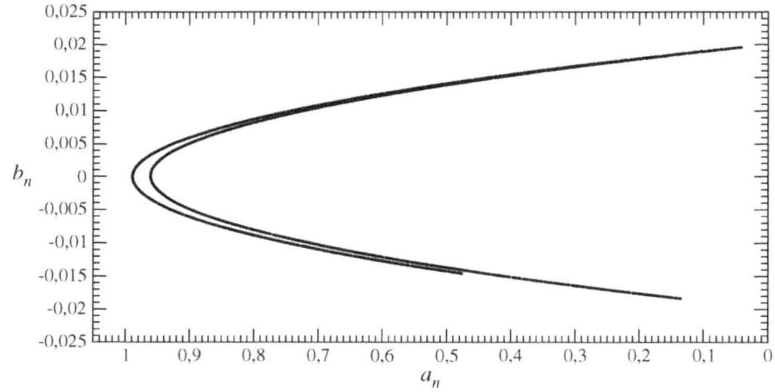

(c) 20,000 iterates from one initial conditions

Fig. 2.9 Calculation of Eq. (2.3) in the limit δ approaches zero. Parameters: $\gamma = 3.9$, $\delta = 0$, $\varepsilon = 0.02$. First (**a**) and second (**b**) iterate of the initial box shown in dashed line. The initial box was defined by the extrema of each variable obtained when the 20 000 iterates (**c**) were obtained from initial conditions $a_0 = 0.98$ and $b_0 = -0.0294$

appears (after an initial rotation by exactly 90° and a subsequent change of scales) to be carefully 'shifted', horizontal layer by horizontal layer, in such a way to again fit into the original box.

This 'unnatural' (because physically unrealizable) behavior is, however, confined to the first iterate. Because no similar 90° constraint applies to originally horizontal distances, vertical distances (which after one iteration have become horizontal distances) cease to obtain special treatment beyond the first iteration. The second iterate of Fig. 2.9b therefore looks remarkably similar to that of Fig. 2.7b. (Unfortunately, the somewhat stronger compression in Fig. 2.9a admits only a poor optical resolution of the second iterate in Fig. 2.9b).

The extremal case to Eq. (2.3) that has been presented in Fig. 2.9 is, again, a well-known equation. Equation (2.3) with $\delta = 0$ is equivalent to Hénon's recurrence after a change of scales. Michel Hénon [6] considered the following equation (which he called the Cremona transformation):

$$\begin{cases} p_{n+1} = 1 - \alpha p_n^2 + q_n \\ q_{n+1} = \beta p_n \end{cases} \tag{2.6}$$

where both a and b were assumed greater than zero. (Specifically, the example set $\alpha = 1.4$, $\beta = 0.3$, $p_0 = q_0 = 0.5$ was considered.)

The following scaling transformation

$$\begin{cases} a = \kappa p + \dfrac{1}{2} \\ b = -\kappa q \,, \end{cases}$$

with

$$\kappa = -\frac{1}{4} + \frac{1}{2}\sqrt{\frac{1}{4} + \alpha} \,,$$

changes Eq. (2.6) into

$$\begin{cases} a_{n+1} = \dfrac{\alpha}{\kappa} a_n (1 - a_n) - b_n \\ b_{n+1} = \dfrac{\beta}{2}(1 - 2a_n) \,, \end{cases} \tag{2.7}$$

as can be verified by insertion. Equation (2.7), however, is Eq. (2.3) with $\gamma = \frac{\alpha}{\kappa}$, $\delta = 0$, and $\varepsilon = -\frac{\beta}{2}$.

The above original parameters ('example set') yield $\gamma = 3.569\ldots$, $\delta = 0$, $\varepsilon = -0.15$, with initial value $a_0 = 0.7$ and $b_0 = -0.2$, respectively. What is unexpected is the negative value of ε. Looking again at a rectangular initial box that contains all higher iterates shows what happens. (In this case, ε has to be taken as -0.08 rather than -0.15 if one wants to avoid having to choose a more complicated initial domain.) The first iterate within the box looks just like of Fig. 2.9a, but with the lower arrow (being the image of the right-hand side of the original box) pointing leftward rather than rightward. This means that the analogy to a taffy-puller has broken down.

For the first iterate now has the same orientation as a mirror image of the original domain. Thus, with the present parameter set, the map of Eq. (2.6) is not orientation preserving. It therefore also cannot be used as a model of what happens in a smooth chaotic flow.

However, such 'appropriate' parameter values do also exist, as follows from Fig. 2.10. This figure at the same time represents a numerical calculation of Eq. (2.6), with parameter $\alpha = \kappa\gamma = \frac{\gamma^2}{4} - \frac{\gamma}{2} = 1.71$ and $\beta = -2\varepsilon = -0.062$.

Is, then, Eq. (2.3) or (2.6) preferable as a crude model of what happens in a smooth flow like that of Fig. 2.3? The answer is: it depends. In the first case one has to put up with two quadratic terms Eq. (2.3). In the other, one is studying a map Eq. (2.6) which itself is only a 'half-iterate' to a more realistic map (namely, the second iterate of Eq. (2.6) which this time contains terms up to the fourth degree, however). On the other hand one knows that fractional iterates do preserve much structure [26]. Thus, they are good arguments for either choice.

The simulated taffy-puller of Fig. 2.10 is good for an empirical surprise: when substituting the value 0.03 for the present value of ε (0.031) in oder to have only good-looking constants, one finds that the equation for all initial conditions that one cares to test, soon (that is, after at most a few hundred iterates) locks into a period-five cycle (always the same) which is obviously attracting. This observation has theoretical significance in the context of whether or not the present maps contain a strange attractor in the sense of Smale [25]: they do not (see Sect. 7.2).

Secondly, there is the question, how does one know whether the preceding maps are indeed diffeomorphisms (as they ought to be in order to simulate continous flows)? They are, if (i) the first iterate is nowhere self-overlapping or self-touching in the region of interest, as is easy to check for the case of simple maps—for example, by using to check for the case of simple maps—for example, by using a computer to produce an initial grid (or at least outline as was done in Figs. 2.7 and 2.10) and (ii) the maps are everywhere 'locally affine' in the region of interest. The latter property (which implies the maps being locally diffeomorphic) is guaranteed if the locally linearized maps do nowhere contract volume to zero in one step. Since the local volume-contraction factor is indicated by the Jacobian determinant, it suffices to make sure that the latter is nowhere zero in the region of interest ('inverse function theorem'; [7]). The Jacobian determinant of Eq. (2.3) is

$$\gamma\delta(1-2a)^2 - 2(\delta b - \varepsilon)$$

and hence does not become zero in Figs. 2.7 and 2.10 as it is easy to verify. In Fig. 2.6, where Eq. (2.5) with Jacobian determinant

$$\gamma\delta(1-2a)^2 - 2\varepsilon(\delta b - 1)$$

be used, the line $a = \frac{1}{2}$ properly makes an exception if ε goes to zero.

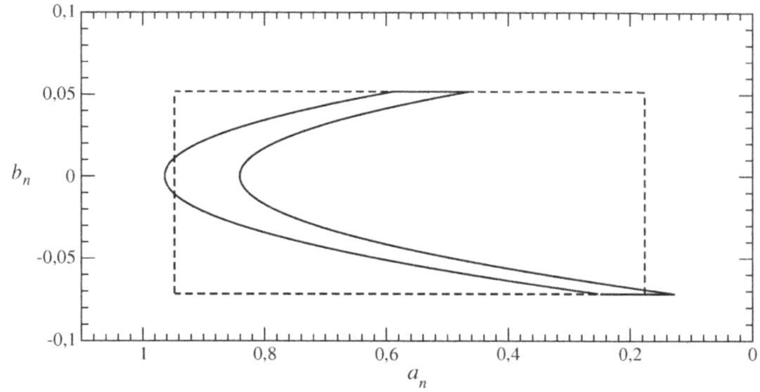

(a) First iterate ($\varepsilon = -0.08$)

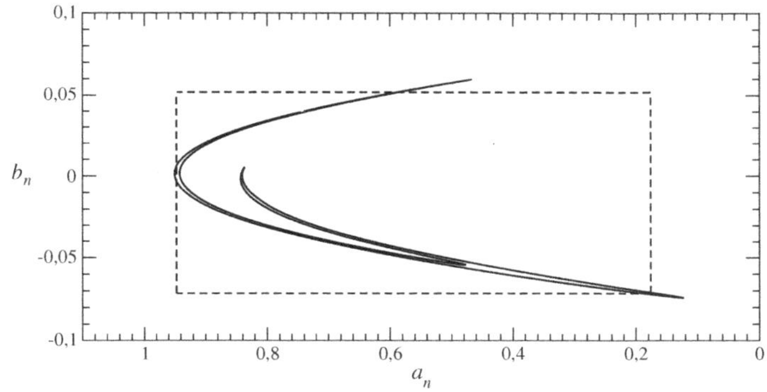

(b) Second iterate ($\varepsilon = -0.08$)

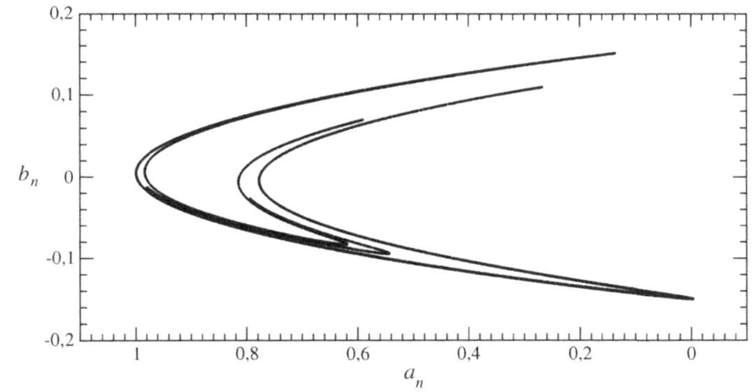

(c) 20,000 iterates from one initial conditions ($\varepsilon = -0.15$)

Fig. 2.10 Calculation of Eq. (2.3) in the "Hénon limit". Parameters: $\gamma = 3.659$, and $\delta = 0$. First (**a**) and second (**b**) iterate of the initial box shown in dashed line. The initial box was defined by the extrema of each variable obtained when the 20 000 iterates (**c**) were obtained from initial conditions $a_0 = 0.98$ and $b_0 = -0.0294$

2.6 Screw-Type Chaos

The two running rinks of Fig. 1.1 can be combined in another, equally simple fashion. The combination shown in Fig. 2.11a is, in a sense, even simpler than that of Fig. 1.1, for there is no spiral-shaped subpath involved. On the other hand, there are more transitions between 'upstairs' and 'downstairs' implied, thereby suggesting the picture of a screw (with folding-over) rather than spiral (with folding-over). Figure 2.11b presents a direct 'translation', first into an equation, and then into a simulation. The simulated equation is, by the way, Eq. (2.1)—with three constants given a different value than in Fig. 2.2. Figure 2.11c finally shows another simulation of screw-type chaos. This time, Eq. (2.2) is encountered again. (Compare Rössler, [17, 19–21] for several more screw-like pictures of screw-type chaos in these and other equations.)

Everything that has been said above about the relationship between primal sketch (Fig. 1.1), its limiting (and close-to-limiting) realization (Fig. 2.2), its empirical 'de-idealization' (Fig. 2.3), and finally its relationship to a folded-over, smoothly invertible map (Fig. 2.6), pertains again. Even the rotating taffy puller (Fig. 2.6a) is applicable again, once Fig. 2.5 has been replaced by its analogue (Fig. 2.12) [21].

Aside from those parallels, the present family of chaotic flows possesses two mathematical advantages. First, there no longer applies a discontinuous transition between the limiting flow and its invertible neighbors: the 'hole plugging' (by a continuous curtain of trajectories) which occurred in Eq. (2.1) for the transition $\varepsilon = 0$ to $\varepsilon > 0$ in the case of spiral chaos (Sect. 2.5) has no analogue this time for, as Fig. 2.11a shows, there is no 'hole' to plug in the first place. This means that the 'limiting transition between maps' that was encountered in the preceding section (Eqs. (2.3) and (2.4); Figs. 2.7 and 2.8) can now be invoked directly as a model of what happens in the flow of Fig. 2.11b in its own limiting transition. Secondly, screw-type chaos lends itself better to a quantitative analysis (see the next-following story).

2.7 An Example with an Explicit Cross Section

Screw-type chaos differs from spiral chaos in another respect (apart from the missing hole): the flow 'downstairs' never makes a full turn. Therefore it is no longer necessary to base the flow downstairs on a spiral. Instead, one can use a family of concentric circles, or ellipses, or even parabolas. At the same time, the flow upstairs might (just as in the spiral case) be replaced by straight lines.

This implication of screw-type chaos was first seen and exploited by Mira [11][4] who proposed the following equation as an alternative to Eq. (2.1):

[4]Otto met Christian Mira for the first in 1978, in Toulouse.

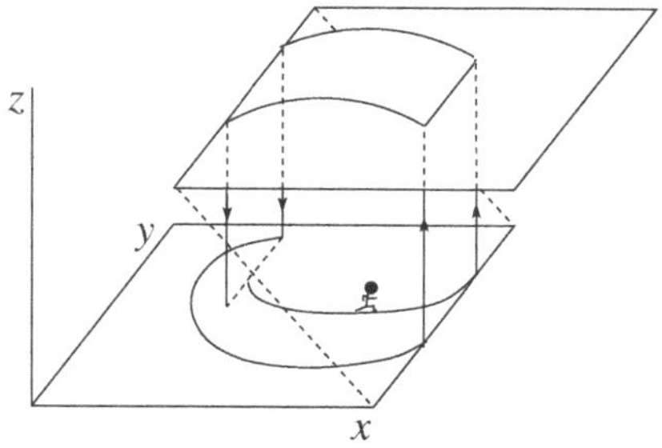

(a) Paper model (compare to Fig. 1.1)

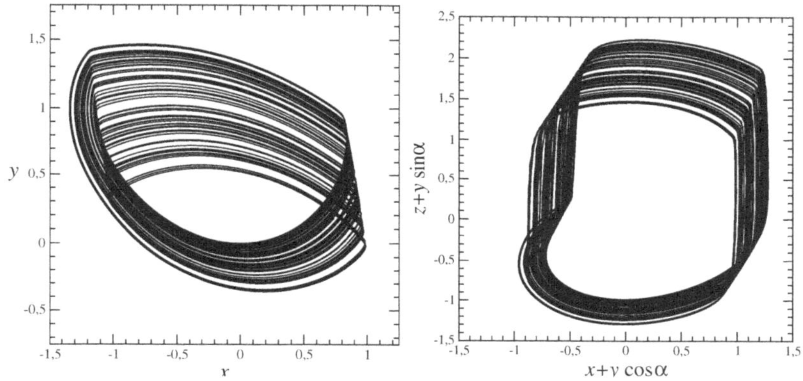

(b) Numerical simulation of Eq. (2.1). (compare to Fig. 2.2)

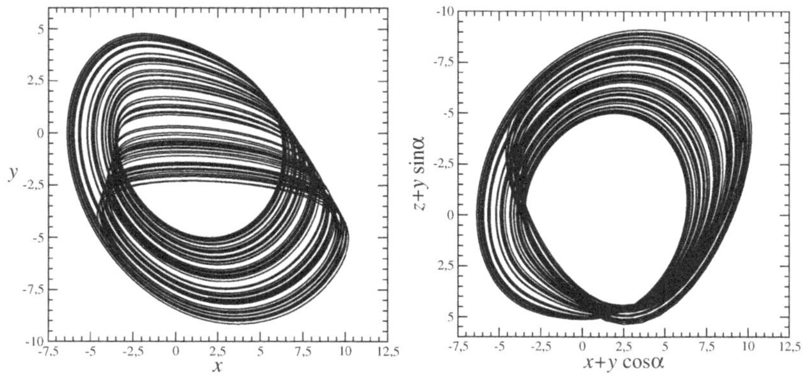

(c) Numerical simulation of Eq. (2.2). (compare to Fig. 2.3)

Fig. 2.11 Screw-type chaos in Eq. (2.1). For the numerical simulations of Eq. (2.1) as written above, 0.95 has been replaced by unity, 0.15 by 0.53, and +1 by zero. Same initial conditions as in Fig. 2.2. Numerical simulations of Eq. (2.2) with parameter values $a = 0.4$, $b = 0.2$ and $c = 2.95$. Initial conditions: $x_0 = -3.5$, $y_0 = z_0 = 0$. $\alpha = \frac{\pi}{3}$

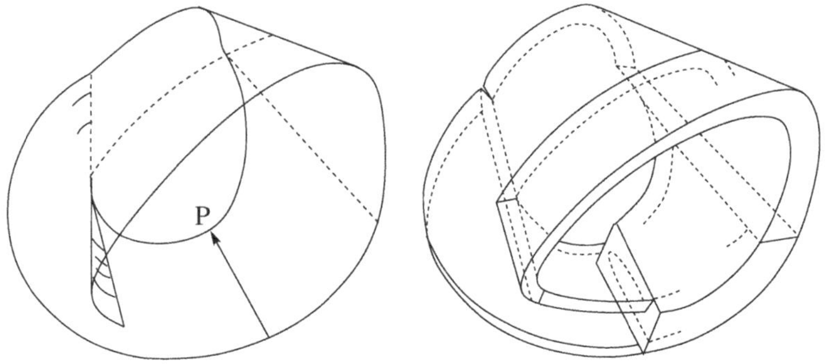

(a) As a 'paper flow' (b) As a 'cake flow'

Fig. 2.12 The flow of Fig. 2.11c rationalized. Compare Fig. 2.5

$$\begin{cases} \dot{x} = (x-b)z + (y+c)(1-z) \\ \dot{y} = yz + (1-z) \\ \varepsilon\dot{z} = z(1-z)(z-1+x) - \delta(z-0.5)\,. \end{cases} \tag{2.8}$$

The third line was originally given as a piecewise linear algebraic constraint of letter Z-shape since only the asymptotic behavior was considered. Equation (2.8) involves three quadratic terms in its first two lines and hence is more complicated than Eq. (2.1). On the other hand, it in the limit still describes two linear subflows in x and y, one holding true if $z = 0$ (right hand block of terms in the first two lines of Eq. (2.8) rendered effective), the other holding true if $z = 1$ (left-hand block effective). These two linear sub-flows are each simpler than those represented in Eq. (2.1): The first (with $z = 0$), representing the equation of a falling stone, possesses a family of displaced parabolas for solutions, the second (with $z = 1$) yields a family of straight lines (two uncoupled autocatalytic variables with equal growth rates).

The resulting overall behavior can be inspected in Fig. 2.13. In this simulation, ε has been given a small finite value (and so has δ).

The advantage of Eq. (2.8) is that a cross section through the limiting flow can be calculated not only 'in principle' as in Eq. (2.1) [17], but in reality and explicitly. At the positions marked P in Fig. 2.13, the next value of y as a function of the last becomes on the basis of simple geometric reasoning [11] :

$$y_{n+1} = -c + \sqrt{2 + c^2 + \frac{2cb}{b-1}y_n + \frac{b^2}{(b-1)^2}y_n^2}\,. \tag{2.9}$$

A graph of this function (with P at $z = 0.5$) is shown in Fig. 2.14b.[5] The parameters chosen are the same as in Fig. 2.13 except for ε and δ which are both set equal to

[5]The first-return map to a Poincaré section of the attractor produced by continuous equations (2.8) was added while editing the present version of this book.

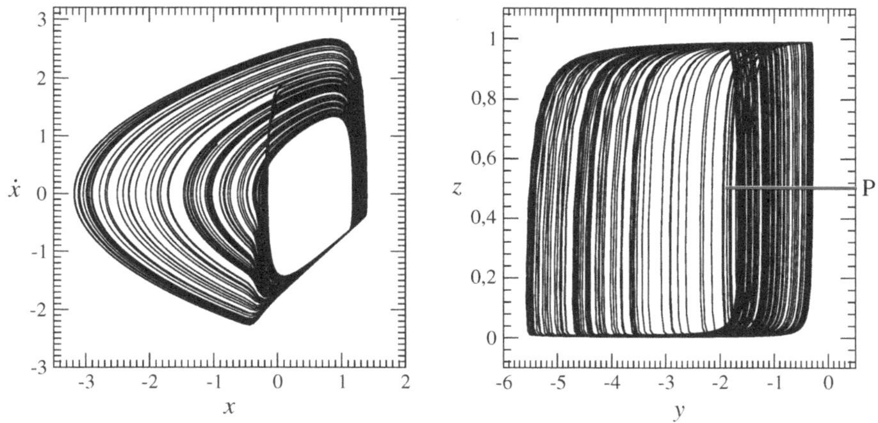

Fig. 2.13 Screw-type chaos in Mira's equation (2.8). Parameters: $b = 1.8$, $c = 3.3$, $\varepsilon = \delta = 0.03$. Initial conditions: $x_0 = z_0 = 0$, $y_0 = -1.8$

zero (corresponding to the limit of ε and δ as well as $\frac{\varepsilon}{\delta}$ approaching zero in Eq. (2.8), and also except for b which had to be reduced as the hysteresis loop of z is now no longer both smoothed and widened—as it is if ε differs perceptibly from zero as in Fig. 2.13. The present value ($b = 1.53$)[6] indeed reestablishes the global appearance of Fig. 2.13 if inserted into Eq. (2.8) with $\varepsilon = \delta = 0.0003$. A numerical calculation of Eq. (2.9) is provided in Fig. 2.14b. Equation (2.9) hereby appears as a close analogue to Eq. (2.4) above, the logistic map, which is the simplest single-maximum smooth one-dimensional map.

[6]There is a displacement in the parameter space between continuous equations and Mira's map (2.9). The bifurcation diagram (see below) clearly shows that a period-1 limit cycle is observed for the original b-value ($b = 1.8$) and not $b = 1.53$ as retained in Fig. 2.9b.

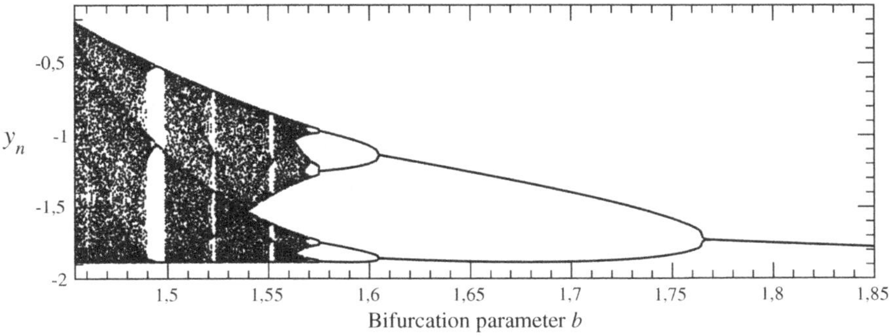

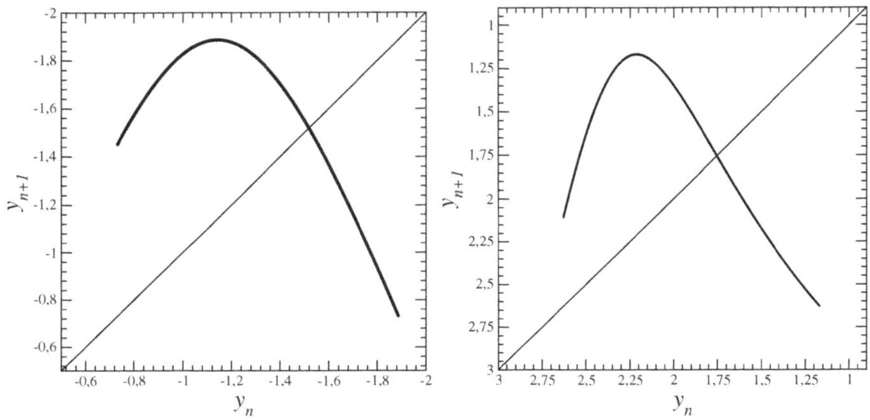

(a) First-return map in the flow (2.8) (b) Numerical iterations of map (2.9)

Fig. 2.14 **a** First-return map to a Poincaré section of the flow (2.8) with parameter values as in Fig. 2.13b. **b** Numerical iterations of Eq. (2.9) of the map to Poincaré section P as shown in Fig. 2.13 in the limit of the underlying differential equation becoming singular. P at $z = 0.5$. 1200 iterates of an arbitrarily picked initial point (lower left corner) computed with 12-digit accuracy. Parameters: $b = 1.53$ and $c = 3.3$. Initial condition: $y_n = -1$

2.8 A Two-Dimensional Embedding

The present context is again basically two-dimensional. It is therefore of interest to ask what happens to Eq. (2.9) if in the underlying continuous system (2.8), ε does not quite reach its zero limit. The unknown two-dimensional equation which describes the cross section in this case must include Eq. (2.9) in the limit of a certain intrinsic parameter ε' becoming zero, but must be invertible (diffeomorphic) as soon as this parameter turns nonzero ('hairpin map to walking-stick map transition' [21]). The intrinsic parameter ε' should be a monotone function of ε in Eq. (2.8).

This very type of transition has been observed already in the preceding chapter (Sect. 2.5). In dependence on its own parameter ε, Eq. (2.3) was either a one-dimensional non-invertible map (endomorphism) or a two-dimensional smoothy invertible map (diffeomorphism). On the other hand, Eq. (2.3) as proposed above served only as an illustration for a whole class of walking-stick maps that all possessed a single-maximum smooth function in their first line. Therefore, the following analogue to Eq. (2.3) suggests itself immediately as an 'embedding' for Eq. (2.9) [22]:

$$\begin{cases} p_{n+1} = -c + \sqrt{2 + c^2 + \dfrac{2cb}{b-1} p_n + \dfrac{b^2}{(b-1)^2} p_n^2} + \varepsilon' q_n \\ q_{n+1} = (d q_n - 1)(1 - r p_n) \,. \end{cases} \tag{2.10}$$

Here r is a constant defined by $r = \frac{1}{\max\left(f(p_n)\right)}$, (with the parameter of Fig. 2.14, r is approximatively equal to 0.2), while d plays the role of the constant δ in Eq. (2.3). Note that Eq. (2.10) has been written in the convention of (2.5) rather than that of Eq. (2.3).

Mira [11] independently made an analogous proposal:

$$\begin{cases} p_{n+1} = f(p_n) - \varepsilon' g(p_n, q_n) \\ q_{n+1} = h(p_n) \, . \end{cases} \tag{2.11}$$

Here $f(p_n)$ is again the right-hand side of Eq. (2.9) and g and h are unspecified continuous functions of their arguments such that Eq. (2.11) is a diffeomorphism. Using estimates on the rapidity of non-uniform convergence in relaxation type dynamical systems [5, 11] specifically proposed ε' to be proportional to $\varepsilon^{2/3}$ in a neighborhood to the limit $\varepsilon = 0$ in Eq. (2.8).

The most general form of an embedding to Eq. (2.9) is, of course,

$$\begin{cases} p_{n+1} = f'\left(p_n, q_n, \varepsilon'\right) \\ q_{n+1} = h\left(p_n, q_n, \varepsilon'\right) \end{cases} \tag{2.12}$$

where f' is assumed to approach f in the limit $\varepsilon' = 0$. If ε' is very small, however, Eq. (2.11) constitutes a first-order approximation to Eq. (2.11) under rather general conditions. (Specifying these conditions may be a worthwhile exercise.)

The argument can formally be extended to a transition Eq. (2.11) $\rightarrow$ Eq. (2.10) whereby the conditions become more and more *ad hoc* with each sub-step. The overall transition from Eq. (2.12) to Eq. (2.10) can, nonetheless be pictured as follows: The true map (2.12) with f' and h unknown has an image that most probably, (i) is non-uniformly 'sheared' (which includes formation of an asymmetrically shaped 'knee' and unequally slanted 'feet') and (ii) has one or more intrinsic 'wriggles'. Both of these unknown properties develop at an unpredictable differential pace as soon as ε' deviates perceptibly from zero. Equation (2.10) with all of its three free parameters (ε', d, r) is, on the other hand, only capable of capturing some simple 'systematic' shearing effects (including an asymmetric knee and two outward-sloped feet). Of these, only the form of the knee can be expected to be adaptable to reality: the correct slopes of the feet will in general require a higher-order polynomial in the second line of Eq. (2.10) in order to be reproducible. The same holds true for the inhomogeneous shearing effects (supposed their correct positions were known), while corrections for a curved and wriggled overall shape can no longer be provided so easily. All of this means that the value of Eq. (2.10) as an approximate model to the 'true taffy puller' has by now been brought into perspective: the equation is completely false but nonetheless arbitrarily good in the limit.

A picture of Eq. (2.10) is presented in Fig. 2.15a. This value of ε' chosen (0.05, corresponding seemingly to a value of $\varepsilon = 0.0112$ in Eq. (2.8)), is much too large in order for Eq. (2.10) to approximatively model the true cross section. This follows already from the fact that at such large finite ε (Fig. 2.13) where ε was 0.03), b must be considerably increased in order for the flow not to 'explode' because of excessive

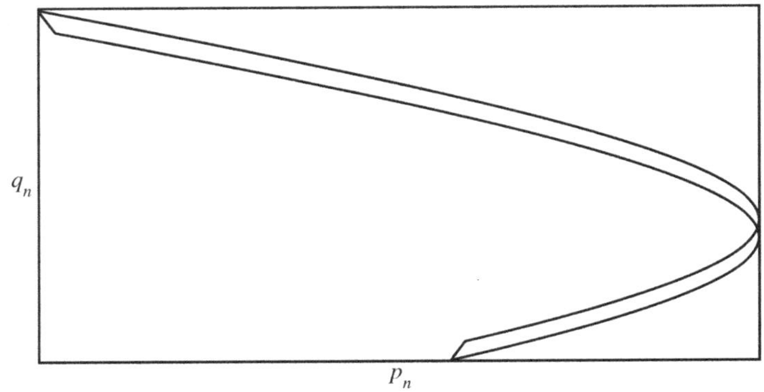

(a) Zeroth and first iterate

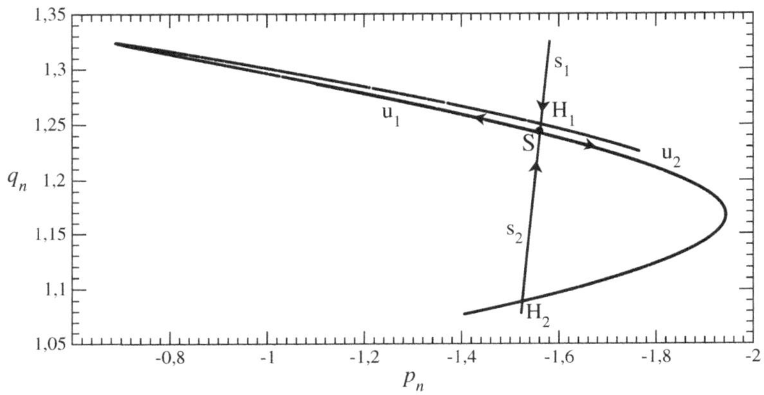

(b) Displaying of 20000 iterates from an initial condition

Fig. 2.15 Numerical calculation of Eq. (2.10). Compare to Fig. 2.7. In (**b**), are plotted the fixed point S ($p_* = -1.56, q_* = 0.994$), the stable and unstable manifolds (s_1, s_2, u_1, u_2), and two homoclinic points (H_1, H_2). Parameters: $\varepsilon' = 0.05, d = 0.04, r = 0.2$; b and c are as in Fig. 2.14

overlap in the direction of P (in Fig. 2.13). Such larger values of b do, however, no longer create a long enough 'handle' of the 'walking-stick' in Fig. 2.15 in order for this map still to generate chaos. For more realistic (that is, smaller) values of ε', only two major changes occur in the appearance of Fig. 2.15, however: (i) the width of the first iterate at the 'knee' decreases proportionally to ε'; (ii) the slopes of the 'feet' of the two legs of the first iterate becomes more and more vertical, approaching infinity as ε' approaches zero (compare to Fig. 2.9). This latter limit is, of course, the same for all diffeomorphisms of the type of Eq. (2.12)—hence the realistic touch of Fig. 2.15.

After having thus succeed in reducing the behavior of at least one chaotic trajectorial flow (that of Fig. 2.13) to a process of the type of Fig. 2.6 (a taffy puller), it is perhaps time for us to look at a slightly more sophisticated picture of the present prototypic map: Fig. 2.13c. One there sees a saddle point and two pairs of singular lines emerging from it or approaching it, respectively. The lines hitting the saddle, two under positive t, two under negative t, were found numerically by trying out different initial conditions at the top and at the bottom of the box until finally only the last digit in the value of p determined whether the saddle point would be passed to the left or to the right under further iteration (similarly for the saddle's unstable manifolds which can be determined by replacing the map of Eq. (2.10) by its time-inverted form—multiplying the right hand sides by -1 and adding $-2p_n$ or $-2q_n$, respectively.). These two pairs of singular lines obviously intersect in two points that are doubly asymptotic to the saddle (namely, for positive t asymptotic on the approaching line, and for negative t asymptotic on the inverted departing line). This 'leaning on' (Greek *klinein*) toward the 'same' (*homos*) object was called *homoclinic* by Poincaré [13]. Poincaré also gave the first discussion of the implications of the existence of a homoclinic point in a cross section (see Sect. 7.1). Here it suffices to note that the 'dollar-shaped' ($) set of invariant lines of Fig. 2.13c is a general characteristic of two-dimensional orientable diffeomorphisms of walking-stick shape producing a connected chaotic regime [23].

References

1. A.A. Andronov, A.A. Vitt, S.E. Chaikin, *Theory of Oscillators* (Pergamon Press, Oxford, 1966)
2. S.E. Chaikin, Continuous and discontinuous oscillations. Zhurnal Prikladnoi Fiziki **7**, 6–21 (1930)
3. S.S. Chern, S. Smale, Global analysis, in *Proceedings of the Symposia in Pure Mathematics*, vol. 14 (AMS, 1970)
4. D.R.J. Chillingworth, *Differential Topology with a View to Applications*, vol. 9, Research Notes in Mathematics (Pitman, London, 1976)
5. I. Gumowski, C. Mira, Sensitivity problems related to certain bifurcations in the nonlinear recurrence relations. Automatica **5**, 303–317 (1969)
6. M. Hénon, A two-dimensional mapping with a strange attractor. Commun. Math. Phys. **50**, 69–78 (1976)
7. M.W. Hirsch, S. Smale, *Differential Equations, Dynamical Systems, and Linear Algebra* (Academic Press, New York, 1974)
8. F.C. Hoppensteadt, J.M. Hyman, Periodic solutions of a logistic difference equation. SIAM J. Appl. Math. **32**, 73–81 (1977)
9. W. Jentsch, *Digitale Simulation Kontinuierlicher Systeme* (R. Obldenburg Verlag, Munich, 1969)
10. N. Minorski, *Nonlinear Oscillations* (R. E. Krieger, Huntington, New York, 1974)
11. C. Mira, Dynamique complexe engendrée par une équation différentielle d'ordre 3, in *Proceedings Equadiff'78*, eds. by R. Conti, G. Sestini, G. Villari (Florence, 1978), pp. 25–37
12. R.E. O'Malley, *Introduction to Singular Perturbation* (Academic Press, New York, 1974)
13. H. Poincaré, *Les méthodes nouvelles de la mécanique céleste*, vol. III (Gauthier-Vilars, 1899)
14. N. Rashevsky, über Hysterese-Erscheinungen in Physikalish-chemischen Systemen. Zeitschrift für Physik **53**, 102–106 (1929)
15. O.E. Rössler, Chaos, in *Structural stability in Physics*, eds. by W. Güttinger, H. Eikemeier (Springer, Berlin, 1979), pp. 290–309
16. O.E. Rössler, Chaotic behavior in simple reaction systems. Zeitschrift für Naturforschung A **31**, 239–264 (1976)

17. O.E. Rössler, Chemical turbulence. Synopsis, in *Synergetics. A Workshop*, ed. by H. Haken (Springer, Berlin, 1977), pp. 184–197
18. O.E. Rössler, Chemical turbulence: chaos in a simple reaction-diffusion system. Zeitschrift für Naturforschung A **31**, 1168–1172 (1976)
19. O.E. Rössler, Continuous chaos, in *Synergetics. A Workshop*. ed. by H. Haken (Springer, Berlin, 1977), pp. 174–183
20. O.E. Rössler, Different types of chaos in two simple differential equations. Zeitschrift für Naturforschung A **31**, 1664–1670 (1976)
21. O.E. Rössler, Quasi-periodic oscillation in an abstract reaction system (abstract). Biophys. J. A **17**, 281 (1977)
22. O.E. Rössler, Chaotic oscillations: an example of hyperchaos. Lect. Appl. Math. **17**, 141–156 (1979)
23. O.E. Rössler, Chaos in abstract kinetics. Two prototypes. Bull. Math. Biol. **39**, 275–289 (1979)
24. S. Smale, Dynamical systems and the topological conjugacy problem for diffeomorphisms, in *Proceedings of the International Congress of Mathematics* (Stockholm, 1962) (Institut Mittag-Leffler, Djursholm, 1963), pp. 440–496
25. S. Smale, Differentiable dynamical system. Bull. Am. Math. Soc. **73**, 747–817 (1967)
26. G. Targonski, *Topics in Iteration Theory* (Vandenhœck & Raprecht, Gottingen, 1981)
27. R. Thom, *Stabilité structurelle et Morphogénèse* (Interéditions, Paris, 1972)

Chapter 3
The Lorenzian Paradigm

3.1 Lorenz Chaos

It would be astonishing if the just described 'royal way' to chaos was the only one. And it would also be astonishing if the next alternative—Lorenzian chaos—would not at the same time reveal a second principle—a whole second world. Lorenzian chaos is based on a 'doors in state space' principle. We will see it in operation in the Lorenz equation itself, and then in another similarly acting equation.

First the original equation. It was found empirically by Lorenz [1]—by Saltzman [2] who however did not look at its behavior—in an attempt to understand a particular hydrodynamic system (Rayleigh–Bénard[1] convection) described by particular differential equations, the Navier–Stokes equations of fluid motion (see [3]). Proceeding in the usual way—going to a Hilbert space representation with an infinity of modes, and then carefully selecting three—, these authors came up with the following simple quadratic three-variable differential equation:

$$\left\{ \begin{array}{l} \dot{x} = \sigma y - \sigma x \\ \dot{y} = Rx - y - xz \\ \dot{z} = -bz + xy \, . \end{array} \right. \tag{3.1}$$

The three constants σ, R, b represent the well-known hydrodynamic parameters: Prandtl number, dimensionless Rayleigh number and an aspect ratio, respectively. It should be added that the values of these constants chosen to reveal the 'astonishing persistence' [1] of the main phenomenon of turbulence, nonperiodicity are not quite realistic physically [4]. Moreover, the 'naturalness' of the particular mode selection chosen by Saltzman and Lorenz was a problem until recently (when Curry [5] first looked at a much larger number of modes and still found chaos). Worth noting is

[1]In fact, Saltzman already observed irregular oscillations, but did not pay attention because its model was too rough for providing realistic solutions for the large Rayleigh number at which these irregular oscillations were observed.

O. E. Rössler and C. Letellier, *Chaos*, Understanding Complex Systems,
https://doi.org/10.1007/978-3-030-44305-4_3

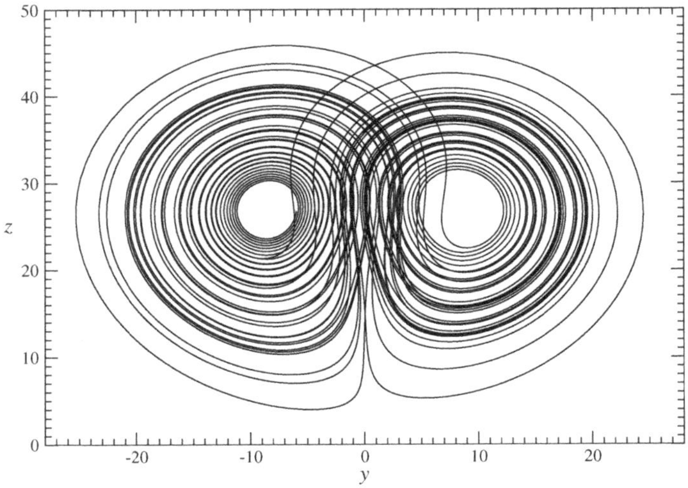

Fig. 3.1 Lorenzian chaos in the Lorenz equation (3.1). Parameter values: $\sigma = 10$, $R = 28$, and $b = \frac{8}{3}$

also that in plasma turbulence, a single mode (involving three variables in this case) suffices for chaos [6].

Figure 3.1 shows the appearance in state space of Lorenz type chaos, with Lorenz' original parameters [1]. The system obviously is highly symmetrical—as a look at Eq. (3.1) confirms.[2] While formerly (in Eq. (2.2), say) we had a *single* spiral that was carrying overcritical amplitudes back toward its own side (Fig. 2.3) we now have two spirals that each send their overcritical amplitudes toward the sides of the other. However, there is also a deeper difference: while formerly, the 'overcritical amplitude' was determined by a (smooth) folding over, we now have a sharp fracture instead: there is a saddle point present which separates the two unstable spirals. (If we had waited long enough in the simulation of Fig. 3.1, some trajectories would have almost hit the saddle, which lies at the origin of the state space.) This saddle point in three-dimensional space acts like a trap door that carries trajectories crossing it into a different garden, so to speak. In the present case, the other garden looks like a mirror image of the first.

[2]There is in fact an order-2 rotation symmetry around the z-axis: the system is said to be equivariant, meaning that $\Gamma \cdot f(\boldsymbol{x}) = f(\Gamma \cdot \boldsymbol{x})$ where $f(\boldsymbol{x})$ is the Lorenz vector field and Γ a 3×3 matrix defining the rotation symmetry according to

$$\Gamma = \begin{bmatrix} -1 & 0 & 0 \\ 0 & -1 & 0 \\ 0 & 0 & +1 \end{bmatrix}.$$

See [7] for details.

3.2 An Analogue to the Lorenz Equation

If one wants to completely understand a three-dimensional system, one has to look for a perturbation method. Apart from singular perturbation (underlying the reinjection principle), ordinary perturbation can also be tried. The following equation was found in this way. When looking at Eq. (3.1) and trying to get a 'feeling' for what really goes on dynamically between its variables, one is struck by the high symmetry of interaction between the first two variables. If these variables were, for example, chemical substances, one could call them 'cross-activating,' because they would jointly form an autocatalytic system. This suggests to 'collaps' them into a single variable and see what happens.

The proposed condensation leads to

$$\begin{cases} \dot{y} = ay - yz \\ \dot{z} = y^2 - bz\,, \end{cases} \tag{3.2}$$

where y now stands for the (contracted) two former variables x and y. The behavior of the contracted system is presented in Fig. 3.2. One sees a saddle point that separates two stable foci. (This may be confirmed with analytical methods; cf. Rosen [8], or Hirsch and Smale [9].[3]) More interestingly, the two horizontal unstable manifolds of the saddle point both come close to the descending stable manifold (and each other) before finally spiralling into their ipsilateral[4] focus.

Enter now a slight asymmetry-generating perturbation. Obviously, such a perturbation will be sufficient to push one of those trajectories that lie close to the separating middle line over into the other basin. After the trajectory has then returned toward the middle line from the other side, the time-dependent perturbation gets another chance to throw the trajectory back into the first basin. This throwing back and forth may go on several times in a non-repetitive fashion if small periodic perturbation is assumed to be added to the right-hand side of the first line of Eq. (3.2). It now appears possible that in the original, three-variable version of (3.1), an analogous time-dependent perturbation is somehow generated endogenously—perhaps by the lack of a perfect symmetry between the first two variables.

The wild hypothesis may be checked by adding (or removing) terms on the right hand side of Eq. (3.1) such that finally perfect symmetry is reached, and then experimentally re-introducing the minimum amount of asymmetry needed to reobtain the

[3] System (3.2) has one singular point located at the origin of the state space: it is associated with the eigenvalues $\lambda_1 = a$ and $\lambda_2 = -b$, thus corresponding to a saddle point. There are two other singular points ($y_\pm = \pm\sqrt{ab}$, $z_\pm = a$) associated with a pair of complex conjugated eigenvalues

$$\lambda_\pm = \frac{-b \pm \sqrt{b^2 - 8ab}}{2}\,.$$

These two symmetry-related points are stable focus type points when $b < 8a$; there are saddle points otherwise.

[4] Ipsilateral: pertaining to, situated on, or affecting the same side of the body.

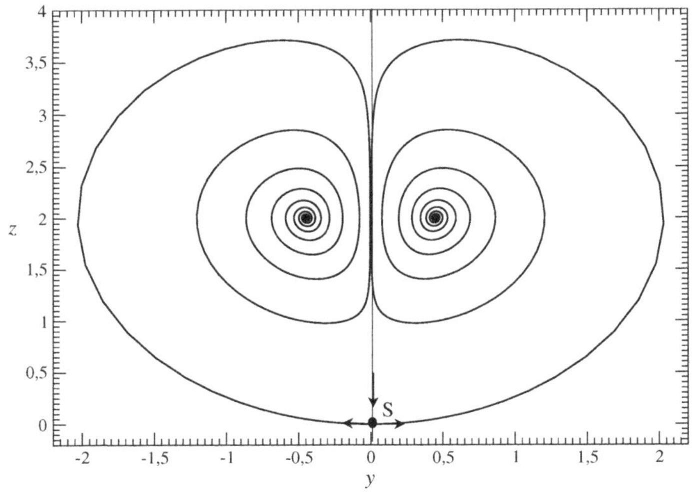

Fig. 3.2 Trajectorial flow in Eq. (3.2). Parameter values: $a = 2$, and $b = 0.1$. Initial conditions: $y_0 = -10^{-10}$, $z_0 = 0$ (left half picture); $y_0 = 10^{-10}$, $z_0 = 0$ (right half picture). S = saddle point

irregular switching behavior. An alternative easier way is to take Eq. (3.2) and simply add a third variable as an ad hoc perturbation [10]:

$$\begin{cases} \dot{x} = ay - dx + e \\ \dot{y} = y - cx - yz \\ \dot{z} = y^2 - bz \, . \end{cases} \tag{3.3}$$

The first variable follows y with a time-lag. If c is sufficiently small x can be considered as an 'endogenous perturbation.' A simulation is shown in Fig. 3.3.

3.3 Two Internal Blue-Sky Catrastrophes

Figure 3.4 shows how the mode of action of Eq. (3.3) could be rationalized in a perturbation context. Assuming a two-dimensional limit cycle sub-system of the type shown in the middle of Fig. 3.4, it should be possible to find a set of parameter values that causes one of the two limit cycles to grow until Abraham's 'blue sky catastrophe' [11] occurs (so that all trajectories formerly to the right of the saddle are bound to switch over to the left, owing to the disappearance of the corresponding limit cycle).[5] The whole three-dimensional flow, with the variable x infinitely weakly coupled to y, then yields to a non-periodic flow of Lorenz type (although with infinitely long 'inter-transition times').

[5] We were not able to find such a scenario in Lorenz-like system.

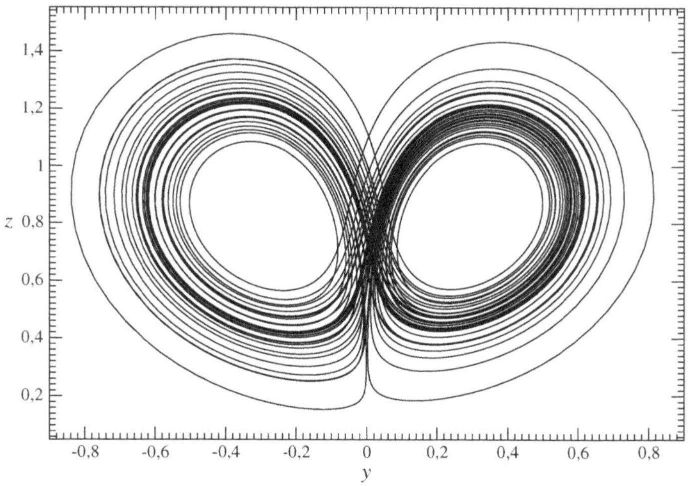

Fig. 3.3 Lorenzian chaos in Eq. (3.3). Parameter values: $a = 1$, $b = 0.1$, $c = 0.08$, $d = 0.44$, and $e = 0$. Initial conditions: $x_0 = 0.125$, $y_0 = 0.2$, and $z_0 = 0.4$

Fig. 3.4 Proposed mode of action of Eq. (3.3), idealized. At two different parameter values, a stable limit cycle coalesces with the saddle and disappears 'into the blue.' If the variable x is (arbitrarily weakly) coupled to y, these two bifurcations should occur 'internally,' generating Lorenz chaos

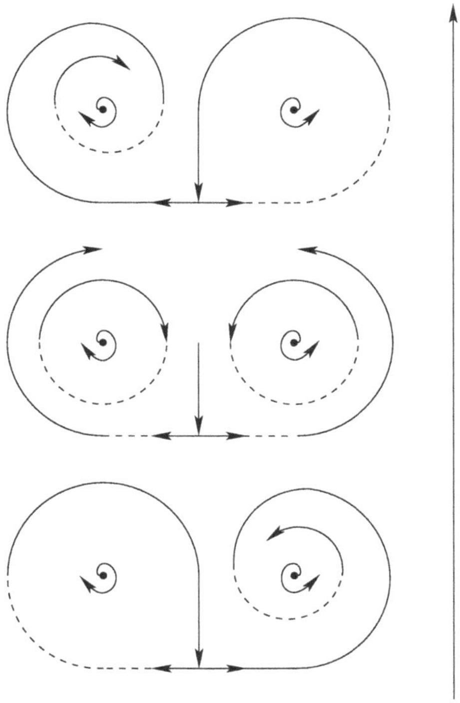

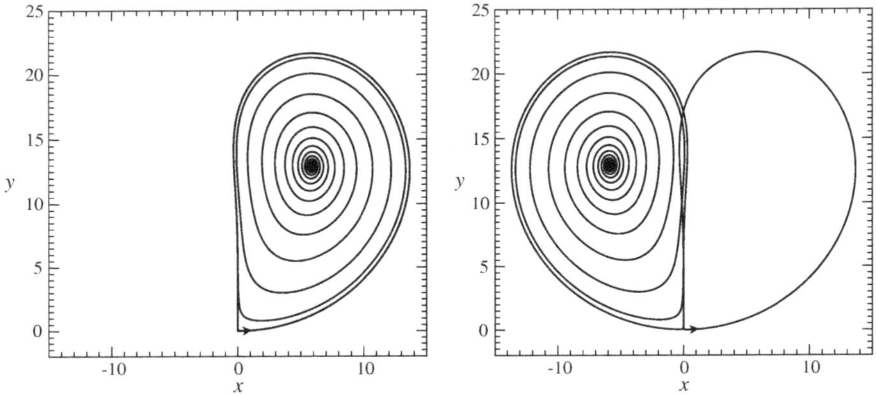

(a) $R = 13.926557$ (b) $R = 13.926558$

Fig. 3.5 Homoclinic connection to the singular point located at the origin of the phase space associated with the Lorenz system (3.1). Other parameter values: $\sigma = 10$ and $b = \frac{8}{3}$. Initial conditions: $x_0 = 0$, $y_0 = 10^{-4}$, and $z_0 = 0$. Time step: $\mathrm{d}t = 5 \cdot 10^{-3}$ s

Now Eq. (3.2) does not contain a stable limit cycle on either side (only a stable focus). But adding variable x should generate such a limit cycle on one side. Since x is relatively fast-changing, the fact that on the other side there is no full-fledged limit cycle (only a stable focus) waiting 'does not matter' since the latter soon loses stability anyhow.[6] In the original Lorenz equations (3.1), a homoclinic connection can be actually found but there is not limit cycle that disappears during that connection [12]. For instance, for $R = R_{\rm H} \approx 13.926557$, such a connection is observed (Fig. 3.5). In this example, the unstable orbit that exists is not ejected to infinity for disappearing in the blue sky.

A transition of the type idealized in Fig. 3.4 was in fact observed in the perturbed van der Pol oscillator [13], written in the Liénard form [14]

$$\left\{ \begin{array}{l} \dot{x} = -y + C \\ \dot{y} = \kappa x + \alpha(\beta - x^2)y \end{array} \right. \tag{3.4}$$

when parameter C was varied [15]. This system has three singular points

$$\mathrm{S}_0 = \left| \begin{array}{l} x_0 = 0 \\ y_0 = C \end{array} \right. , \text{ and } \mathrm{S}_\pm = \left| \begin{array}{l} x_\pm = \frac{1}{2} \left[\dfrac{\kappa}{\alpha C} \pm \sqrt{\left(\dfrac{\kappa}{\alpha C}\right)^2 + 4\beta} \right] \\ y_\pm = C \end{array} \right. ,$$

[6]These features were not found in the system (3.3). The subsequent part of this section and the next one were rewritten for the present edition of the original manuscript, trying at our best to preserve the original views.

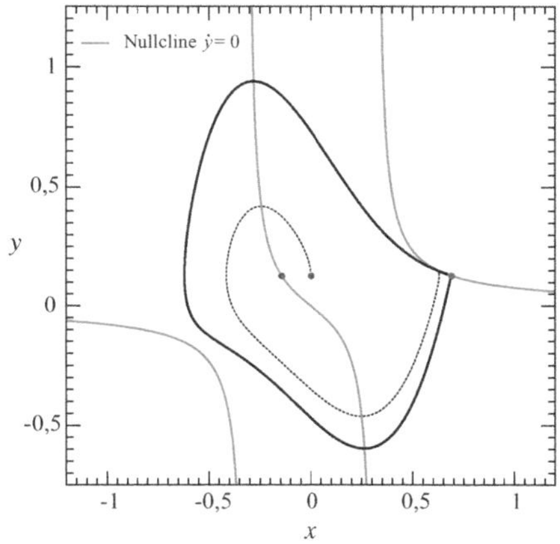

(a) $C = 0.1288$: limit cycle

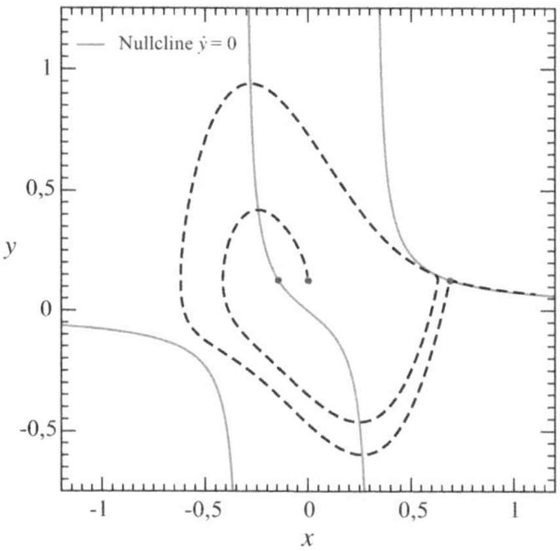

(b) $C = 0.1289$: ejection to the blue sky

Fig. 3.6 Blue sky catastrophe observed in the van der Pol oscillator perturbed by Abraham and Stewart. Other parameter values: $\kappa = 0.7$, $\alpha = 10$, and $\beta = 0.1$

Varying C from 0.10 to 0.14, there is a homoclinic saddle connection to point S_+, and there is a catastrophic disappearance of the limit cycle into the "blue sky" (Fig. 3.6). This is not exactly true since the limit cycle is destroyed at the homoclinic connection and the trajectory is then sent to the blue sky.[7]

This transition (which is shown in Fig. 3.6) is a simpler analogue to the blue sky catastrophe. Just as the van der Pol oscillator 'automatically traverses the homoclinic saddle connection' [16], so does the analogous, more complicated scheme of Fig. 3.4.

There are, of course, differences. Especially, there is the nonperiodicity which has to be assured. Today a blue sky catastrophe—one of the main bifurcations of periodic orbits—is associated with one orbit whose period and length become infinite before the bifurcation through which it disappears. In the problem 37 from the list established by Palis and Pugh [17], a blue sky catastrophe corresponds to an orbit whose time period becomes infinite at the bifurcation point without approaching any of the singularities of the vector field, thus implying an infinite length. This is not the case in the example previously discussed because only its time period becomes infinite while the orbits reaches the neighborhood of the saddle singular point located at the origin of the state space.

A blue sky catastrophe can be actually found in a Lorenz-like system as the one proposed by Leonov [18]

[7] Such homoclinic connection (the unstable manifold of a saddle connects with its stable manifold) is sometimes (wrongly?) associated with the blue sky catastrophe. Otto E. Rössler is quoted by Ralph Abraham and Bruce Stewart to be the first to have observed a blue sky catastrophe in differential equations. But for Abraham and Stewart, blue sky catastrophe meant [15]: "*an infinitesimal change in a* [...] *parameter eradicates a chaotic attractor from the phase portrait.*".

$$\begin{cases} \dot{x} = y \\ \dot{y} = x - \lambda y - xz - x^3 \\ \dot{z} = x - \beta z - \alpha y^2 \end{cases} \tag{3.5}$$

where the parameters can be expressed in terms of a single one, s, as

$$\begin{cases} \alpha = \sqrt{1-s} \\ \beta = 1 + 2\sin\left[k\pi(1-s)\right] \\ \lambda = \dfrac{s}{\sqrt{1-s}} \,. \end{cases} \tag{3.6}$$

This allows to investigate the dynamics of this system versus a single parameter s. This system has a rotation symmetry as the Lorenz system. The different behaviors are shown in Fig. 3.4. This system has three singular points

$$S_0 = \left| \begin{array}{l} x_0 = 0 \\ y_0 = 0 \\ z_0 = 0 \end{array} \right. , \text{ and } S_\pm = \left| \begin{array}{l} x_\pm = \pm 1 \\ y_\pm = 0 \\ z_\pm = 0 \end{array} \right. .$$

Points $S_\pm$ are associated with complex conjugated eigenvalues for $-6 - 4\sqrt{3} < s < -6 + 4\sqrt{3} = 0.928$; the real part is negative for positive values of s within this interval.

With $k = 2$ (kept constant during this analysis) and for $s = 0.8564$, the two singular points $S_\pm$ become unstable (*via* another bifurcation than a Hopf bifurcation) and there is a resulting limit cycle around each of them. These two symmetry-related limit cycles grow in size up to reaching the point S_0, leading to a homoclinic connection ($s \approx 0.8420$ as shown Fig. 3.7a). After this bifurcation, there is a single symmetric limit cycle (Fig. 3.7b). There are then few different successive limit cycles up to the period-10 limit cycle returning in the neighborhood of the saddle point S_0, leading to a new homoclinic bifurcation inducing a symmetry breaking and, then, two symmetry-related period-12 limit cycles ($s = 0.7730$, Fig. 3.4d). Once the symmetry of the solutions is broken, there is a period-doubling cascade leading to a homoclinic chaos (Fig. 3.7e), which then develops into a "Burke and Shaw attractor"[8] ($s = 0.6620$, Fig. 3.7f). After various bifurcations, there is a symmetric period-4 limit cycle (Fig. 3.7g), close to a homoclinic bifurcation. Then, there is a period-2 limit cycle ($s = 0.60195$, Fig. 3.7h) and, a new homoclinic bifurcation, breaking the symmetry, leads to two symmetry-related period-1 limit cycles (Fig. 3.7i). Another homoclinic bifurcation leads to two period-1 limit cycles, one surrounding each singular points, S_- and S_+, respectively, as observed for $0.8564 > s > 0.8420$. Nevertheless, these limit cycles are not the same in the two situations. A last homoclinic bifurcation leads to a symmetry period-1 limit cycle ($s = 0.064335$, Fig. 3.7k), which

[8]This type of chaotic attractor was observed in a system introduced by Bill Burke and Robert Shaw [19] and later topologically characterized [20].

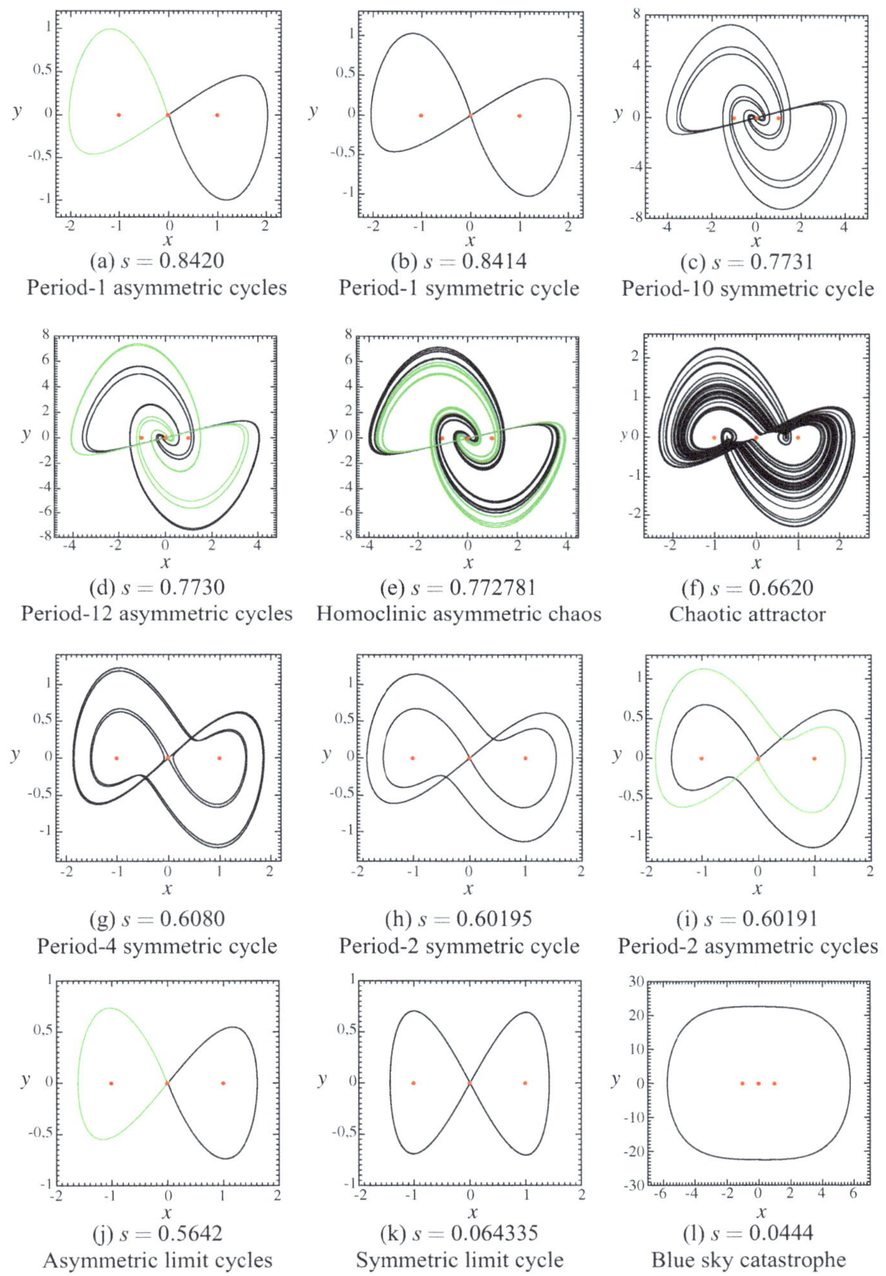

(a) $s = 0.8420$ Period-1 asymmetric cycles

(b) $s = 0.8414$ Period-1 symmetric cycle

(c) $s = 0.7731$ Period-10 symmetric cycle

(d) $s = 0.7730$ Period-12 asymmetric cycles

(e) $s = 0.772781$ Homoclinic asymmetric chaos

(f) $s = 0.6620$ Chaotic attractor

(g) $s = 0.6080$ Period-4 symmetric cycle

(h) $s = 0.60195$ Period-2 symmetric cycle

(i) $s = 0.60191$ Period-2 asymmetric cycles

(j) $s = 0.5642$ Asymmetric limit cycles

(k) $s = 0.064335$ Symmetric limit cycle

(l) $s = 0.0444$ Blue sky catastrophe

Fig. 3.7 Blue sky catastrophe in Eq. (3.5). After few homoclinic bifurcations, a limit cycle grows in size and coalesces with an unstable periodic orbit and disappears 'into the blue sky.' Other parameter value are provided by Eq. (3.6), and $k = 2$

starts to grow in size. When this limit cycle is sufficiently large, it has its stability, most likely *via* a boundary crisis with an unstable periodic orbit, and the trajectory is ejected to infinity, the limit cycle disappearing in the 'blue sky'.[9]

3.4 A Twin System

The findings of the last Section point to the existence of an underlying principle. The 'trapdoor principle' in its most general form says that saddle points (and their generalizations) can act like 'sluices' in state space through which trajectories can enter another basin (which then, too, may after a while turn out to possess a hatch and hence not be a genuine basin).

Stated in this form, the trapdoor principle is closely related to the reinjection principle. It has the advantage that it is not derived from a lower-dimensional representation, and the disadvantage that a rigorous (perturbation) formulation is not easy.

Heuristically, a restricted version of the principle is helpful if the finding of a new chaotic system is at stake. The procedure goes like this: take an arbitrary two-variable system containing two basins separated by the stable manifold of a saddle, and add a third variable that weakly perturbs the two in such a way that the (now three-dimensional) boxes cease to be air-tight. Usually, only one of the two boxes formed becomes leaky, but in symmetrical systems a symmetrical leakiness is also quite easy to obtain. This mutual leakiness then allows for a 'shuttle service', that is, for either periodic (of more or less complicated type) or even chaotic trajectories.

One such example may be obtained from the original Lorenz equations (3.1) by rewriting the Lorenz equations in a jerk form starting from variable x [21].[10] The equations are

[9] All this scenario indicates that, most likely, it would be very difficult to observe a blue sky catastrophe occurring in a Lorenz-like system by a crisis with a singular point as observed by Abraham and Stewart in their modified van der Pol oscillator. We kept here this section about the blue sky catastrophe to preserve the original table of contents.

[10] The equations here reported are not those initially intended by Otto Rössler, but numerical simulations were performed and it appeared while editing this book that the original equations were not producing the expected dynamics. We therefore replaced the original equations by the rational system obtained from the Lorenz system using the change of coordinates

$$\left| \begin{array}{l} x = X \\ y = X + \dfrac{Y}{\sigma} \\ z = (R-1) - \dfrac{(\sigma+1)Y}{\sigma X} - \dfrac{Z}{\sigma X} \,. \end{array} \right.$$

Variable x was chosen because it provides the simplest and closest form to what Otto initially wrote. The text was left unchanged as much as possible.

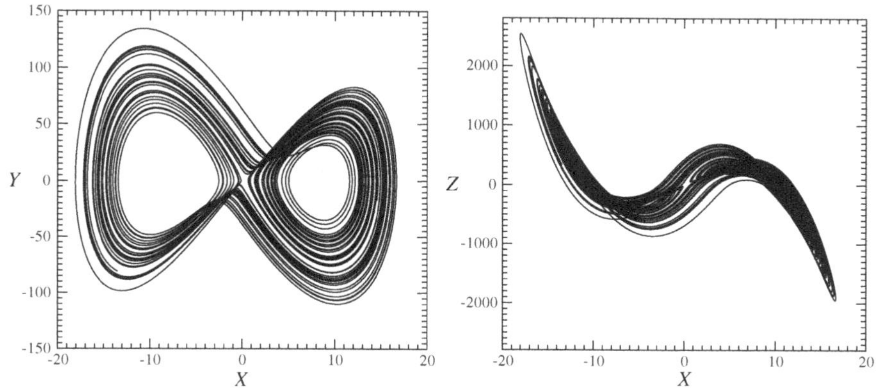

(a) 'Top-down' view (b) 'Side' view

Fig. 3.8 Trajectorial flow in the 'twin system' (3.7) of Eq. (3.1). Two different views of the same flow are presented. Parameter values: $R = 28$, $\sigma = 10$, and $b = \frac{8}{3}$

$$\begin{cases} \dot{X} = Y \\ \dot{Y} = Z \\ \dot{Z} = (R-1)b\sigma X - b(\sigma+1)Y - (b+\sigma+1)Z - \sigma X^3 - X^2 Y \\ \qquad + (\sigma+1)\dfrac{Y^2}{X} + \dfrac{YZ}{X} \end{cases} \tag{3.7}$$

This is an analogue to the original Lorenz system (3.1). Fig. 3.8 gives a simulation in two different views. A flow like this has, apparently, not yet been described. Other, more complicated examples (based on three basins, or with less symmetry) can be expected to be easy to find also.[11]

Focusing on the picture of Fig. 3.8a, one sees that the present chaotic flow is not actually identical with that of Fig. 3.3. Instead of a 'letter B' type underlying structure (rotated by $-\frac{\pi}{2}$), we now have a 'number 8' type one (also rotated by $-\frac{\pi}{2}$). More specifically, the two 'spirals' involved (one on either side) circle not in opposite directions (as in Fig. 3.3), but in the same (clockwise) direction.

The whole flow therefore can be described by the schematic picture of Fig. 3.9 rather than that of Fig. 3.4. (There are again two homoclinic connections generated internally). However, this time it is possible to understand even more.

Equation (3.7) thus is even 'more ideal' than Eq. (3.1) is. It is not a further member of the 'zoo' of Lorenz-like equations, however, but illustrates a new—though related—class.[12] This is because the two spiralling subregimes involved (which each show a homoclinic connection at a certain parameter value, see one of them in Fig. 3.9) do not circle like in a mirror-inverted fashion (letter-B pattern),

---

[11]This was in fact done by Rick Miranda and Emily Stone, in 1993 [31].

[12]This was developed only recently by Claudia Lainscsek [22].

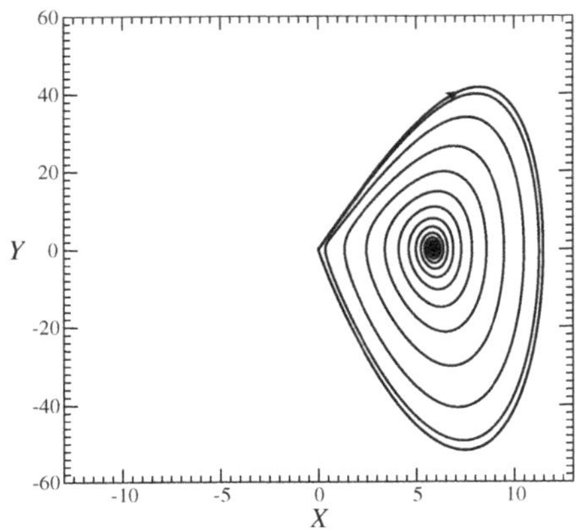

(a) $R = 13.926557$

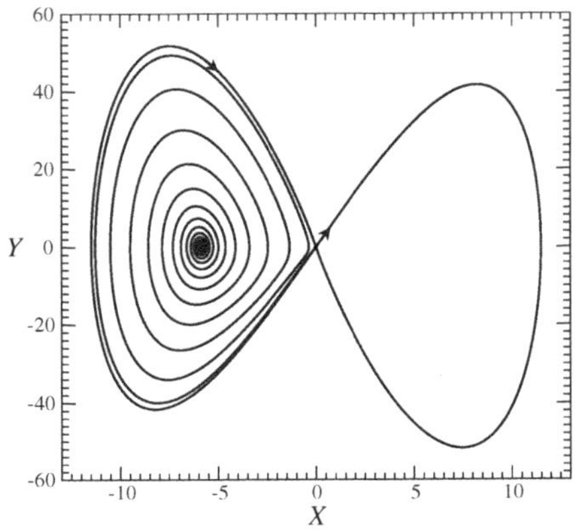

(b) $R = 13.926558$

Fig. 3.9 Mode of action of Eq. (3.7). Same parameter values and initial conditions as in Fig. 3.5

but rather possess the *same* rotation sense,[13] that is, are point-symmetric (figure-8 pattern).

The qualitative behavior is nonetheless evidently 'the same'. In fact, Eq. (3.7) can be understood as an idealized prototype (in the same sense as Eq. (3.3) is an idealized prototype), yet not to the Lorenz [1] equation (3.1), as Eq. (3.3) is, but to the Moore and Spiegel [24] equation. This equation was (independently of the Lorenz equation) found when its authors tried to model a certain mechano-hydrodynamic gadget (a 'balloon' diver' floating in a narrow, fluid-containing tube with convection generated by an external heat gradient) which in turn served as a crude model of solar flares. It has the form [24]

$$\begin{cases} \dot{x} = y \\ \dot{y} = z \\ \dot{z} = -az - by - cx - dx^2y\,. \end{cases} \tag{3.8}$$

where $a = 1$, $b = R - T$, $c = T$ and $d = R$, R and T being the two parameters used by Moore and Spiegel [24]. Like Eq. (3.7) it contains but a single cubic term on the right-hand side. A simulation is provided in Fig. 3.10a, with 'aperiodicity' generating values of the parameters indicated by Moore and Spiegel [24].

There are some differences to Fig. 3.8. The two 'tubes' do not grow thicker before ending (switch-over), but become narrower. As a consequence, what this equation shows is not actually Lorenzian behavior (cf. next Section), but a hybrid between Lorenzian behavior on the one hand and the simpler kind of folded-over chaos discussed previously (cf. Sect. 6.4). It thereby combines the disadvantage of both: 'non-diffeomorphicity' and 'pollutedness' (to be discussed later). Nonetheless, as Eq. (3.3)

[13]The twin system (3.7) is invariant under an inversion symmetry and not a rotation symmetry as the Lorenz system and the Leonov system (3.5) are. This is only in plane projection that an apparent mirror symmetry in Lorenz-like attractor can be observed. See for instance [23] for details.

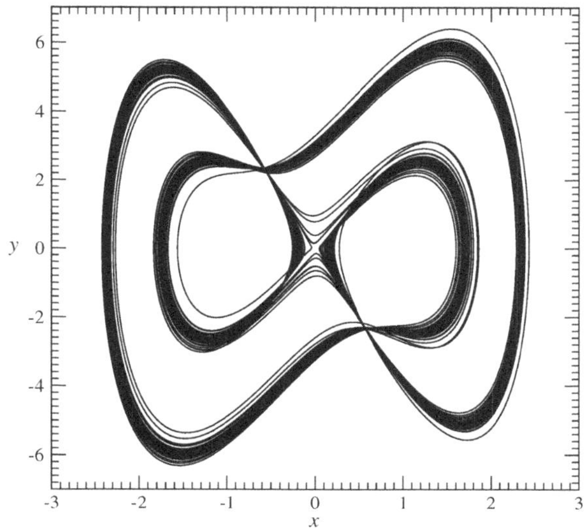

Fig. 3.10 Aperiodic behavior produced by the Moore and Spiegel equation (3.8). Parameter values $R = 19.2$ and $T = 6$

shows, it belongs into one ballpark with the 'twin-like' alternative to Lorenzian behavior that was illustrated (almost ideally) by Fig. 3.8.

3.5 Understanding Lorenzian Flows

Figure 3.11 shows three paper models designed to explain Lorenzian behavior, two taken from the literature [25, 26] and one devised ad hoc to explain the flow of Fig. 3.8a. So far, no example of Lorenzian behavior was yet found for the paper model of Fig. 3.11a (see, however, next Section).

The flow shown in Figs. 3.1 and 3.2 can be approximately understood with the aid of the paper model of Fig. 3.11b which in similar form was first indicated by Lorenz [1] and later by Williams [26]. In contrast, the flow of Fig. 3.8 leads to the paper model of Fig. 3.11c. Evidently, by folding down the upper half of Fig. 3.11c, the paper model of Fig. 3.11b is reobtained. Thus, both systems basically generate the same type of flow—a so-called Lorenz attractor. This is seen most easily by looking at the (smile-shaped) dashed line in Fig. 3.11b (which was first proposed as a valid cross section for Lorenzian flows by Guckenheimer [27]).

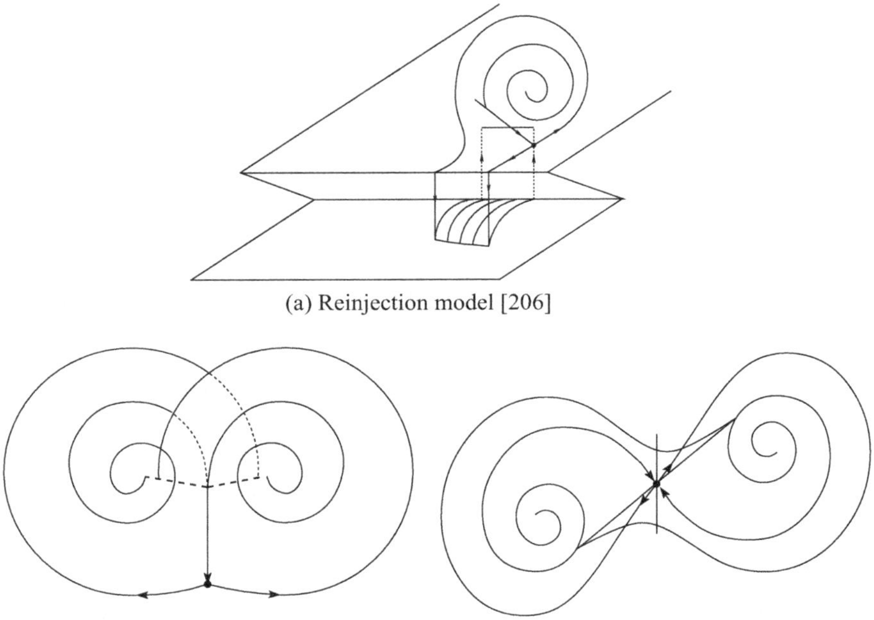

(a) Reinjection model [206]

(b) Classical paper model [226,118]

(c) Paper model for the flow in Figure3.8

Fig. 3.11 Paper models with Lorenzian behavior

This cutting line determines a next-return (Poincaré) map (Fig. 3.12b) that has the qualitative form seen in Fig. 3.12b.[14] It certainly is a chaos-generating one-dimensional map. It differs from the maps of Sects. 2.7 and 2.8 by its lacking a smooth maximum or minimum, as well as by the fact that the slope is everywhere larger than unity.[15] It therefore generates a type of chaos that certainly lacks any attracting trajectory: every pair of adjacent initial points will diverge exponentially under the iteration.

Attractors with this clear-cut chaotic property were first considered by Smale [28], although in the context of invertible maps (in fact, diffeomorphisms, that is, smooth maps that are smoothly invertible), whereby three dimensions rather one were needed. Later, Ruelle and Takens [29] introduced the term 'strange attractor' for the attactors generated by such maps and flows. The Lorenz attractor formed in Fig. 3.1 was the first example that could be conjectured to actually arise in an explicit differential equation (the Lorenz equation).

In reality, of course, both the flow of Figs. 3.1 and 3.8 are invertible for they are generated by differential equations with continuous right-hand sides (which implies

[14] Otto never computed numerically a first-return map to a Poincaré section and always drawn them amazingly accurately from the stereoscopic views he was using to investigate the flow.

[15] The map idealized from the numerically computed one has a slope that is equal to ± 2.

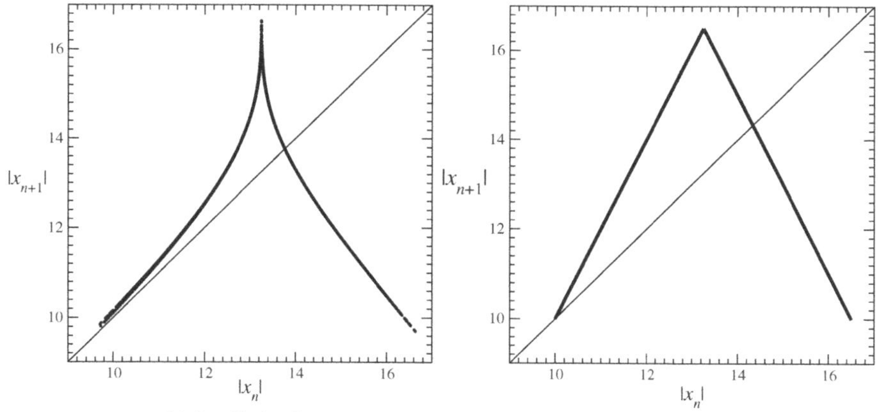

(a) Realistic shape (b) Idealized shape

Fig. 3.12 Computed and approximate Poincaré maps for Lorenzian flows

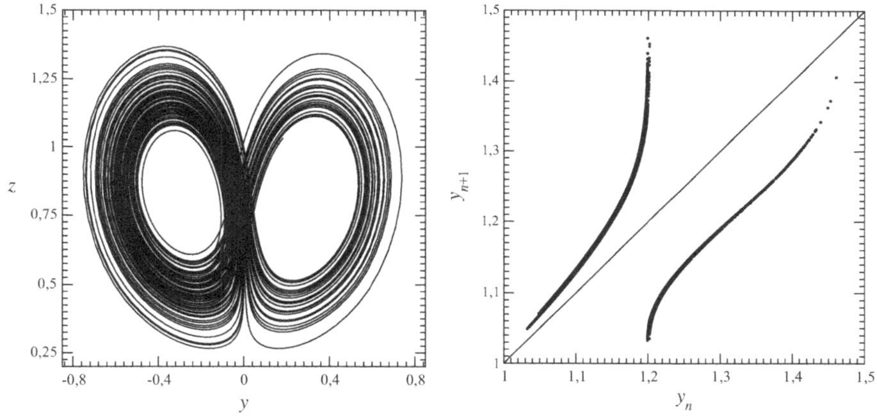

(a) Flow (b) Sandwich map

Fig. 3.13 Conjectured shape of a cross section through an asymmetric Lorenzian flow. Sandwich (**b**), first squeezed and then cut with dull knife (explaining the name 'sandwich map' for this map). Parameter values: $a = 0.08$, $b = 0.1$, $c = 1$, $d = 0.42$, and $e = 0.0015$

their solutions obeying a Lipschitz condition, cf. Hirsh and Smale [9]). The actually applying cross section therefore is two-dimensional and almost everywhere smoothly invertible. Its conjectured shape is shown in Fig. 3.13.

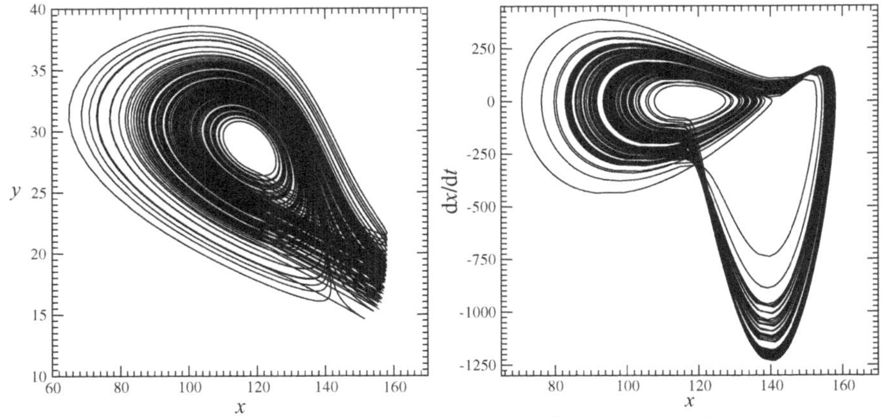

(a) x-y plane projection (b) x-$\dot{x}$ plane projection

Fig. 3.14 Unimodal cut chaotic attractor solution to system (3.9). Parameter values: $a = 33$, $b = 150$, $c = 1$, $d = 3.5$, $e = 4815$, $f = 410$, $g = 0.59$, $h = 4$, $j = 2.5$, $k = 2.5$, $l = 5.29$, $m = 750$, $K_1 = 0.01$ and $K_2 = 0.01$

3.6 A Lorenzian Flow Arising Under Less Symmetric Conditions

So far, the behavior of the Lorenz equation has been looked at without reference to the reinjection principle (Sect. 1.3).

$$\begin{cases} \dot{x} = ax + by - cxy - \dfrac{(dz + e)x}{x + K_1} \\ \dot{y} = f + gz - hy - \dfrac{jxy}{y + K_2} \\ \dot{z} = k + lxz - mz \end{cases} \tag{3.9}$$

We can conclude this Section with a more general statement of a qualitative style. Reinjection along "merging" two-dimensional stable manifolds of oscillatory saddle points[16] appears to be no less successful in producing chaotic flows in both abstract and natural systems, than in reinjection based on one-dimensional thresholds in two-dimensional singular-perturbation subsystems (or at least one such component) of the spiral and screw examples of dissipative chaos. Both principles are equally natural, although the above "touching class" is—owing to the mutuality of touching that is required—more complicated in terms of both the overall architecture of the flow ("doublewingedness") and the minimum of nonlinear terms required in the underlying equations: at least two. A third general "compository recipe" for generating

[16]Or unstable periodic orbits around them; in preparation.

chaotic flows was the—again simpler—"breathing oscillatory principle" (see Eq. (5) in [30]).

It goes without saying that "distorsions" can be introduced in a wide variety of ways, without doing damage to the complexity and beauty of the dissipative chaotic flows created, as the last picture (Fig. 3.14) shows as one case in point.

References

1. E.N. Lorenz, Deterministic nonperiodic flow. J. Atmos. Sci. **20**, 130–141 (1963)
2. B. Saltzman, Finite amplitude free convection as an initial value problem-I. J. Atmos. Sci. **19**, 329–341 (1962)
3. D.D. Joseph, Factorization theorems and repeated branching of solutions at a simple eigenvalue. Ann. N. Y. Acad. Sci. **316**, 150–167 (1979)
4. J.B. McLaughlin, P.C. Martin, Transition to turbulence of a statically stressed fluid system. Phys. Rev. Lett. **33**, 1189–1192 (1974)
5. J.H. Curry, A generalized Lorenz system. Commun. Math. Phys. **60**(3), 193–204 (1978)
6. P.K.C. Wang, Nonperiodic oscillations of Langmuir waves. J. Math. Phys. **21**, 398–407 (1980)
7. * C. Letellier, P. Dutertre, G. Gouesbet, Characterization of the Lorenz system taking into account the equivariance of the vector field. Phys. Rev. E **49**(4), 3492–3495 (1994)
8. R. Rosen, *Dynamical System Theory in Biology* (Wiley, New York, 1970)
9. M.W. Hirsch, S. Smale, *Differential Equations, Dynamical Systems, and Linear Algebra* (Academic, New York, 1974)
10. O.E. Rössler, Chemical turbulence: chaos in a simple reaction-diffusion system. Zeitschrift für Naturforschung A **31**, 1168–1172 (1976)
11. R. Abraham, J.E. Marsden, *Foundations of Mechanics*, 2nd enlarged edn. (Benjamin/Cummings, Reading, 1978)
12. * A.F. Vakakis, M.F.A. Azeez, Analytic approximation of the homoclinic orbits of the Lorenz system at $\sigma = 10$, $b = 8/3$ and $\rho = 13.926\ldots$. Nonlinear Dyn. **15**(3), 245–257 (1998)
13. B. van der Pol, On "relaxation-oscillations". Philos. Mag. **7**(2), 978–992 (1926)
14. A. Liénard, Etudes des oscillations entretenues. Revue Générale de l'Electricité **23**(901–912), 946–954 (1928)
15. * R.H. Abraham, H.B. Stewart, A chaotic blue sky catastrophe in forced relaxation oscillations. Phys. D **21**, 394–400 (1986)
16. O.E. Rössler, A synthetic approach to exotic kinetics, with examples, in *Lecture Notes in Biomathematics*, vol. 4 (1974), pp. 546–582
17. J. Palis, C.C. Pugh, Fifty problems in dynamical systems, in *Lecture Notes in Mathematics*, vol. 468 (1975), pp. 343–353
18. * G.A. Leonov, Cascade of bifurcations in Lorenz-like systems: Birth of a strange attractor, blue sky catastrophe bifurcation, and nine homoclinic bifurcations. Dokl. Math. **92**(2), 563–567 (2015)
19. * R. Shaw, Strange attractor, chaotic behavior and information flow. Zeitschrift für Naturforschung A **36**, 80–112 (1981)
20. * C. Letellier, P. Dutertre, J. Reizner, G. Gouesbet, Evolution of multimodal map induced by an equivariant vector field. J. Phys. A **29**, 5359–5373 (1996)
21. * G. Gouesbet, C. Letellier, Global vector field reconstruction by using a multivariate polynomial L_2-approximation on nets. Phys. Rev. E **49**(6), 4955–4972 (1994)
22. * C. Lainscsek, A new class of Lorenz-like systems. Chaos **22**, 013126 (2012)
23. * C. Letellier, R. Gilmore, Covering dynamical systems: two-fold covers. Phys. Rev. E **63**, 16206 (2001)

24. D.W. Moore, E.A. Spiegel, A thermally excited non-linear oscillator. Astrophys. J. **143**(3), 871–887 (1966)
25. F. Takens, Implicit differential equations: some open problems, in *Lecture Notes in Mathematics*, vol.535 (1976), pp. 237–253
26. R.F. Williams, The structure of Lorenz attractors. Publications Mathématiques de l'Institut des Hautes Études Scientifiques **50**(1), 73–99 (1979)
27. J. Guckenheimer, G.F. Oster, A. Ipaktchi, Periodic solutions of a logistic difference equation. J. Math. Biol. **4**, 101–147 (1976)
28. S. Smale, Differentiable dynamical system. Bull. Amer. Math. Soc. **73**, 747–817 (1967)
29. D. Ruelle, F. Takens, On the nature of turbulence. Commun. Math. Phys. **20**, 167–192 (1971)
30. O.E. Rössler, Continuous chaos - four prototype equations. Ann. N. Y. Acad. Sci. **316**, 376–392 (1979)
31. * R. Miranda, E. Stone, The proto - Lorenz system, Physics Letters A, **178**, 105–113, 1993

Chapter 4
Hyperchaos

4.1 An Equation for 'Hyperchaos'

As mentioned at the beginning, the reinjection is not confined to planar sub-systems. Reinjection processes between three-dimensional sub-systems, initiated along two-dimensional cross sections, are still quite easy to conceptualize and to generalize. The fact that two-dimensional reinjection processes can be drawn easily in two dimensions stands not alone. For example, Fig. 1.1 needed not have been drawn in a three-dimensional convention, with the two two-dimensional layers separated from each other like two different floors, but could as well have been generalized. In that case, it would only have been necessary to distinguish graphically between trajectories meant to lie on the lower plane, and those belonging to the upper plane. Dashing one of the two sets (the one with the shorter segments) would have been sufficient in order to obtain all necessary information about the flow of Fig. 1.1. In the same vein it is possible to represent a three-dimensional flow which at a two-dimensional threshold changes its structure in four space, by simply letting new—dashed—trajectories emerge from the same threshold. This time, one is using 'three-dimensional paper'—or jelly—to draw the two sets of trajectories, so to speak.

Just as in the two-dimensional setting, we are now confronted with the challenge of finding the simplest possibilities in the next higher dimension. A first candidate for new behavior is the flow depicted in Fig. 4.1. The underlying equation

$$\begin{cases} \dot{x} = -y - z \\ \dot{y} = x + 0.25y + w \\ \dot{z} = 2.2 + xz \\ \dot{w} = -0.5z + 0.05w \end{cases} \tag{4.1}$$

is identical to Eq. (2.2), with the variable x shifted by a constant in order to get rid of the third term in the third line, and with another linear variable added, w. The linear variables, x, y, w now form a three-dimensional flow, that has the form of an expanding screw. The remaining 'switching' variable, z, defines a horizontal

O. E. Rössler and C. Letellier, *Chaos*, Understanding Complex Systems,
https://doi.org/10.1007/978-3-030-44305-4_4

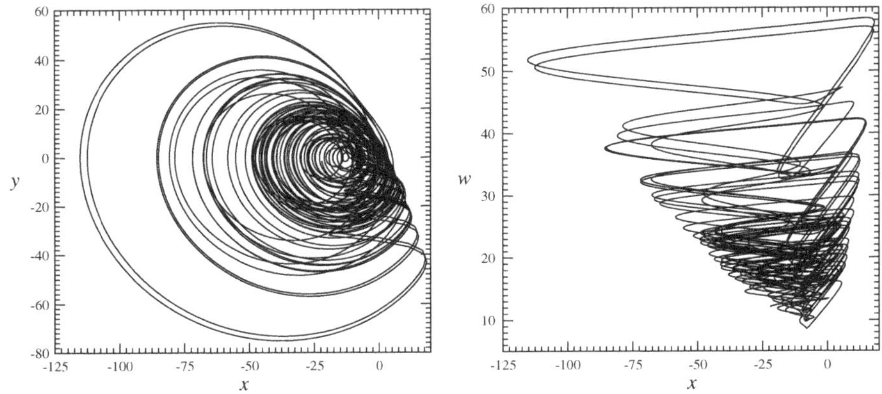

(a) x-y plane projection (b) x-w plane projection

Fig. 4.1 Hyperchaos in Eq. (4.1). Initial conditions: $x_0 = -10$, $y_0 = -6$, $z_0 = 0$, and $w_0 = 10.1$

threshold in x. Trajectories reaching it becomes dashed, meaning that they are now on the other three-dimensional sheet on which they slant backwards until they hit another (near parallel) threshold that is suspended inside the screw. Thereafter, the non-dashed flow takes over again, and so forth.

There is no doubt, that in the two-dimensional projection of Fig. 4.2, the same folding over-type reinjection takes place (despite a continual variation of size). However, there is the possibility of a second two-dimensional projection now as shown in Fig. 4.2. While Fig. 4.2 is not itself realizable in three dimensions, it is related to the three-dimensional flow depicted in Fig. 4.1. The latter shows (once more) the fact that the reinjection along the windings of a screw leads to chaos if during the reinjection, the interval between one winding is mapped back onto more than one winding interval. An analogous (though differently realized) 'screw-like reinjection' can be said to apply to the flow of Fig. 4.1.

Thus, in principle a new kind of chaos is possible in four dimensions: one composed of two 'ordinary' chaotic regimes being present simultaneously, so to speak. In Fig. 3.3, both 'chaos within windings' and 'chaos between windings' of the screw appear to be present simultaneously [1]. Of course, Eq. (4.1) is 'too' simple to be easily understood in detail. More complicated, but better 'decomposable,' examples are needed.

The simplicity of Eq. (4.1), on the other hand, allows to implement it in a rapid analogue computer and watch different projections on screen while listening to $z(t)$ via an amplifier and loudspeaker. The time behavior of z in Eqs. (2.2) and (4.1) is displayed in Fig. 4.3. The corresponding noises are those of an idling motorcycle (a), a snore (b), and a 'nasty snore' (c), respectively. Time-dependent spectra (sonograms) of the three have yet to be prepared. Farmer et al. [2] found that (a) contains both a broad component and a peak in its spectrum (compare with Swinney et al. [3], for a similar experimental result), while (b) is peak-free; (c) has yet to be analyzed. A

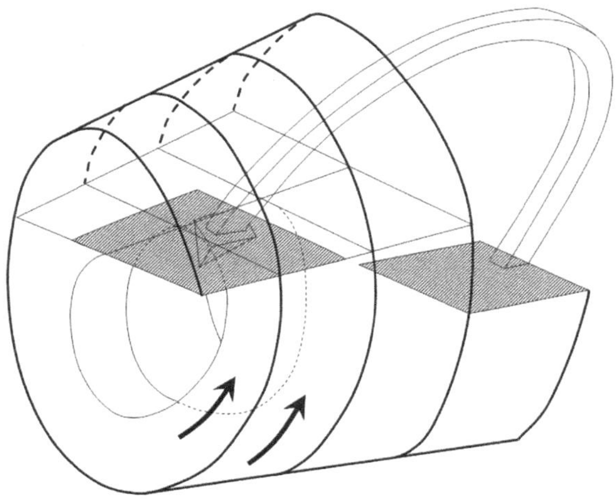

Fig. 4.2 Schematic explanation of the flow of Fig. 4.1

super-8 sound movie displaying Eqs. (2.2), (3.3) and (4.1) has been prepared [4, 5].[1] The flow of Eq. (4.1), by the way, possesses two positive characteristic exponents.[2]

4.2 Hyper Chaos—An Explicit Example

Visualizing all possible kinds of reinjection between three-dimensional flows will take its time. Equation (4.1) of the last Section was heuristically found and presented as such—allowing for a breathing pause. Returning now to geometry and 'pure' examples, I have to admit that, even though the job is easy in principle (suspending two kinds of wires between two plates), so far only clear-cut example has been visualized and, as a consequence, found. One way to proceed is to start out with two 'two-dimensional' reinjections already known to produce chaos and then try to preserve the behavior in the combined flow. It is possible that in this way, one of the most nontrivial three-dimensional possibilities can be hit up directly. On the other hand it is also possible that this is the surest way to miss the 'really new thing' that happens in four dimensions, whatever it is.

Figure 4.4 shows the two simplest 'two-dimensional' components to try: two identical 'Ø' flows, each consisting of a combination of straight lines (dashed) and parabolas (full lines) as originally presented, in full three-dimensional view in Fig. 2.5. The two 'specks' visible in Fig. 4.4 each correspond to a top-down view on the flow of Fig. 4.4, in the limit of ε in Eq. (2.8) tending to zero. The problem now is, simply, to

[1] See the movie made by Otto and Reimara Rössler in 1979 and available at https://www.youtube.com/watch?v=Tmmdg2P1RIM.

[2] Norman Packard, personal communication, 1979.

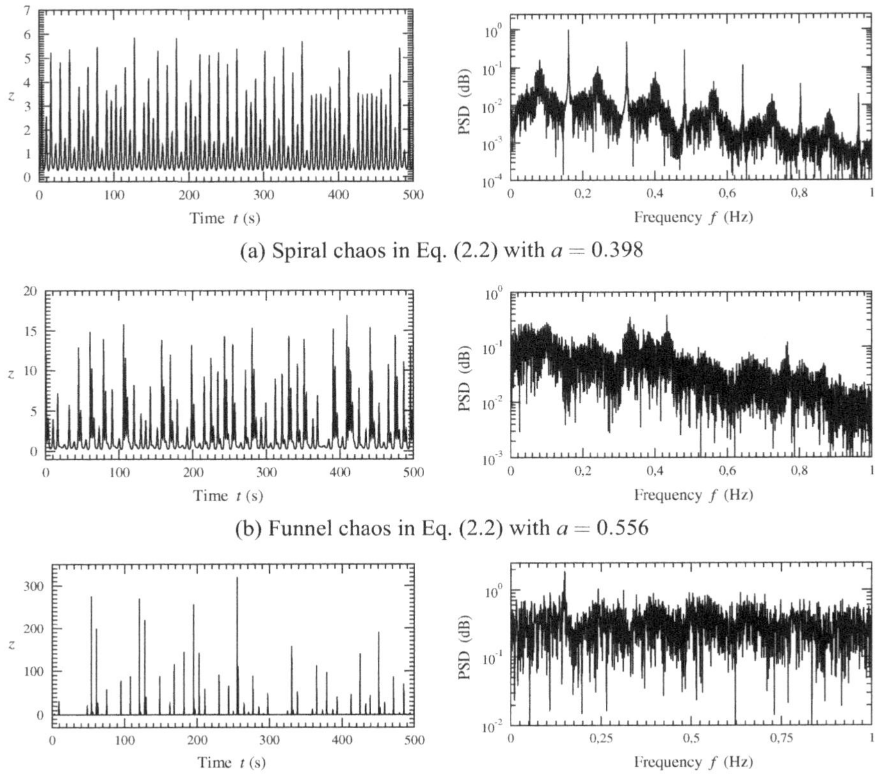

(a) Spiral chaos in Eq. (2.2) with $a = 0.398$

(b) Funnel chaos in Eq. (2.2) with $a = 0.556$

(c) Hyperchaos in Eq. (4.1)

Fig. 4.3 Time series and the corresponding power spectral densities of variable z of Eq. (2.2) for the spiral (**a**) and funnel (**b**) type of chaos, and of variable z of Eq. (4.1) (**c**). Other parameter values in Eq. (2.2): $b = 2$ and $c = 4$

put these two together in three-dimensional space in an appropriate 'symmetrical' fashion.

Thinking of each 'Ø' as a letter, two of them can be put together like a "P" and a "d", yielding the symmetrical pair Pd. To avoid interference, they should be positioned at right angles yielding $\widehat{\text{Pd}}$ (whereby the 'hat' symbolizes the orthogonality). Using paper and pencil, a pair of scissors, and glue, one comes up with Fig. 4.5.

The two 'wooden blocks' thus prepared can indeed be put together at right angles, one upside down, if mutual interpenetration is allowed to occur at some places. By making cuts into the sides of the two paper blocks, Fig. 4.5c is obtained. The latter can be translated back into an equation of the form[3]

[3]This equation was published in [6].

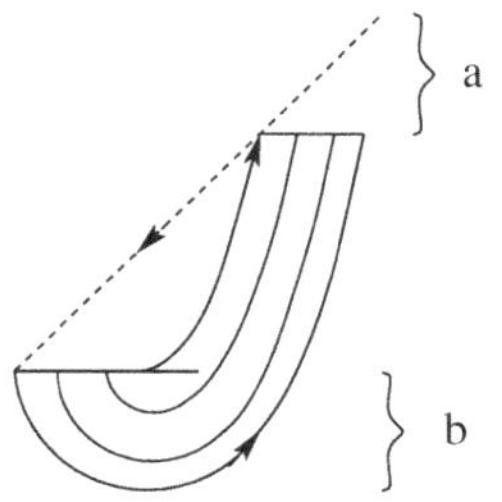

Fig. 4.4 Two specimens of the flow of Fig. 2.5, presented in a two-dimensional ('z-free') projection as described in Sect. 2.7. Schematic drawing. The two flows are to be put together in a behavior-preserving manner (see text). a, b = non-overlapping portions of flows

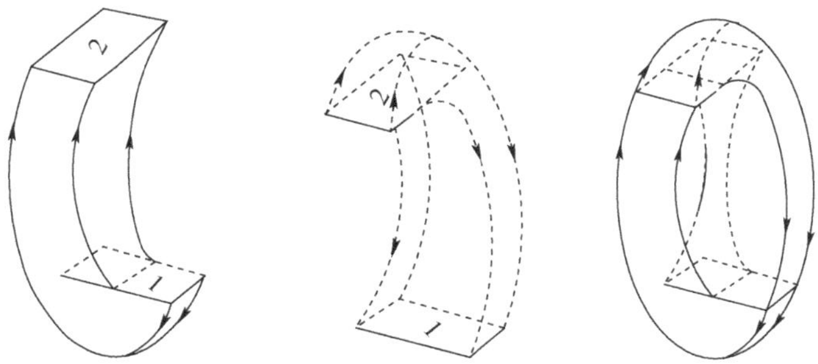

(a) 'Dashed' sub-flow (b) 'Non-dashed' sub-flow (c) Combination

Fig. 4.5 Two three-dimensional building blocks having only the four (dashed and non-dashed, respectively) shapes of Fig. 4.4 as their normal projections. 'Dashed' (**a**) and 'non-dashed' (**b**) three-dimensional subflow, respectively. A possible combination (**c**) in which the two 'central' portions of the two blocks (located between the two threshold planes 1, 2) overlap in three-dimensional space

$$\left\{\begin{array}{ll} \dot{x} = -(y+c)w & +(z+c)(1-w) \\ \dot{y} = -(1-a-x)w & +y\dfrac{z+c}{x+b-1}(1-w) \\ \dot{z} = -zw\dfrac{y+c}{x-b} & +(a-x)(1-w) \\ \varepsilon\dot{w} = w(1-w)(w-1+x) & -\delta(w-0.5)\,. \end{array}\right. \tag{4.2}$$

This equation consists of two equations like Eq. (2.8), only slightly modified (since when actually playing with the paper blocks, one finds that the parabolas have to be replaced by segments of concentric circles, as the next-simplest possibility, in order to meet the mechanical constrain that the distance b in Fig. 4.4 must not exceed the distance a. The modified Eq. (2.8), represented twice in Eq. (4.2), is

$$\left\{\begin{array}{l} \dot{x} = (x-b)w + (z+c)(1-w) \\ \dot{y} = zw \qquad\qquad + (a-x)(1-w) \\ \varepsilon\dot{w} = w(1-w)(w-1+x) - \delta(w-0.5) \end{array}\right. \tag{4.3}$$

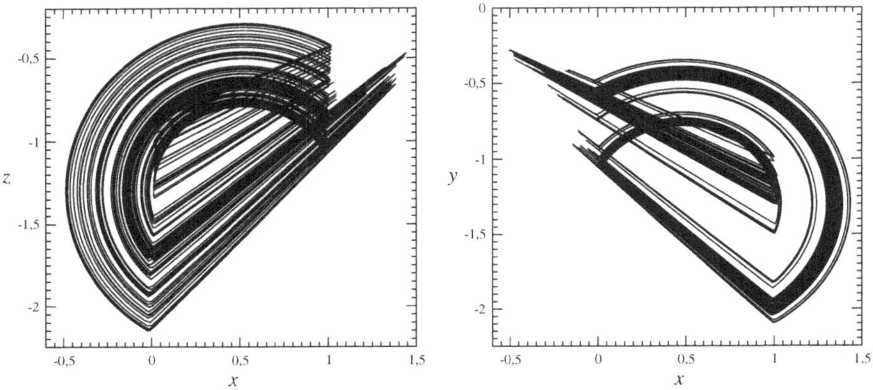

(a) x-z plane projection (b) x-y plane projection

Fig. 4.6 Higher chaos in Eq. (4.2). Parameters: $a = 0.52$, $b = 2$, $c = 1.3$, $\delta = 0.001$, and $\varepsilon = 0.001$. Initial conditions: $x_0 = 0$, $y_0 = -1$, $z_0 = -1.5$, and $w_0 = 0.01$. Time step: $\mathrm{d}t = 10^{-4}$ s

and

$$\begin{cases} \dot{x} = (x + b - 1)(1 - w) + (y + c)w \\ \dot{y} = y(1 - w) \qquad -(1 - a - x)w \\ \varepsilon\dot{w} = w(1 - w)(w - 1 + x) - \delta(w - 0.5) \end{cases}, \tag{4.4}$$

respectively. The latter is the former, mirrored at the planes $x = \frac{1}{2}$ and $w = \frac{1}{2}$. Numerical simulation with Eq. (4.2) is presented in Fig. 4.6. It is much like Fig. 4.5.

A limiting cross section through Eq. (4.3) reads, similarly to Eq. (2.9) [7],

$$y_{n+1} = -c + \sqrt{2a - 1 + (2y_n + c)^2}, \tag{4.5}$$

and a cross section through Eq. (4.4) accordingly

$$z_{n+1} = -2c + 2\sqrt{2a - 1 + (z_n + c)^2}, \tag{4.6}$$

The limiting cross section through Eq. (4.1) therefore becomes

$$\begin{cases} y_{n+1} = -c + \sqrt{2a - 1 + (2y_n + c)^2} \\ z_{n+1} = -2c + 2\sqrt{2a - 1 + (z_n + c)^2}. \end{cases} \tag{4.7}$$

The resulting behavior is shown in Fig. 4.7.

When introducing a non-zero ε in Eq. (4.2), the cross section can, in analogy to Eq. (2.11), in a neighborhood be approximately described by[4]

[4]The equation suggested in the original manuscript of this book was

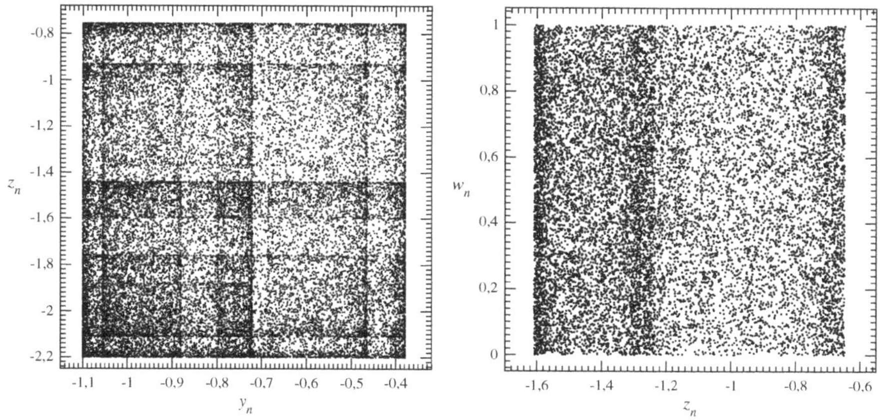

(a) Numerical iterations of map (4.7)

(b) Poincaré section of the flow (4.2)

Fig. 4.7 **a** Numerical simulation of Eq. (4.7) for the same parameters and initial conditions as in Fig. 4.6, except for $\delta = \varepsilon = 0$. **b** Poincaré section—defined by $x_n = 0.5$ and $\dot{x}_n < 0$—of system (4.2) for the same parameter and initial conditions as in Fig. 4.6

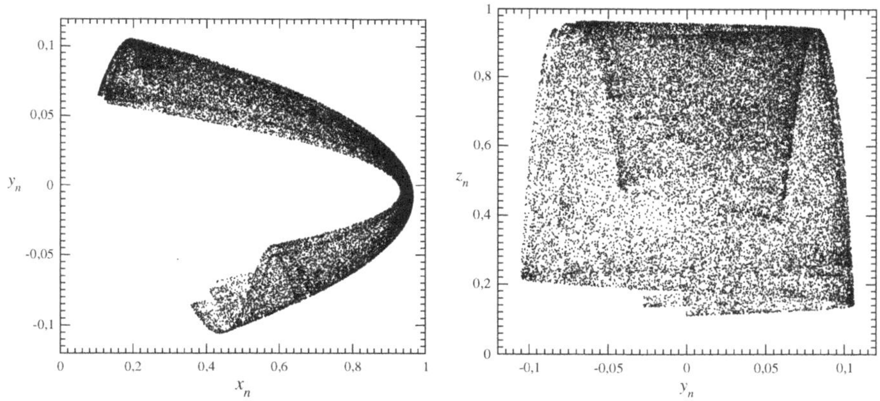

(a) x-y plane projection

(b) x-z plane projection

Fig. 4.8 Folded towel map. Numerical simulation of Eq. (4.8). Parameter values: $a = 0.52, b = 1.9$ $c = 1.3, d = 0.05$, and initial conditions: $x_0 = 0.1$, $y_0 = 0.1$, $z_0 = 0.1$

$$\begin{cases} x_{n+1} = 2bx_n(1 - x_n) - d(y + 0.35)(1 - 2z_n) \\ y_{n+1} = 0.1\,[(y_n + 0.35)(1 - 2z_n) - 1]\,(1 - bx_n) \\ z_{n+1} = 3.78z_n(1 - z_n) + 0.2y_n\,, \end{cases} \tag{4.8}$$

but all our trials lead to a trajectory ejected to infinity. We therefore reported the equations for the folded towel map published in [8].

$$\begin{cases} x_{n+1} = 0.1\,[(x_n + 0.35)(1 - 2z_n) - 1]\,(1 - 2y_n) \\ y_{n+1} = -c + \sqrt{2a - 1 + (2y_n + c)^2} - 0.05(x_n + 0.35)(1 - 2x_n) \\ z_{n+1} = -2c + 2\sqrt{2a - 1 + (z_n + c)^2} + 0.2x_n \end{cases}$$

A numerical simulation of Eq. (4.8), a 'folded towel map' is presented in Fig. 4.8.

References

1. O.E. Rössler, Chaotic oscillations: an example of hyperchaos, in *Lectures in Applied Mathematics*, vol. 17 (1979), pp. 141–156
2. J.D. Farmer, J.P. Crutchfield, H. Frœling, N.H. Packard, R.S. Shaw, Power spectra and mixing properties of strange attractors. Ann. N. Y. Acad. Sci. **357**, 453–472 (1980)
3. H.L. Swinney, P.R. Fenstermacher, J.P. Gollub, Transition to turbulence in a fluid flow, in *Synergetics, a Workshop*, ed. by H. Haken (Springer, Berlin, 1977), pp. 60–69
4. O.E. Rössler, Chemical turbulence: chaos in a simple reaction-diffusion system. Zeitschrift für Naturforschung A **31**, 1168–1172 (1976)
5. O.E. Rössler, C. Kahlert, Winfree meandering in a 2-dimensional 2-variable excitable medium. Zeitschrift für Naturforschung A **34**, 565–570 (1979)
6. O.E. Rössler, The chaotic hierarchy. Zeitschrift für Naturforschung A **38**, 788–801 (1982)
7. C. Mira, Dynamique complexe engendrée par une équation différentielle d'ordre 3, in *Proceedings Equadiff'78*, ed. by R. Conti, G. Sestini, G. Villari, Florence (1978), pp. 25–37
8. O.E. Rössler, Chaos, in *Structural Stability in Physics*, ed. by W. Güttinger, H. Eikemeier (Springer, Berlin, 1979), pp. 290–309

Chapter 5
The Gluing-Together Principle

5.1 Chaos in Single-Loop Feedback Systems

So far, a single design principle has proved sufficient to generate a whole world of nontrivial flows in three and higher dimensions. The resulting limiting equations appeared rather 'robust' in the sense that the limiting parameter ε could be increased from zero up to the order of unity (cf. Eq. (2.2)); and also in view of the fact that no more than a single quasi-threshold (an 'L' instead of a 'Z'-shaped slow manifold) with a single quadratic term was required (Eqs. (2.2) and (4.1)). However, this empirical observation could not be directly transformed into hard analytical evidence. Similarly for the second class of systems about which existence statements could be made (infinitesimally small chaotic regimes, and chaotic regimes with infinitesimally small cross sections): again there were chaotically behaving systems in a large neighborhood (to judge from computer results). In one case the same Eq. (3.4) could be linked to either limiting case.

While chaos thus is apparently possible in a wide range of 'ordinary' (that is, not near-singular) systems, one would still like to have at least one completely non-singular example, residing squarely within the class of analytically unreachable systems, for which the presence of chaos could be demonstrated analytically. The gluing-together principle provides a class of examples.

The principle consists in the 'gluing-together', in an non-overlapping fashion, of two fully n-dimensional flows. In the simplest (three-dimensional) case, the two subflows are glued together smoothly along a two-dimensional plane. Both subflows may be linear. This leads to the class of piecewise linear systems as subclass.

One of the simplest piecewise linear 3-variable systems is the following equation:

$$\begin{cases} \dot{x} = -x + f(z) \\ \dot{y} = x - y \\ \dot{z} = y - z \,, \end{cases} \tag{5.1}$$

O. E. Rössler and C. Letellier, *Chaos*, Understanding Complex Systems,
https://doi.org/10.1007/978-3-030-44305-4_5

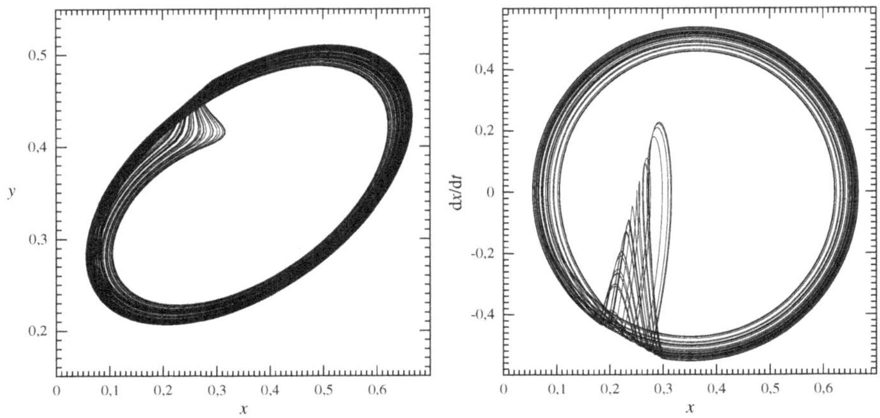

(a) x-y plane projection (b) x-$\dot{x}$ plane projection

Fig. 5.1 Spiral-type chaos in Eq. (5.1). Note the topological similarity to the flow of Figs. 2.3 and 2.4. Initial conditions: $x_0 = 0.1$, $y_0 = 0.3$, and $z_0 = 0.4$

where $f(z)$ is a continuous, piecewise linear function involving two pieces only. Specifically, choose the 'V'-shaped function [8][1]

$$f(z) = -0.25 + (3.6 - 8.4z) - 20(3.6 - 8.4z)\mathscr{H}\left(z - \frac{9}{21}\right), \qquad (5.2)$$

where $\mathscr{H}(\zeta)$ is the Heaviside function $\mathscr{H}(\zeta) = 1$ if $\zeta \geq 0$ and $\mathscr{H}(\zeta) = 0$ if $\zeta < 0$. A numerical solution is provided in Fig. 5.1.

The flow has a familiar appearance (compare Fig. 2.3). The asset of Eq. (5.1) is that it can in principle be solved analytically, whereby an infinite sequence of linear pieces have to be patched together smoothly. An approximate analytical calculation, based on inserting the analytical expressions into a computer, has been performed by Uehleke [11]. It is also possible to explicitly calculate the shape of invariant lines in the two subtransformations formed at the boundary between the two linear regimes. The Poincaré map in the present case is clearly the product of two submaps each mediated by a linear flow. Several analytic approaches to Eq. (5.1) can be found in Sparrow's recent paper [10].

The system (5.1) is a special case of a biological feedback equation, proposed in 1956 as a model of thyroid hormone control by Danziger and Elmergreen [3]. In the original version,[2] the right-hand part of $f(z)$ was flat (zero) instead of ascending. The

[1] The simpler function

$$f(z) = \left| \begin{array}{ll} Gz - G & \text{for } z \leq 1 \\ Fz - G & \text{for } z > 1 \end{array} \right.$$

was later found and published with Bernahrd Uehleke in [11, 12]. For instance, spiral chaos was found for $F = 143$ and $G = 10$.

[2] The original system

original equation was discussed extensively by Rashevsky [6] and Cronin-Scanlon [2]. Cronin-Scanlon pointed out that the limit cycle claimed to exist by the authors has never been demonstrated rigorously. All that had been shown was the existence of an attracting (Poincaré–Bendixson type) toroidal region. The expectation, explicitly stated by Cronin-Scanlon [2], that a more general type of recurrence might be possible in Eq. (5.1) is vindicated by the above numerical and Sparrow's and Uehleke's analytical results.

It is an interesting question to ask whether an equation of the same form as Eq. (5.1), but prolonged by a third variable in the linear chain of phase-shifting variables, will produce the next higher type of chaos (and so on). The question is not easy to tackle because in order to solve it, one would have to deal carefully with two pieces of four-dimensional linear flows, to be composed smoothly along a three-dimensional threshold (and so on).

Equation (5.1) was orginally found heuristically along a roundabout way which is of interest in its own right [7]. The equation from which to start out was an arbitrary single-variable endomorphism capable of generating chaos, like the logistic map (Eq. (2.3) with $\varepsilon = 0$) or like a piecewise linear letter V-shaped function. All these equations can be written in the form

$$\varepsilon\dot{x} = -x(t) + f(x(t-1)), \tag{5.3}$$

where $f(x)$ is the endomorphic function and $\varepsilon \equiv 0$. Equation (5.3), then, describes not just a single chaos-generating delay differential equation, but a whole continuum of such equations, all uncoupled. This is probably the most complicated kind of (hyper) chaos conceivable. By introducing a finite value of ε, however small, much of the incredible complexity of this equation should be destroyed—presumably in a graded fashion depending on ε. Makey and Glass [4] described an equation which can be brought to the form of Eq. (5.3) and which indeed shows (ordinary) chaos if $\varepsilon = 0.3$. By raising ε even further, the chaotic motion could be made disappear.

Since delay-type (so-called 'functional') differential equations like Eq. (5.3) are known to be equivalent to an infinite system of ordinary differential equations of the form of Eq. (5.1), but with n equal to infinity rather than 3 as considered by Roger Nussbaum [5],[3] it was straightforward to ask for the minimum n in Eq. (5.1) which

---

$$\begin{cases} \dot{x} = -gx + (c - hy)\mathscr{H}\left(\frac{c}{h} - z\right) \\ \dot{y} = mx - ky \\ \dot{z} = ay - bz \end{cases}$$

accounts for the observed sustained oscillations and includes a specific mechanism by which the pituitary stimulates the thyroid [3]. In this system, x is the concentration of thyrotropin, y is the concentration of activated enzyme, z the concentration of thyroid hormone. b, g and k are loss constants, a, h and m are constants expressing the sensitivity of the glands to simulation or inhibition and c is the rate of production of thyrotropin in the absence of thyroid inhibition.

[3] The dimension of the system made of the first equation of system (5.1) with n additional ordinary differential equations is $n + 1$. The corresponding last two equations would be

$$\begin{cases} \dot{x}_{n-1} = x_{n-2} - x_{n-1} \\ \dot{x}_n = x_{n-1} - x_n \,. \end{cases}$$

still allows for chaos. Sparrow [9] found chaos numerically with $n = 50$; an der Heiden [1] found chaos with $n = 15$; and Glass[4] found chaotic solution with $n = 9$. An explanation for the differing numbers of n found empirically can be seen in the varying degrees of 'smoothness' of the respective $f(z)$ functions assumed. In the case of Eq. (5.1), the non-differentiable function of Eq. (5.2) can be replaced with the smooth (C^∞) function

$$f(z) = -0.25 - 9(3.6 - 8.4z) + \sqrt{100}(3.6 - 8.4z)^2 + \varepsilon\,, \tag{5.4}$$

where $\varepsilon = 10^{-4}$. The pointed V-shaped function of Eq. (5.2) is now replaced by a (more or less 'sharp') branch of hyperbola. If ε is only ten times larger, chaos was already disappeared. Thus, polynomial of a rather high order seem to be required if $n = 3$.

The open problem whether indeed the whole 'hierarchy of chaos' can be displayed by Eq. (5.1) as a function of n has already been mentioned. The important question of applications of Eq. (5.1), with varying n, will be taken up below (Sect. 10.6).

References

1. U. an der Heiden, *Analysis of Neural Networks*, vol. 35, Lecture Notes in Biomathematics (Springer, Berlin, 1980)
2. J. Cronin-Scanlon, A mathematical model for catatonic shizophrenia. Ann. N. Y. Acad. Sci. **231**, 112–130 (1974)
3. L. Danziger, G.L. Elmergreen, The thyroid-pituitary homeostatic mechanism. Bull. Math. Biophys. **18**, 1–13 (1956)
4. M.C. Mackey, L. Glass, Oscillation and chaos in physiological control systems. Science **197**, 287–289 (1977)
5. R.D. Nussbaum, Periodic solutions of some nonlinear autonomous functional differential equations. J. Differ. Equ. **14**, 360–394 (1973)
6. N. Rashevsky, *Some Medical Aspects of Mathematical Biology* (Charles C. Thomas, Springfield, 1964)
7. O.E. Rössler, Continuous chaos, in *Synergetics. A Workshop*, ed. by H. Haken (Springer, Berlin, 1977), pp. 174–183
8. R. Rössler, F. Götz, O.E. Rössler, Chaos in endocrinology (abstract). Biophys. J. **25**, 261 (1979)
9. C. Sparrow, Bifurcation and chaotic behaviors in simple feedback systems. J. Theor. Biol. **83**, 93–105 (1980)
10. C. Sparrow, Chaos in a three-dimensional single loop feedback system with a piecewise linear feedback. J. Math. Anal. Appl. **82**, 275–291 (1981)
11. B. Uehleke, Chaos in einem stückweise linearen System: Analytische Reultate. Ph.D. Thesis, University of Tübingen, 1982
12. * B. Uehleke, O.E. Rössler, Analytical results on a chaotic piecewise-linear O.D.E. Zeitschrift für Naturforschung A **39**, 342–348 (1983)

[4] 1979, personal communication.

Chapter 6
Chaos in Toroidal Systems

6.1 The Bonhoeffer-Van der Pol Equation

The oldest example of complicated behavior found in an unconstrained (non-conservative) dynamical system is, as mentioned in the Introduction, that found in the periodically forced van der Pol oscillator [1]. Cartwright and Littlewood's results were based on painstaking analytical arguments (see the elaboration in [2]). Is it, at the same time, also possible to understand their observations in terms of the simple reinjection principle? The answer is yes. One can even say that these authors were the first to make use of the reinjection principle in order to find nontrivial behavior in a simple dynamical system.

Using Liénard's familiar nonlinear transformation (cf. [3, 4])

$$y = -x + \frac{x^3}{3} + \frac{\dot{x}}{\mu}, \tag{6.1}$$

the van der Pol oscillator

$$\ddot{x} - \mu\dot{x}(1 - x^2) + x = 0, \tag{6.2}$$

can be written in the form

$$\begin{cases} \varepsilon\dot{x} = x - \dfrac{x^3}{3} + y \\ \dot{y} = -\varepsilon x \end{cases} \tag{6.3}$$

where $\varepsilon = \frac{1}{\mu}$. Cartwright and Littlewood considered Eq. (6.3) under the condition of very small values of ε [1] but periodically forced. Another simple model of such periodically forced van der Pol oscillator consists of a Bonhoeffer-van der Pol (BVP) excitable system as investigated by FitzHugh [5] and whose equations are

O. E. Rössler and C. Letellier, *Chaos*, Understanding Complex Systems,
https://doi.org/10.1007/978-3-030-44305-4_6

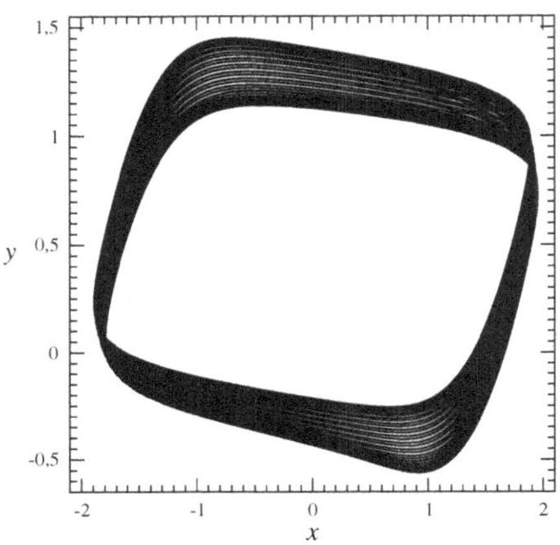

(a) Quasi-periodic regime

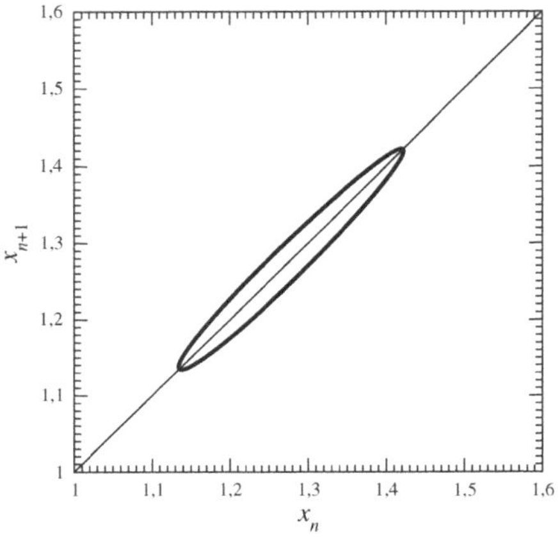

(b) First-return map

Fig. 6.1 Quasi-periodic regime produced by the periodically forced Bonhoeffer-van der Pol oscillator. Numerical solution to Eq. (6.5). Parameters: $a = 0.467$, $b = 0.8$, $\varepsilon = 1/3$, $A = 0.1$, $\zeta = -0.4$ and $\omega = 0.021$. Initial conditions: $x_0 = 0.4$ $y_0 = 0$, $z_0 = A$, and $w_0 = 0$

$$\left\{\begin{aligned} \varepsilon\dot{x} &= x - \frac{x^3}{3} + y + z \\ \dot{y} &= -\varepsilon(-a + x + by) \end{aligned}\right. \tag{6.4}$$

where terms z, a and b were added by Bonhoeffer to explain the behavior of passivated iron wires [6–8]. This BVP-model can be periodically forced by using $z = A \sin \omega t$ as investigated in [9].[1] The periodically forced BVP-model is thus

$$\left\{\begin{aligned} \varepsilon\dot{x} &= x - \frac{x^3}{3} + y + \zeta + A \sin \omega t \\ \dot{y} &= -\varepsilon(-a + x + by) \end{aligned}\right. \tag{6.5}$$

where we added a constant input ζ. The solution of this periodically driven Bonhoeffer-van der Pol oscillator is a quasi-periodic regime as exemplified in Fig. 6.1.

In the context of the reinjection principle, one may ask whether Eq. (6.4) can be transformed into an autonomous (non-forced) equation. An obvious possibility is the following four-variable autonomous system

[1] The parameter values indicated in this latter work did not allow us to recover a simple chaotic attractor. The chaotic attractor found in the Bonhoeffer-van der Pol oscillator (6.4) is investigated in the Appendix C.7. We thus adapted this section and the next one to a better example whose parameter values were proposed in [10]. This example is more in the spirit of Otto's original work than the first observation of chaos in a driven van der Pol system published by Ulrich Parlitz et al. [11].

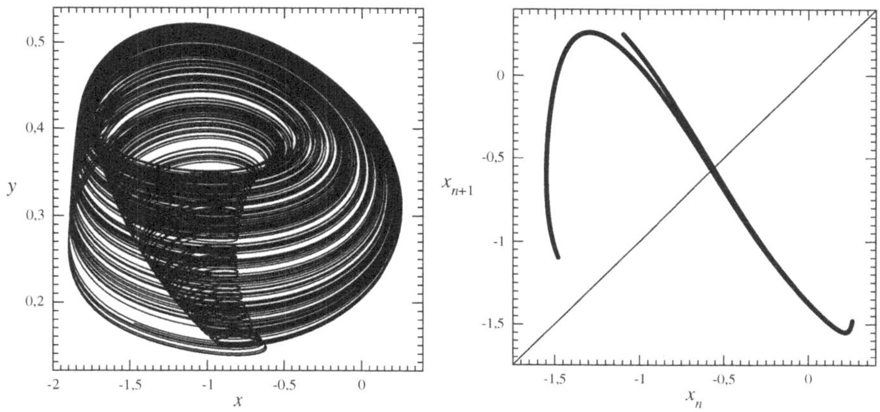

(a) Flow (b) First-return map

Fig. 6.2 Chaotic regime produced by the periodically forced Bonhoeffer-van der Pol oscillator. Numerical solution of the four-dimensional system (6.7). Parameter values: $a = 0.695$, $b = 0.8$, $\varepsilon = 0.1$, $\omega = 1$, and $A = 0.74$

$$\left\{ \begin{array}{l} \varepsilon \dot{x} = x - \dfrac{x^3}{3} + y + \zeta + z \\ \dot{y} = -\varepsilon(-a + x + by) \\ \dot{z} = w \\ \dot{w} = -\omega^2 z \,, \end{array} \right. \tag{6.6}$$

where $z_0 = A$ and $w_0 = 0$. Here z and w together form an autonomous sub-system generating $z(t) = A \sin \omega t$.

In order to have a simple chaotic attractor produced by the Bonhoeffer-van der Pol oscillator, we therefore prefered the modified BVP-oscillator [10]

$$\left\{ \begin{array}{l} \dot{x} = x - \dfrac{x^3}{3} + y + z \\ \dot{y} = -\varepsilon(a + x + by) \\ \dot{z} = w \\ \dot{w} = -\omega^2 z \,. \end{array} \right. \tag{6.7}$$

With appropriate parameter values, a non symmetric attractor issued from a period-doubling cascade is obtained (Fig. 6.2). The first-return map is clearly a smooth unimodal map as obtained after period-doubling cascade.

However, four variables (and hence four dimensions) do not yet allow one to gain a visual understanding of what happens, since the reinjection is not taking place between two-dimensional surfaces. Therefore, it is fortunate that a three-variable system reproducing the behaviour of x exists also, namely,[2]

$$\begin{cases} \dot{u} = \dfrac{u}{r}\left(r - \dfrac{r^3}{3} + y + u\right) + \omega v \\ \dot{v} = -\omega u + \dfrac{v}{r}\left(r - \dfrac{r^3}{3} + y + v\right) \\ \dot{y} = -\varepsilon(a + r + by) + C \end{cases} \tag{6.8}$$

where $r = \sqrt{u^2 + v^2} = x$, C is a positive constant (for example, $C = 0.15$),[3] and $u_0 = A$.

The first and the second lines of Eq. (6.8) describe a rotation-symmetrical oscillator which in polar coordinates reads

$$\begin{cases} \dot{r} = r - \dfrac{r^3}{3} + y + u \\ \dot{\theta} = \omega \,, \end{cases} \tag{6.9}$$

with $w \equiv \theta = \theta_0 \cos \omega t$ and $r = \theta_0$. Here r is the 'amplitude' (actual radius) of the u-v system: $r = \sqrt{u^2 + v^2}$, as it is easy to verify by calculating $\dot{r} \equiv r - \frac{r^3}{3} + y + u$ explicitly using the chain rule (see [12] for a similar equation with r replaced by r^2). Choosing $r \equiv x$, one obtains

$$\dot{x} \equiv \dot{r} = x - \frac{x^3}{3} + y + A \cos \omega t \;;$$

that is to say, one has re-obtained the first line of Eq. (6.5) from the first and second lines of Eq. (6.8).

For bounded initial conditions, Eq. (6.5) indeed has only bounded solutions for bounded values of ε [13] and bounded values of A [1]. Equation (6.8) therefore is a valid substitute of Eq. (6.5). A numerical simulation of these equations is shown in Fig. 6.3. There is a singularity in the attractor as clearly evidenced in the u-y plane projection: different trajectories goes through a very thin tube and cannot be distinguished, rendering difficult the characterization of the dynamics.

There is another possibility for constructing a periodically driven van der Pol oscillator: it is based on the original equations written in 1980 which were modified for finding a trajectories with a rather similar shape to what was initially drawn. The equations, in the four-dimensional form, are

---

[2]The idea is to replace the w-z oscillator whose amplitude is driven by the second equation of system (6.6).

[3]Constant c is introduced to make sure that the amplitude does not shrink to zero.

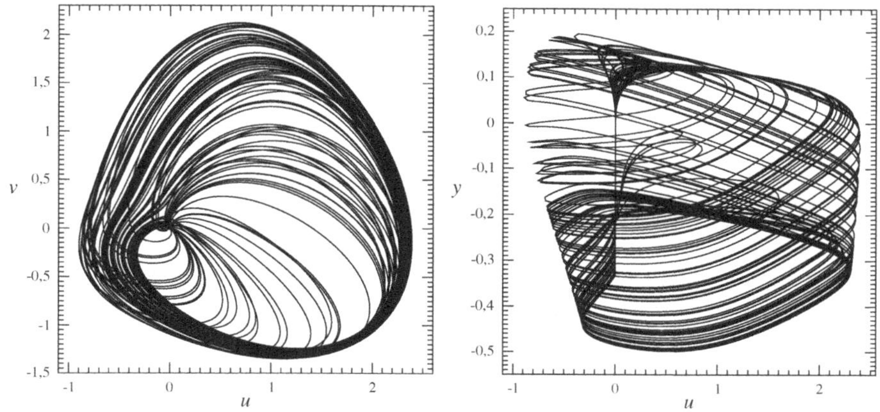

(a) u-v plane projection (b) u-y plane projection

Fig. 6.3 Chaotic regime of the periodically forced Bonhoeffer-van der Pol oscillator rewritten in the three-dimensional form (6.8). Same parameter values as in Fig. 6.2

$$\begin{cases} \dot{x} = c\left[x - \dfrac{x^3}{3} + (y-d) - \mu z\right] \\ \dot{y} = -\dfrac{-a + x + b(y-d)}{ec} \\ \dot{z} = w \\ \dot{w} = -\omega^2 z\,. \end{cases} \tag{6.10}$$

As shown in Fig. 6.4, Eq. (6.10) now allows inspection of what happens in the region of the limiting regime of the periodically forced van der Pol oscillator.

6.2 Chaos in the Bonhoeffer-Van der Pol Equation

The complicated flow of Fig. 6.4 can be analyzed in the same way as the complicated flow of Fig. 2.2) could. The first possibility is to look at a Poincaré cross section. When attempting to do this, one immediately encounters an 'unnecessary' complication, however. It concerns the fact that two folding processes occur in the flow of Eq. (6.8) in a 'symmetrical' fashion, one being produced upstairs and the other downstairs. This is due to the great symmetry of the underlying Liénard equation [13]. A straightforward way to circumvent this complication is to simply take only the lower half of the flow of Fig. 6.4b and combine it with the upper half of the flow of Fig. 6.4a. This procedure yields the 'modified' Cartwright–Littlewood equation

$$\begin{cases} \varepsilon\dot{x} = x - x^3 - y \\ \dot{y} = x + (a\cos\varepsilon b't)\mathscr{H}(-x)\,, \end{cases} \tag{6.11}$$

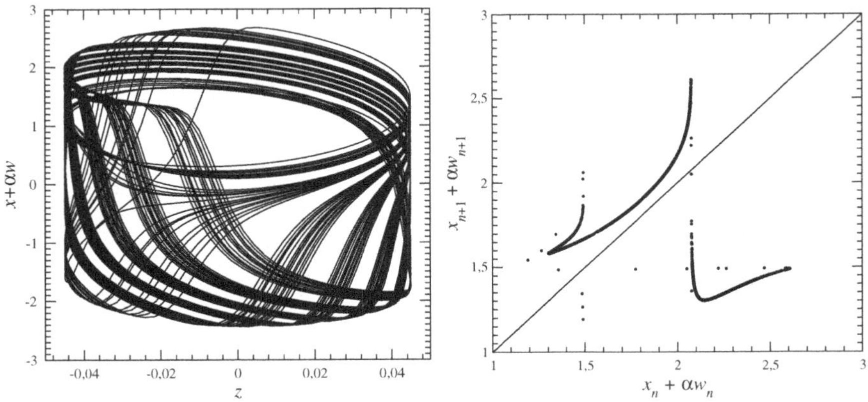

(a) Chaotic regime (b) First-return map

Fig. 6.4 Reinjection in the periodically forced van der Pol oscillator of Eq. (6.10). Parameters: $a = 0.44$, $b = 0.8$, $c = 1$, $d = 1.5$, $e = 9$, $A = 0.045$ and $\omega = 0.2$. Initial conditions: $x_0 = -1.5$, $y_0 = 1.5$, $z_0 = 0$ and $w_0 = A\omega$. Axes: $x + \alpha w$ with $\alpha = 80$ is plotted in ordinates

where $\mathscr{H}(-x)$ is the Heaviside function (being zero for negative values of its argument and unity otherwise). Note that also b has been replaced by $\varepsilon b'$ for convenience. The corresponding autonomous 3-dimensional system is

$$\left\{ \begin{aligned} \varepsilon \dot{x} &= x - x^3 - r + c \\ \dot{z} &= -\varepsilon b' w + \frac{z}{r}\left(x + \frac{aw}{r}\mathscr{H}(-x)\right) \\ \dot{w} &= \varepsilon b' z + \frac{w}{r}\left(x + \frac{aw}{r}\mathscr{H}(-x)\right) . \end{aligned} \right. \tag{6.12}$$

A top-down view of the same flow is presented in Fig. 6.5. For simplicity, only the flow 'downstairs' is shown. The flow 'downstairs' merely connects to the two arrows P_1 and P_2 (of exit and reentry, respectively) in a linear, one-to-one fashion.

When comparing Fig. 6.5 with Fig. 2.2, one sees that the two are virtually identical. Indeed, Eq. (6.11) can produce both spiral and screw-type chaos, just as Eq. (2.1) could (see Figs. 2.2 and 2.8). The only 'disadvantage' of Eq. (6.11) is that the thresholds involved are curvilinear rather than straight lines.

Returning to the 'full' Cartwright–Littlewood equation (Eqs. (6.3) and (6.8)), it is evident that the same process which occurs once in the modified equation occurs twice in the original one. Hereby no qualitatively new features are to be expected (compare [14], for other examples of 'combined' chaotic flows).

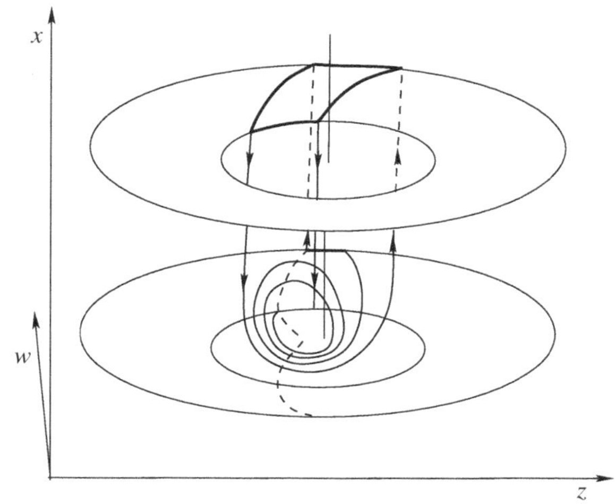

Fig. 6.5 Top-down view on Cartwright–Littlewood chaos. The y-axis has been contracted to a point

6.3 A Related Prototype

The limiting flow seen in Fig. 6.5 is similar to one generated by the following, even simpler equation

$$\begin{cases} \dot{s} = s\,, & s \bmod (1 + a\cos\pi bt) \\ \dot{\theta} = 1\,. \end{cases} \tag{6.13}$$

A corresponding two-variable autonomous system[4] is

$$\begin{cases} \dot{x} = \dfrac{x}{r} + bz \\ \dot{y} = -bx + \dfrac{y}{r} \end{cases} \tag{6.14}$$

[4] As explained in [15], the general equivalence comes from the two-dimensional system

$$\begin{cases} \dot{x} = \dfrac{x}{r} f(x) + \omega y \\ \dot{y} = -\omega x + \dfrac{y}{r} f(x) \end{cases}$$

in cartesian coordinates which is rewritten as a system in polar coordinates according to

$$\begin{aligned} \dot{s} = \dot{r} &= \frac{1}{2} \frac{2x\,\mathrm{d}x + 2y\,\mathrm{d}y}{\sqrt{x^2 + y^2}} \\ &= \left(\frac{x^2}{r} + \frac{y^2}{r} \right) \cdot f(s) + \underbrace{\frac{\omega xy - \omega xy}{r}}_{=0} \\ &= f(s)\,. \end{aligned}$$

This equation defines the evolution of the radius and it should be associated with a constant angular velocity. In the present case, $f(s) = 1$.

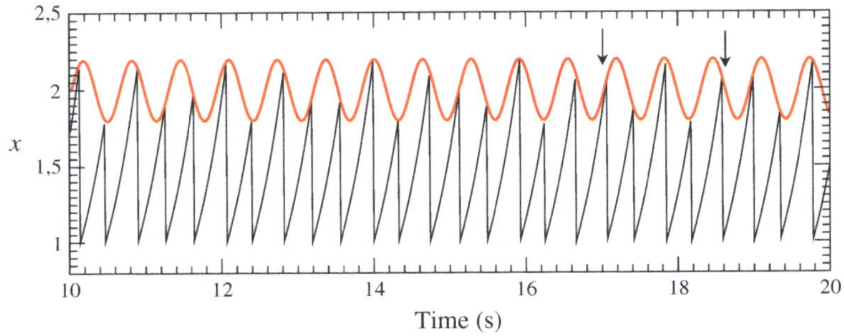

Fig. 6.6 Quasi-periodic regime in Eq. (6.13) obtained by numerical simulation. Parameters: $a = 0.2$, $b = \pi$ and $\mu = 1.8$. Time step: $\delta t = 0.001$ s

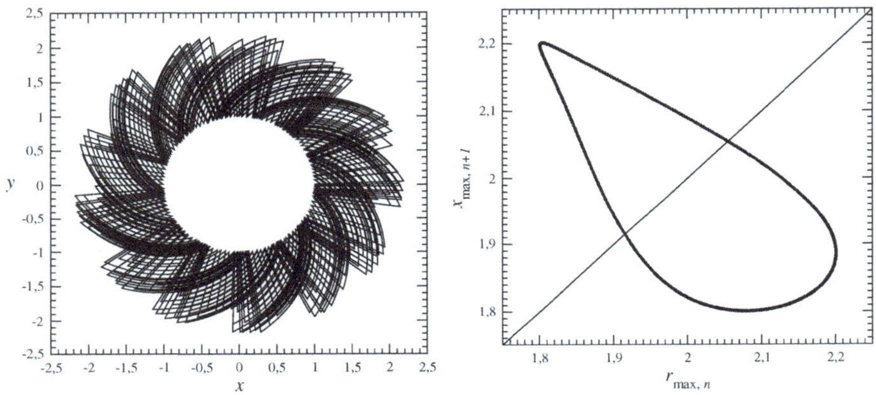

(a) Numerical simulation of Eq. (6.14)

(b) The corresponding first-return map

Fig. 6.7 Quasi-periodic regime in Eqs. (6.13) and (6.14) obtained by numerical simulations. The first-return map is built on the maximum of radius $r_{\max}$. Compare Fig. 6.5. Parameters: $a = 0.2$ and $b = \pi/4$. Initial conditions: $x_0 = 1$ and $y_0 = 0$

where $(r - 1) \bmod (1 + a \cos \pi bt)$ (for example). For the simulation shown in Fig. 6.6, this meant that whenever $r = \sqrt{y^2 + z^2}$ surpassed $2 + a \cos \pi bt$ from below, r is replaced with $r - (1 + a \cos \pi bt)$. With a linear growth of the radius, the dynamics produced by Eq. (6.14) can only be quasi-periodic.

In addition to the x-y flow, the time behavior of x itself is also shown in Fig. 6.7. The two oscillators are not pulsating with the same period as especially seen at the two points marked by an arrow. Equation (6.13) therefore gives a quick glimpse of the basic mechanism of the Bonhoeffer-van der Pol equation itself. The close relation between Eqs. (6.3) and (6.13) has also been seen by Hoppensteadt.[5]

When comparing the relaxation oscillator of Eq. (6.14) with that of Eq. (6.9) which shows essentially the same behavior, one difference becomes apparent: there

[5]Personal communication, 1979.

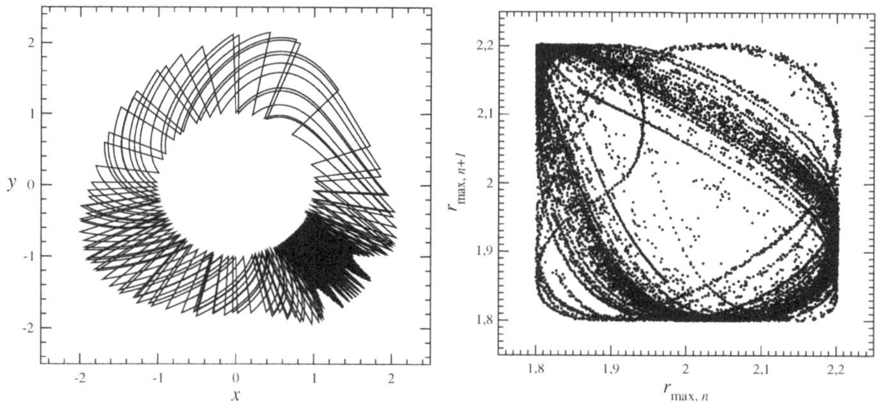

(a) Numerical simulation of Eq. (6.16) (b) The corresponding first-return map

Fig. 6.8 Toroidal Chaos in Eqs. (6.13) and (6.16) obtained by numerical simulations. The first-return map is built on the maximum of radius r_{max}. Compare to Fig. 6.5. Parameters: $a = 0.2$, $b = \pi/4$ and $\mu = 1.1$. Initial conditions: $y_0 = 1$ and $z_0 = 0$

is no ε in Eqs. (6.13) and (6.14). They represent typical 'single threshold relaxation oscillators' and as such do not readily translate into neighboring' non-degenerate systems. Several examples of non-degenerate electronic and reaction kinetic systems possessing such degenerate equations as their ideal (though not limiting) equations were presented earlier [16].

In fact, only periodic or quasi-periodic regimes can be produced with a linear growth of the radius as involved in Eq. (6.14). In order to produce chaos, the radius evolution needs to be nonlinear. We thus introduced nonlinear terms in Eq. (6.14) to obtain the following equations

$$\begin{cases} \dot{x} = \mu \dfrac{x(1-y)}{r} + bz \\ \dot{y} = -bx + \mu \dfrac{y(1-x)}{r} \end{cases} \tag{6.15}$$

where $(r - 1) \bmod (1 + a \cos \pi bt)$ as in the previous example. The radius evolution is now chaotic as evidenced by the first-return map built on the maxima of the radius (Fig. 6.8b).

An avatar of Eq. (6.15) can be written in a slightly different way. The principle is the same as in the two previous examples but the equations are simpler since there is no longer rational terms

$$\begin{cases} \dot{x} = \mu x(1-y) - by \\ \dot{z} = bx + \mu y(1-x) \, . \end{cases} \tag{6.16}$$

These equations produced the toroidal chaotic regimes shown in Fig. 6.9.

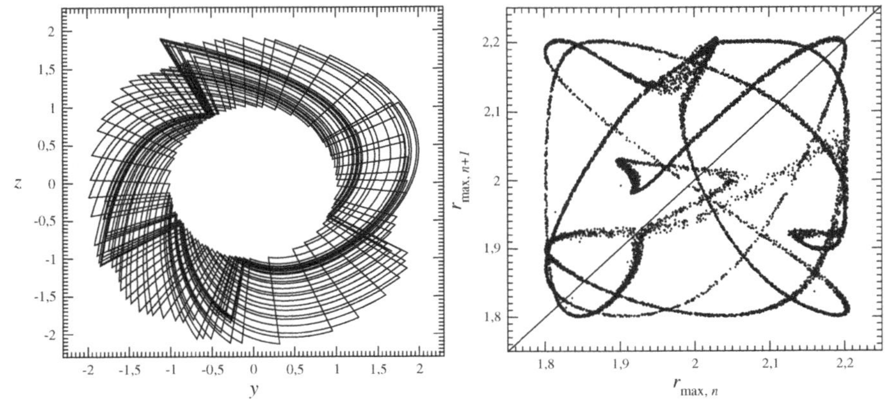

(a) Numerical simulation of Eq. (6.16) (b) The corresponding first-return map

Fig. 6.9 Toroidal chaos in Eq. (6.13) and (6.16) obtained by numerical simulations. The first-return map is built on the maximum of radius $r_{\max}$. Compare to Fig. 6.8. Parameter values: $a = 0.2, b = \pi$, and $\mu = 1.2$. Initial conditions: $y_0 = 1$, and $z_0 = 0$

Taking such a non-degenerate equation and subjecting it to a periodic forcing again yields perfectly invertible three-dimensional chaotic flows. And, again, the three-variable system can be simplified by omitting some of the terms necessary to decouple the harmonic sub-oscillator from the rest of the system. For a 3-variable reaction-kinetic example based on a non-degenerate single-threshold relaxation oscillator, yielding screw-type chaos [17].

Another example is Eq. (2.2), above, with the simulation already shown in Fig. 2.3. The time behavior of its 'slow' variable z (corresponding to x in Eq. (6.13)) was shown in Fig. 2.4c. This time behavior and that of Fig. 6.6 indeed show a striking similarity. It is therefore possible to say that Eq. (6.13) is an 'ideal equation' to a number of more realistic equations and systems.

6.4 An Autonomous 'One-Liner'

A simple autonomous single-variable equation also showing 'toroidal chaos' is

$$\dddot{y} - b\ddot{y} + \dot{y} + (a - b)y - ay^2 = 0\,. \tag{6.17}$$

Two numerical simulations are presented in Fig. 6.10. The first (with $b = 0$) shows a flow on an invariant torus, the second (Fig. 6.10b) shows a once folded-over attracting chaotic flow. In fact, the regime is quite unstable and only transient chaos is observed.[6]

[6]Transient chaos was later called *metastable chaos*, see [18].

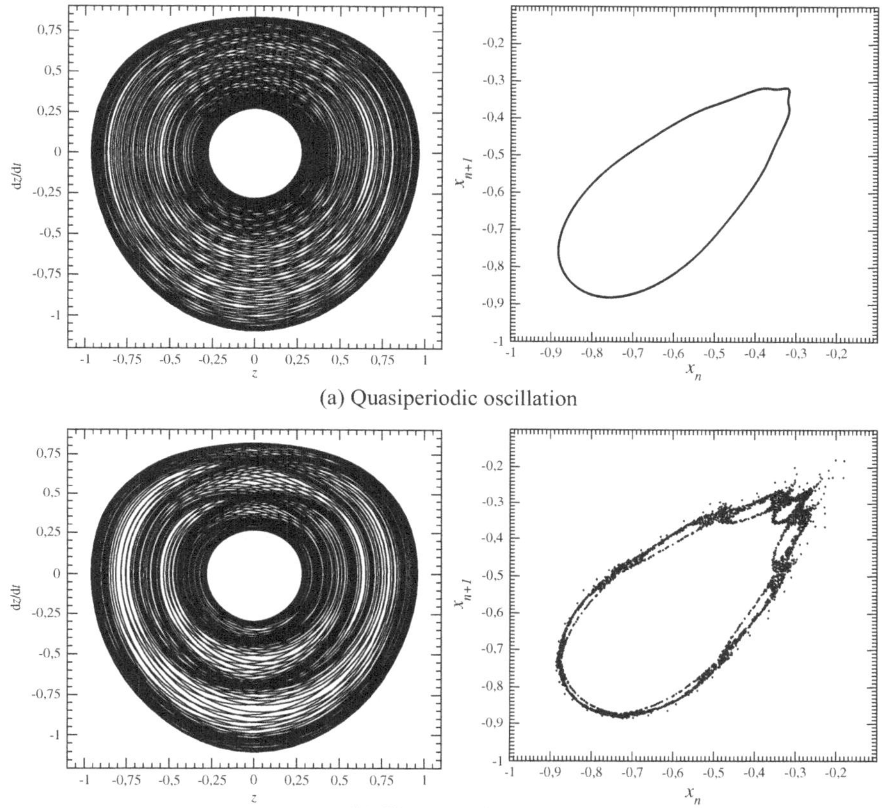

(a) Quasiperiodic oscillation

(b) Chaotic oscillation

Fig. 6.10 Quasi-periodic oscillation (**a**) and (metastable) chaos (**b**) in a simple third order autonomous system. Numerical simulation of Eq. (6.17). Parameters: $a = 0.2$, $b = 0.0$ (a) and $a = 0.219$, $b = 0$ (**b**). Initial conditions: $y_0 = 0.29$, $\dot{y}_0 = 0$ and $\ddot{y}_0 = -0.25$

Equation (6.17) is a linear transformation of the following three-variable system [19]

$$\begin{cases} \dot{x} = -y - z \\ \dot{y} = x \\ \dot{z} = ay(1 - y) - bz\,. \end{cases} \tag{6.18}$$

The variable y is the same as in Eq. (6.17), as it is easy to verify by substitution. This simple equation can, if $b = 0$ and a sufficiently small, be considered as a simplified version of the equation[7]

[7]This equation is the only analytically certain case. Here $r = \sqrt{x^2 + y^2}$ is the amplitude of an harmonic oscillator, in x and y. The subsystem

$$\begin{cases} \dot{x} = -\dfrac{xz}{r} - y \\ \dot{y} = x - \dfrac{yz}{r} \\ \dot{z} = a(1-r) \end{cases} \tag{6.19}$$

with $r = \sqrt{x^2 + y^2}$. In this system, z interacts only with the amplitude r of the x-y sub-oscillator, forming a second, higher order linear oscillator (z-r).

Equation (6.19) is closely related to Eq. (6.8) above. The difference lies only in the higher-order' two-variable system. The latter no longer is the Cartwright–Littlewood equation (6.3) in the present case, but (if $a = 1$) simply the linear oscillator

$$\begin{cases} \dot{r} = -z \\ \dot{\theta} = 1 \\ \dot{z} = a(1-r)\,, \end{cases} \tag{6.20}$$

having concentric circles as solutions if $a = 1$.

Equation (6.19) thus illustrates a general 'building block principle' for the design of toroidal oscillators. Specifically, the principle allows to 'blow up' any variable in a given system (like r in Eq. (6.8) or y in Eq. (6.3)) by replacing it by a two-variable harmonic oscillator whose 'amplitude' $r = \sqrt{x^2 + y^2}$ obeys the same differential equation as the former single variable. In this way, $n - 1$ oscillators can be implemented simultaneously in n-dimensional state space. At most two of the original variables—those forming the highest-order sub-oscillator—are hereby allowed to be nonlinear, the rest being harmonic oscillators. This 'linearity constraint' can, of course, be dropped once the system has been set up: Eq. (6.18) is a 'nonlinear' analogue to the 'linear' Eq. (6.19).

The nice fact about this building-block principle is that there are no constrains for the relative magnitudes of the rotation frequencies of the individual sub-oscillators involved. Another feature which makes Eq. (6.19) and (6.8) and their higher-dimensional analogues an attractive toy to play with is that the resulting 'ideal' systems can be 'distorted' by changing (that is mostly, dropping) parameters. In this way, any of the $\frac{n-1}{2}$ rotation symmetries involved can be relaxed.

Returning to Eq. (6.18) and its relationship to Eq. (6.19), it is empirically to replace in Eq. (6.19) the term $\frac{x}{r}$ by unity in the first line and by zero in the second line, and the term r by r^2, and then by y^2 (one of the two summands of r^2), without changing the qualitative behavior of Eq. (6.19)—presence of a family of invariant foci—over a certain range of amplitudes at least. The only empirical condition for this similarity is that a stays sufficiently small, for example, $c = 0.05$.

The constant a was so far assumed to be zero in Eq. (6.20). This means that the damping term $-k_i x_i$ absent in any line except the third line is zero. Accordingly, volumes in state space are preserved (cf., for example, Hirsch and Smale [20]). It is,

---

$$\begin{cases} \dot{r} = -z \\ \dot{z} = r - \text{ constant}\,, \end{cases}$$

is just an harmonic oscillator too.

therefore, not surprising that the system of Eq. (6.18) mimicks a Hamiltonian system. Hamiltonian flows are divergence-zero flows on a three-dimensional (in the simplest case) smooth manifold and are therefore only locally three-dimensional (cf. [21]). Equation (6.18) may therefore be considered as an unduly simplified Hamiltonian system.

Indeed, just as Hamiltonian systems can produce chaos (in fact, a homoclinic point in a cross section [12, 22]), so does Eq. (6.18) with $a = 0$, or at least it appears so numerically. By empirically increasing c one apparently reaches a 'threshold of distortion' of the torus that was defined by the initial conditions one had chosen, where the trajectories no longer lie on a torus (that is, on a closed curve if looking at a cross section), but begin to wander around chaotically. This threshold is the higher the smaller the cross section of the torus, that is, the closer the initial conditions lie to the zero amplitude torus (Fig. 6.11a). If $a = 0.4$, even that 'torus' can no longer be found.

The chaotic solutions found with $a = 0.4$ and $b = 0.0$ all escape to infinity after an unpredictable number of recurrent returns to the cross section (Fig. 6.11b). A numerical study of a cross section through this flow can be expected to show the typical combination of regions governed by homoclinic points (and chaotic behavior) and regions governed by concentric ellipses (and quasi-periodic behavior) characteristic of area-preserving map with random behavior [23, 24]. Figure 6.11c shows an analogous, but bounded chaotic regime.

Taking a small finite value of a in Eq. (6.18) does not cause the chaotic behavior to disappear. By increasing a, and at the same time readjusting b so that chaos persists, one finally reaches a single chaotic attractor as shown in Fig. 6.11c.

Equation (6.17) can be interpreted as a damped linear oscillator subjected to a time-dependent potential [25]. These authors considered also other nonlinear terms besides y^2, but did not look at the case $b = 0$. Interestingly, two signs differ in their presentation (the last two ones) so that the relationship to Eq. (6.18) was not obvious.

Another nonlinear equation which, like Eq. (6.18) with a small, produces 'small divergence chaos' (if such a term is allowed) was studied numerically by Ueda et al. [26]; for a recent account, see [27]. Ueda's equation is the periodically forced two-variable oscillator

$$\begin{cases} \dot{x} = y \\ \dot{y} = \mu(1 - \gamma x^2)\dot{x} - x^3 + B\cos(\omega t) \end{cases} \tag{6.21}$$

similar to the Cartwright–Littlewood oscillator but containing two third-order nonlinearities.

The simplicity of Eq. (6.17) and (6.18), which exceeds that of Eq. (2.2), may render it an appropriate 'guinea pig' for further numerical and analytical studies. Recently, Guckenheimer [28] suggested to look at the bifurcations of rotation-symmetric flows between two saddle-foci oriented like those of Eq. (6.17)—a saddle spiralling outward and another spiralling inward, both connected as if forming an onion (which is the case in Eq. (6.17) for the parameters chosen in Fig. 6.11a). Maybe, even an algebraic criterion for the emergence of an infinitesimally small 'onion' can be found.

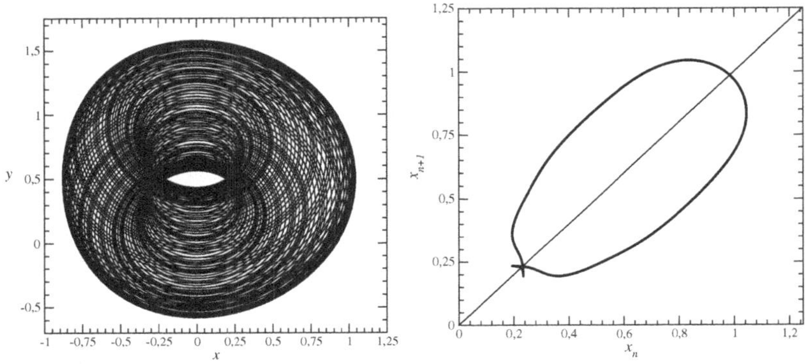

(a) Attracting torus: $a = 0.2$, $b = 0$
Initial conditions: $x_0 = 0.3$, $y_0 = 0$, and $z_0 = -0.25$

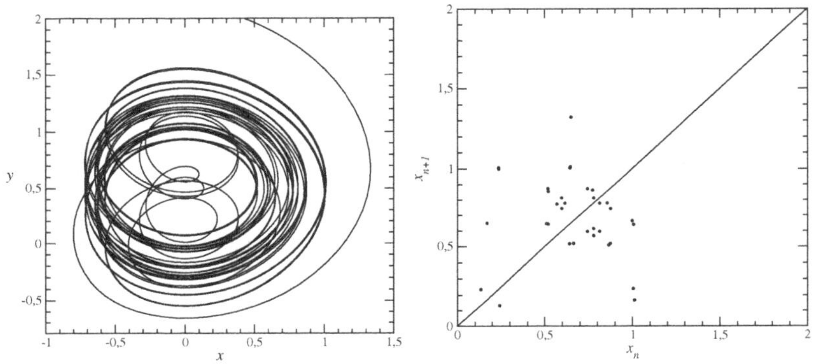

(b) Conservative metastable chaos: $a = 0.4$, $b = 0$
Initial conditions: $x_0 = 0.3$, $y_0 = 0$, and $z_0 = -0.25$

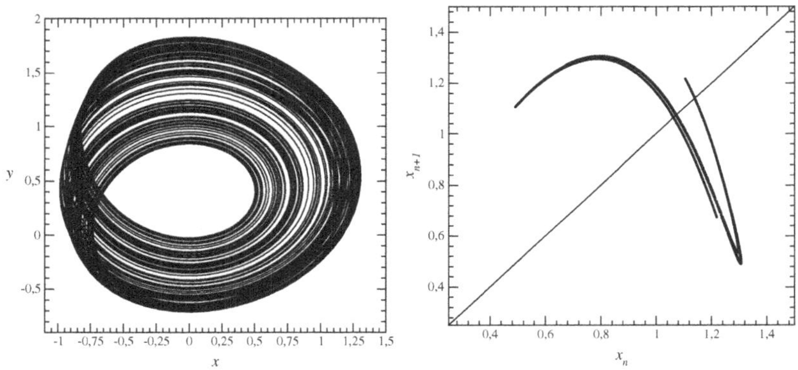

(c) Chaos: $a = 0.386$, $b = 0.2$
Initial conditions: $x_0 = 0.4$, $y_0 = -0.4$, and $z_0 = -0.7$

Fig. 6.11 Recurrent trajectories solution to Eq. (6.18)

6.5 Higher-Order Toroidal Chaos

The simplest torus-generating three-variable system of Eq. (6.18), with $b = 0$, admits a straightforward generalization to the analogous four-variable system

$$\begin{cases} \dot{x} = -y - z - w \\ \dot{y} = x \\ \dot{z} = a\left(y - y^2\right) - bz \\ \dot{w} = c\left(\frac{z}{2} - z^2\right) - dw \end{cases} \tag{6.22}$$

which produces now a 'hypertorus.'

Here the amplitude s of the 'higher-order' oscillator (r, z) forms, with w as the second variable, an analogous 'even higher order' oscillator (s, w). A numerical simulation is presented in Fig. 6.12. The result is a 'periodically breathing torus' or, rather, a whole family of such tori (organized in a 'hyperonion' which has yet to be sliced [29]).

Again, an underlying decomposable equation can be indicated. This four-variable version of Eq. (6.22) contains three sub-oscillators now

$$\begin{cases} \dot{x} = -y + 1 + \frac{x}{r}\left(-\overline{a}z - \frac{rw}{s}\right) \\ \dot{y} = x + \frac{y-1}{r}\left(-\overline{a}z - \frac{rw}{s}\right) \\ \dot{z} = \overline{a}r - \frac{zw}{s} \\ \dot{w} = c(s - 0.5)\,, \end{cases} \tag{6.23}$$

where $r = \sqrt{x^2 + (y-1)^2}$, $s = \sqrt{r^2 + z^2}$, $\overline{c} = \sqrt{c}$. Equation (6.22) stands in the same relationship to Eq. (6.23) as Eq. (6.18) does to (6.19).

Incidentally, there is another possibility for obtaining a four-variable decomposable oscillator containing three uncoupled sub-oscillators:

$$\begin{cases} \dot{x} = -y + \frac{x}{r}(0.5 - s) \\ \dot{y} = x + \frac{y}{r}(0.5 - s) \\ \dot{z} = -cw + \frac{z}{s}a(r - 1) \\ \dot{w} = cz + \frac{w}{s}a(r - 1)\,, \end{cases} \tag{6.24}$$

with $r = \sqrt{x^2 + (y-1)^2}$, $s = \sqrt{z^2 + w^2}$. Here the amplitudes of two sub-oscillators (r, s) interact, so that only a 'first-order' higher-order oscillation is formed. This more symmetric higher analogue of Eq. (6.19) may also possess a pruned (simplified) version analogous to Eq. (6.22). As Eq. (6.23) show, 'pruning' has so far been effective only down to two quadratic terms. Whether a single quadratic term may be sufficient

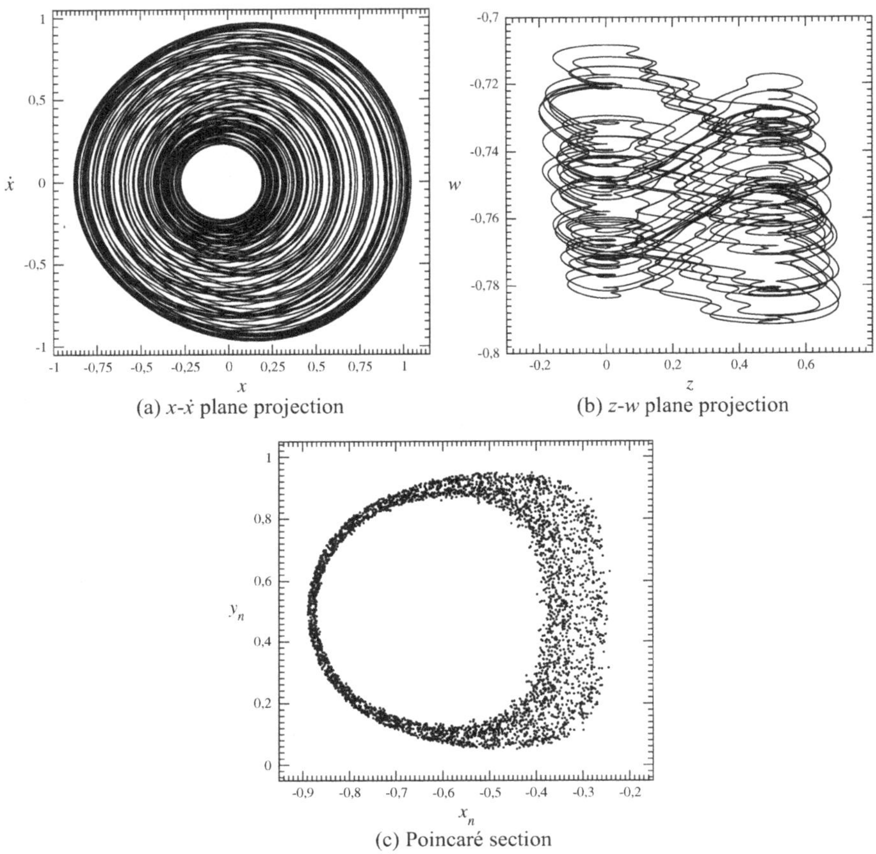

Fig. 6.12 Hypertoroidal behavior in Eq. (6.22). In (a), the solution is seen lying on a 'breathing' three-dimensional torus. In (b), another three-dimensional projection lying on a 'torus without a hole', is shown. In (c) the Poincaré section defined by $\dot{x}_n = 0$ is shown. Parameters: $a = 0.2, b = 0$, $c = 0.04$, and $d = 0$. Initial conditions: $x_0 = 0$, $y_0 = 0.78$, $z_0 = 0.2$ and $w_0 = -0.75$

for generating a hypertoroidal motion in a four-variable system starting, for example, from Eq. (6.24) is presently an open question.[8]

Another remark concerns the symmetries implicit in Eqs. (6.23) and (6.24): each possesses a number of 'stereoisomers' so to speak. The group-theoretical properties of all possible equations of the type of Eq. (6.19) may be amusing to look at.

We now come to the question of higher chaos in any one of the above three equations (6.22)–(6.24). No serious attempts at finding it has been made yet. However, the fact that an equation even simpler than Eq. (6.22), namely, Eq. (4.1), was already sufficient, makes it appear likely that, by tuning parametric 'knobs', any of the pre-

[8]To the best of our knowledge, this is still an open question.

ceding three equations is capable of a transition from higher order toroidal behavior toward higher order chaotic behavior.

So far, only one hypertoroidal system with hyperchaotic behavior has been indicated; it is even more complicated than any of the above equations, but has the asset of being reducible to the reinjection principle. For curiosity, the whole equation may be presented briefly:

$$\begin{cases} \dot{x} = \omega y + \frac{x}{r}\left(-z + F(s) + \frac{x}{r}G(z) - wH(z)J(s)\right) \\ \dot{y} = -\omega x + \frac{y}{r}\left(-z + F(s) + \frac{x}{r}G(z) - wH(z)J(s)\right) \\ \dot{z} = \mu\,(s + K(z)) \\ \dot{w} = \nu\left(s^2 + L(w)\right), \end{cases} \tag{6.25}$$

where $r = \sqrt{x^2 + y^2}$, $s = r = 3$, and

$$\begin{aligned} F(s) = {} & s + p(s+1) - \sqrt{p^2(s+1)^2 + \varepsilon} + p(s-1) + \sqrt{p^2(s-1)^2 + \varepsilon} - \delta p(s+2) \\ & + \sqrt{\delta^2 p^2 (s+2)^2 + \varepsilon} - \delta p(s-2) - \sqrt{\delta^2 p^2 (s-2)^2 + \varepsilon} \\ & - \sqrt{[\delta p(s+1) + p]^2 + \varepsilon} + \sqrt{\delta^2 p^2 (s+1)^2 + \varepsilon} + \sqrt{[\delta p(s-1) - p]^2 + \varepsilon} \\ & - \sqrt{\delta^2 p^2 (s-1)^2 + \varepsilon} \\ G(z) = {} & 1.5z\left[-0.1 + \sqrt{(z+0.1)^2 + \varepsilon} - \sqrt{z^2 + \varepsilon}\right] \\ H(z) = {} & 2.5\left[0.1 - \sqrt{(z-0.1)^2 + \varepsilon} + \sqrt{z^2 + \varepsilon}\right] \\ J(s) = {} & s + \sqrt{s^2 + \varepsilon} \\ K(z) = {} & 0.375\left(z - \sqrt{z^2 + \varepsilon}\right) \\ L(w) = {} & -3 + w + \sqrt{w^2 + \varepsilon} \end{aligned}$$

Ideally, $\delta \to \infty$, $\varepsilon \to 0$, $p = 1.25$, $\mu \to 0$, and $\frac{\nu}{\mu} \to 0$.

This equation is less 'messy' than it looks. For example, $F(s)$ simply determines a 'double Z' shaped function which allows for two attracting and one intermediate repelling limit cycle in the s-z sub-oscillator of relaxation type. A second relaxation sub-oscillator exists between w and the (bistable) amplitude of the s-z system. A third sub-oscillator (x,y) produces a periodic forcing for the s-z system. The nonlinear function G, H and J, as well as those in the second and third line of Eq. (6.25), only play a similar 'simplifying' role as the function H in Eq. (6.11) does. A simulation of Eq. (6.25) has been performed (Fig. 6.13) [15]. For the limiting flow μ and $\frac{\nu}{\mu}$ approaching zero, a two-dimensional cross section can be calculated in principle. Nonetheless, the much greater simplificity of Eq. (4.2) above with Eq. (4.7) as its cross section makes Eq. (6.25) appear superseded. Nonetheless, it still has the asset of being capable of an analytically accessible hypertoroidal motion as well.

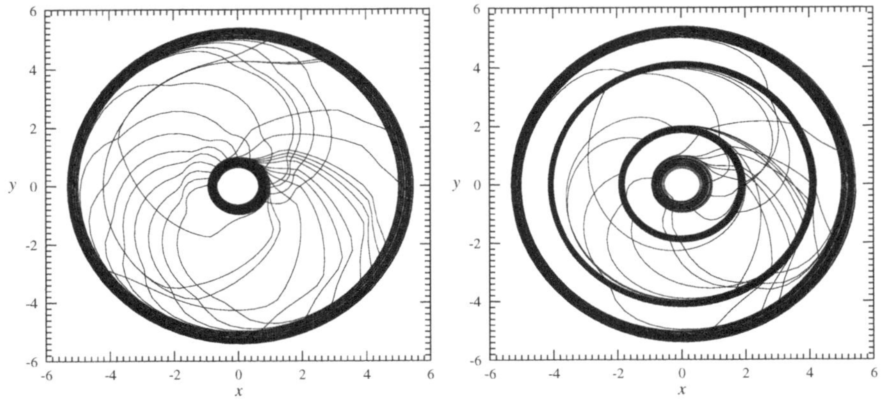

(a) Lower-level chaos ($\mu = 0.05$ and $\nu = 0.0$) (b) Hyperchaos ($\mu = 0.1$ and $\nu = 0.05\mu$)

Fig. 6.13 Chaos in equation (6.25). Parameter values: $\omega = 0.85$, $\varepsilon = 10^{-3}$, $\delta = 4$, $p = 2$. Initial conditions: $x_0 = 4$, $y_0 = 0$, $z_0 = 0.8$, and $w_0 = 0$ for (**a**) and (**b**) except $x_0 = 3.973$ and $z_0 = -1.81$ for the latter

6.6 Near-quasi-Periodic Chaos

Equations (6.19), (6.23) and (6.24) are typical examples of 'quasiperiodic flows' with two or three periodic components. Their flows are confined to tori (T^2, if there are two components or T^3, if there are three). Of course, if all the frequencies involved stand in a finite integer relationship, the compound oscillation is not quasiperiodic but periodic. But this rational special case, which of measure zero, makes no difference in what follows.

Historically, one of the first proposals to search for Poincaré type (homoclinic-point type) motions in ordinary (non-conservative) time-continuous systems—like chemical reaction systems described by differential equations—was made with explicit reference to the class of quasi-periodic (toroidal) systems. Ruelle and Takens [30] proposed that any dynamical system involving four uncoupled sub-oscillators (not all of them linear) should, after the introduction of the slightest amount of coupling, almost always possess a 'strange attractor'.

A strange attractor in the sense of Ruelle and Takens [30] is an axiom $\mathscr{A}$ attractor in the sense of Smale [31]. Briefly, it is an attracting chaotic regime that

1. occurs in a diffeomorphism (or flow with diffeomorphic cross section) and
2. possesses only hyperbolic trajectories (being attractive in one direction and repelling in another; see Sect. 7.2).

Later, the handy notion became gradually synonymous with 'attracting chaotic regime lacking any internal periodic attractor.'

The idea behind this proposal was straightforward. Peixoto [32] had shown that any system involving **two** uncoupled sub-oscillators, not all of them linear, should after being subjected to an infinitesimal coupling almost always possess a *periodic attractor*. The reasoning is easy to follow: The fully uncoupled motion is a quasi-periodic flow on a torus; any coupling (being unidirectional as the simplest possibility) renders one of the two motions dependent on the phase of the other. Thus the torus is going to be 'distorted' in a functional sense. A pair of adjacent trajectories passing through the distorted region will in general be affected differently. That is, overall frequencies will be changed differentially as a function of initial conditions. Even if the coupling is infinitesimal, this smooth infinitesimal change is bound to generate a periodic trajectory (with a rational winding ratio) somewhere, and so with an arbitrarily long periodicity. Neighboring trajectories, being either less or more strongly affected, will spiral either away from or towards the periodic trajectory. Thus in general a pair of periodic trajectories, one attracting and the other repelling, are generated by the coupling. The same argument can also be made in terms of the cross section alone, using the theory of diffeomorphisms on a circle [32]. Again, the 'attractor-free' state is unstable with respect to almost all possible parametric perturbations.

Peixoto's limit cycle can be called 'near quasi-periodic', since distinguishing it empirically from a quasi-periodic motion requires near-infinite observation time. Ruelle and Takens [30] considered the case after next, so to speak: four rather than two oscillators. Their argument has since been sharpened to apply to three oscillators already [33], which is the next case right after Peixoto's example.

Three functionally uncoupled oscillators, one of them a limit cycle oscillator, generate a torus-shaped attracting cross section, the latter being a diffeomorphism onto itself. Without coupling, this cross section is being punctured along quasi-periodic trajectories—since the system then is identical to a two-oscillator quasi-periodic system whose state happens to be observed stroboscopically (with the frequency of the third oscillator).

The slightest coupling between only two sub-oscillators recreates Peixoto's situation, stroboscopically observed. As soon as a second unilateral coupling is introduced, however, coming from the third oscillator used for observation this means that two 'distorting influences' take place simultaneously now, one acting on one direction along the torus, as before, while the other does the same thing in another direction.

The whole toroidal diffeomorphism will therefore again develop fixed points (periodic motions) at some high iterate. In fact, the absence of such fixed points is again 'non-generic' (almost never occurring) in toroidal diffeomorphisms. Their minimum number is now four (compression and expansion along two axes): two saddle points, a sink and a source.

It is known that a toroidal diffeomorphism which contains but a single fixed point of saddle type consists wholly of homoclinic points [34]. This is because the two manifolds of the saddle wrap around the torus in a quasi-periodic way, thereby intersecting densely. If there are other fixed points besides a saddle, however, as is the case here, in principle an ordinary (homoclinicity-free) phase portrait is possible. In this case, the phase portrait will, however, look very different from a simple 'criss-crossing' pattern, as it is the hallmark of the uncoupled situation. (If a quasi-periodic trajectory, wrapping a torus like yarn, is being observed stroboscopically, the points lie on two families of curves simultaneously.) If the emerging fixed points are now to affect this pattern only imperceptibly at first, this means that there will be homoclinic crossings in abundance again—much as in the single-saddle case. Birkhoff's [12] and Smale's [31] theorems then guarantee the presence of both an infinite number of periodic solutions of differing periodicities and an uncountable number of periodic solutions (see Sect. 8.1).

This derivation was far from being rigorous. It may nonetheless suffice to give a 'feeling' for the chaotic potentialities of infinitesimally coupled oscillators in general. Newhouse et al. [33] actually built a much stronger case, providing evidence for the chaotic regime to be i) the only attractor formed and ii) of axiom $\mathscr{A}$ type in the sense of Smale [31], that is, free of 'internal' periodic attractors. Specifically, the formation of Plykin's [35] attractor, involving nine fixed points in a rather rigidly specified pattern (see Sect. 7.4), was postulated.

No matter what the precise nature of the chaotic solutions formed in a neighborhood of the uncoupled state, however, they are near-quasi-periodic. That is to say, they are physically unobservable in that region of parameter space in which their existence is assured.

The present situation is apparently different from one encountered earlier (Sect. 2.6) where only two oscillators were infinitesimally coupled, but with one of them possessing a limit continuum. There, the slightest coupling sufficed to generate a chaotic motion (of Lorenzian type) whose properties were markedly different from those of the uncoupled case, meaning that the chances for its being actually observable were greater from the beginning.

6.7 The 'Bracelet' Hypothesis

The following system of coupled ordinary differential equations was indicated by Hopf [36]:

$$\begin{cases} \dot{x} = -x - y^2 - z^2 \\ \dot{y} = bz + y\left(ax - cx^2\right) \\ \dot{z} = -by + z\left(ax - cx^2\right) . \end{cases} \tag{6.26}$$

It was described as an example of a system that is capable of multiple-level 'Hopf bifurcations,' as they are called today: By changing parameters, it is possible to proceed, both from a damped oscillatory motion (around a steady state) toward a self-maintained one (of infinitesimal amplitude), and from a damped toroidal motion (around a limit cycle) toward a self-maintained one (of infinitesimal toroidal width). Hopf [36] added diffusion terms to the right hand sides of Eq. (6.26) and showed that in this case an infinite hierarchy of analogous, higher bifurcations is possible.

Equation (6.26) is of the same basic structure as Eqs. (6.8) and (6.20) are. The equation

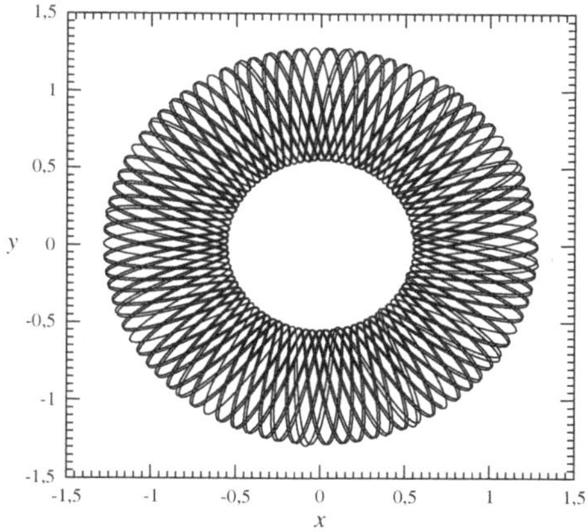

Fig. 6.14 A bracelet-shaped toroidal attractor with numerically lacking periodic attractor. Numerical simulation of Eq. (6.27). Parameter values: $a = b = c = d = 0.1$. Initial conditions: $x_0 = 0.543$, $y_0 = 0$, and $z_0 = 1$

$$\begin{cases} \dot{x} = x - ay - xz \\ \dot{y} = bx + y - yz \\ \dot{z} = x^2 + y^2 + cz - \dfrac{z}{z+d} \end{cases} \tag{6.27}$$

is even closer to Eq. (6.26), and happens to have been analyzed numerically for some asymmetric parameter values ($c \neq c_1$) [14, 37]. If $c = c_1 > 0$, and $a > a_c > 0$, this equation produces an attracting torus (as Eq. (6.8) and (6.26) do). The simulation shown in Fig. 6.14 applies to the case $c_1 = 0$ and b relatively small. The toroidal attractor seen here is just on the brink toward chaos. It has the shape of Zeeman's [38] Möbius-strip analogous 'bracelet'. A cross section through the attracting surface is shaped like a periodic cycloid with three cusps. Similar games can be played with higher-order analogues of Eq. (6.26), like Eq. (6.23) with a term $-ew^3$ added to the fourth line (so that the hypertorus becomes attracting).

The present class of toroidal and hypertoroidal equations is of interest as a possible 'testing ground' for a mathematical controversy. Higher-order Hopf bifurcations were recently postulated to occur easily in realistic systems [39], in contrast to the position taken by Ruelle and Takens [30]. In the same vein, the above equation favor the impression (to be gained from Eqs. (6.18), (6.14), and (6.22), respectively) that the 'weak coupling behaviors' (near-quasi-periodic limit cycle; near-quasi-periodic chaos) of the preceding section have a tendency 'not to appear' if the nonlinearities are all of low order, as is the case in all the above equations. Both proposals may be worth checking numerically (by looking at the spectra of sample trajectories, for example).

References

1. M.L. Cartwright, J.E. Littlewood, On nonlinear differential equations of the second order: I. The equation $\ddot{y} - k(1 - y^2)\dot{y} + y = b\lambda k \cos(\lambda t + \alpha)$, k large. J. Lond. Math. Soc. **20**, 160–189 (1945)
2. M.L. Cartwright, On non-linear differential equations of the second order. Math. Proc. Camb. Philos. Soc. **45**, 495–501 (1949)
3. A. Liénard, Etudes des oscillations entretenues. Revue Générale de l'Electricité **23**, 901–912 and 946–954 (1928)
4. N. Minorski, *Nonlinear Oscillations* (R. E. Krieger, Huntington, 1974)
5. * R. FitzHugh, Impulses and physiological states in theoretical models of nerve membrane. Biophys. J. **1**, 445–466 (1961)
6. * K.F. Bonhoeffer, Über die Aktivierung von passiven Eisen in Salptersäure. Zeitschrift für Electrochemie **47**, 147 (1941)
7. * K.F. Bonhoeffer, Activation of passive iron as a model for the excitation of nerve. J. General Physiol. **32**, 69–91 (1948)
8. * K.F. Bonhoeffer, Modelle der Nervenerrung. Naturwissenschaften **40**, 301–311 (1953)
9. *O.E. Rössler, R. Rössler, H.D. Landahl, Arrhythmia in a periodically forced excitable system (Abstract), in *Proceedings of the Sixth Biopysical Congress, Kyoto (Japan)* (1978), p. 296
10. * S. Rajasekar, M. Lakshmanan, Controlling of chaos in Bonhoeffer-van der Pol oscillator. Int. J. Bifurc. Chaos **2**(1), 201–204 (1992)
11. * U. Parlitz, W. Lauterborn, Period-doubling cascade and devil's staircases of the driven van der Pol oscillator. Phys. Rev. A **36**(3), 1428–1434 (1987)
12. G.D. Birkhoff, On the periodic motions of dynamical systems. Acta Math. **50**, 359–379 (1927)
13. J. LaSalle, Relaxation oscillations. Q. Appl. Math. **7**, 1–19 (1949)
14. O.E. Rössler, Chaos in abstract kinetics. Two prototypes. Bull. Math. Biol. **39**, 275–289 (1979)
15. O.E. Rössler, Chaotic oscillations: an example of hyperchaos. Lect. Appl. Math. **17**, 141–156 (1979)
16. O.E. Rössler, A synthetic approach to exotic kinetics, with examples. Lect. Notes Biomath. **4**, 546–582 (1974)
17. O.E. Rössler, Quasi-periodic oscillation in an abstract reaction system (abstract). Biophys. J. A **17**, 281 (1977)
18. * J.A. Yorke, E.D. Yorke, Metastable chaos: the transition to sustained chaotic behavior in the Lorenz model. J. Stat. Phys. **21**(3), 263–277 (1979)
19. O.E. Rössler, Continuous chaos – four prototype equations. Ann. New York Acad. Sci. **316**, 376–392 (1979)
20. M.W. Hirsch, S. Smale, *Differential Equations, Dynamical Systems, and Linear Algebra* (Academic, New York, 1974)
21. V.I. Arnold, *Ordinary Differential Equations* (The MIT Press, Cambridge, 1978)
22. H. Poincaré, Sur le problème des trois corps et les équations de la dynamique. Acta Math. **13**, 1–271 (1890)
23. G.D. Birkhoff, Recent advances in dynamics. Science **51**, 51–55 (1920)
24. V.I. Arnold, A. Avez, *Ergodic Problems of Classical Mechanics, Benjamin, New York, 1970 – First Edition in French Théorie Ergodique des systèmes dynamiques* (Gauthier-Vilars, Paris, 1967)
25. P. Coullet, C. Tresser, A. Arnéodo, Transition to stochasticity for a class of forced oscillators. Phys. Lett. A **72**(4), 268–270 (1979)
26. Y. Ueda, N. Akamatsu, C. Hayashi, Computer simulation of nonlinear ordinary differential equations and non-periodic oscillations. Trans. Inst. Electron. Commun. Eng. A (IECE, Japan), **56**, 218–225 (1973). English translation in Electron. Commun. Jpn. A, **56**, 27–34 (1973)
27. Y. Ueda, Explosion of strange attractors exhibited by Duffing's equation. Ann. New York Acad. Sci. **357**, 422–434 (1980)
28. J. Guckenheimer, On a codimension two bifurcation. Lect. Notes Math. **898**, 99–142 (1981)

29. O.E. Rössler, Chemical turbulence. Synopsis, in *Synergetics. A Workshop*, ed. by H. Haken (Springer, Berlin, 1977), pp. 184–197
30. D. Ruelle, F. Takens, On the nature of turbulence. Commun. Math. Phys. **20**, 167–192 (1971)
31. S. Smale, Differentiable dynamical system. Bull. Am. Math. Soc. **73**, 747–817 (1967)
32. M.M. Peixoto, Structural stability on two-dimensional manifolds. Topology **1**, 101–120 (1961)
33. S. Newhouse, D. Ruelle, F. Takens, Occurrence of strange axiom $\mathscr{A}$ attractors near quasi-periodic flows on T^m, $m \geq 3$. Commun. Math. Phys. **64**(1), 35–40 (1978)
34. D.V. Anosov, Geodesic flows on closed Riemannian manifolds of negative curvature. Trudy Matematicheskogo Instituta imeni VA Steklova **90**, 3–210 (1967)
35. R.V. Plykin, Sources and sinks of $\mathscr{A}$-diffeomorphisms of surfaces. Math. USSR-Sbornik **23**(2), 233–264 (1974)
36. E. Hopf, A mathematical example displaying features of turbulence. Commun. Pure Appl. Math. **1**, 303–322 (1948)
37. O. Gurel, O.E. Rössler, Bifurcation to toroidal surfaces. Math. Japonica **23**, 491–507 (1979)
38. E.C. Zeeman, *Catastrophe Theory. Selected Papers 1972–1977* (Addison-Wesley, Boston, 1977)
39. H. Haken, Lines of development of synergetics, In *Dynamics of Synergetics Systems*, ed. by H. Haken (Springer, Berlin, 1980), pp. 2–19

Chapter 7
Chaos and Reality

7.1 Some Everyday Examples

If complicated motions indeed occur so easily as has been suggest above, chaos should be among the ubiquitous phenomena of the world. This hypothesis seems not at variance with reality. To start out with a 'far-fetched' example, nonperiodically oscillating astronomical objects might be mentioned. Here the class of Cepheid pulsating stars is especially attractive. As described in 1939 by Wesselink [1], there is an abundance of recordings available. Figure 7.1 presents such a time trace.[1]

Interestingly, the authors explained the behavior of their pulsators in terms of relaxation oscillations produced by a periodically forced van der Pol oscillator, quoting van der Pol [3, 4] and le Corbeiller [5]. They postulated that oscillations of the interior of star are not of the sinusoidal type but of a non-sinusoidal type ('relaxation oscillation') in the stellar envelope (according to Eddington [6]). A pulsating star can therefore be viewed as a periodically forced van der Pol oscillator as discussed in Chap. 6 and, consequently, as we know now, can be of a chaotic nature.

A more 'down-to-earth' example is the *dripping tap*. Whether in the kitchen or elsewhere, the tap makes a nice experimental object. Experience tells that for every tap (supposed it has a sufficiently horizontal and not too small mouth) there is a flow rate (somewhat slower than that yielding a continuously running thread) characterized by nonperiodic dripping [7]. A possible explanation is that at this rate, the budding next droplet is already big enough to suffer noticeable repercussions from the last falling-off event. This induced damped oscillation then influences the threshold of the next falling event, and so forth.

[1] The time series originally selected was not saved and we replaced it by another one. The text was therefore slightly modified. It should be noted that there is still not yet clear evidence that the light curve of the SS Cygni has a chaotic underlying dynamics [2].

O. E. Rössler and C. Letellier, *Chaos*, Understanding Complex Systems,
https://doi.org/10.1007/978-3-030-44305-4_7

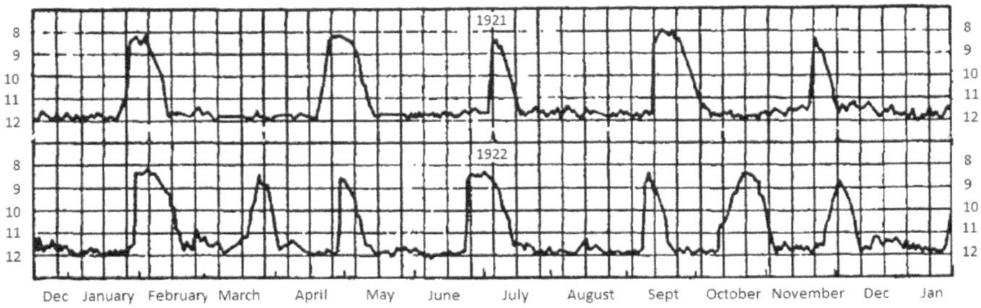

Fig. 7.1 Light-curve of SS Cygni in the years 1921–1922. (Adapted from [1])

Perhaps it would be better to give a time course than an idealized explanation. Farmer et al. [8][2] prepared a whole movie to present the effect. A detailed study in the spirit of Lord Rayleigh may be worth carrying out.

Some years ago, a rather energy-economic gadget was in vogue: the beak dipping-bird, shown in Fig. 7.2. Again, there is a relaxation event of single-threshold type (namely, when upon touching the water the cold beak is suddenly soaked in water of room temperature, so that through the ensuing expansion of the vapor in the anterior part of the device, the bird's head and neck are rather suddenly cleared of fluid again), combined with an oscillatory process (resulting damped oscillation) and a slow motion (the head becoming heavier and heavier under the cooling effect of the soaked—drying—cloth around the beak, hence dipping deeper and deeper...).

There are some more examples of chaos in our everyday environment. Most sounds, for example, are 'impure' (in the sense applied to those 'undesirable' sounds sometimes inadvertently created by child practicing on a flute). These unsettling tones are open to the same analysis to which the former examples were subjected. In many such cases, however, the underlying system turns out to be described by partial differential equations rather than by ordinary ones. Systems in which the spatial extension of the variables plays only a negligible role—like those discussed in the previous examples—are comparatively infrequent. On the other hand, if one listens to the time behaviors of Fig. 4.1, the 'sound of chaos' is not unfamiliar to the ear. One is immediately reminded of the noises made by a 'strittering' idling motor, and of different types of snores, respectively, as was mentioned in the context of Fig. 4.1.

All kinds of hissing[3] and snarling and sizzling sounds are also of chaotic type. So if one wishes, half the consonants of the alphabet represent some kind of chaotic motion. The rest are either of an explosive type or correspond to a resonant filter. (The vowels of course are periodic.) Obviously, a rigorous analysis of these acoustic phenomena will require taking into account partial differential equations (see Sect. 10.9). Other chaotic realizations observed in the real world are briefly described in Appendix C.

---

[2]This lead to few publications by the Santa Fe group [9, 10] where Otto E. Rössler is recognized as being the first to suggest the dripping faucet as a chaotic system.

[3]William Duddell observed some aperiodic oscillations on a musical arc he termed 'hissing' [11].

Fig. 7.2 The beak-dipping bird gadget. The glass bird is suspended at low friction. The body contains a low-boiling fluid. The beak is wrapped in a piece of cloth. The fluid in the beaker is water. The mode of action is explained in the text

7.2 Towards a Definition of Chaos

Chaotic behavior is easy to generate, and also more or less easy to recognize. But how does it fit in with the rest of dynamical phenomena, and what are its most basic features? The best way to characterize it, apparently, to say that it is 'typical three-dimensional behavior'. Dynamical systems in the plane cannot show chaos, but in three-variable systems it occurs quite easily. Still, this is not yet a definition.

Dynamical systems were invented by Poincaré (although that notion was introduced only later by Birkhoff [12]). Poincaré saw that the trajectories formed in systems governed by smooth rate laws are determined by very simple patterning rules: They can show all the configurations—called 'phase portraits' by Poincaré—that are possible in stationary flows of fluids (including gases). Later, dynamical systems were axiomatically defined according to the same analogy. In Bhatia and Szegö's textbook [13], for example, a dynamical is defined as a pair X, π where X is Euclidean space (in the simplest case) and π, the rate law, is a map with simple properties: in dependence on the parameter R (interpretable as time), π assigns a unique image to every point of X. Hereby, a commutativity and transitivity law applies, meaning that a unique thread (streamline) is determined running through every regular point. Dynamical systems have a simple semi-group structure.

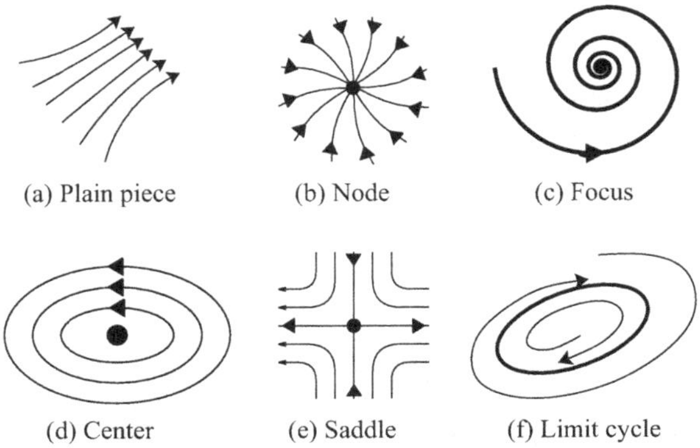

Fig. 7.3 Poincaré's elements of two-dimensional stationary flows. All arrows can be inverted. Plain piece is nowadays called spray [14]

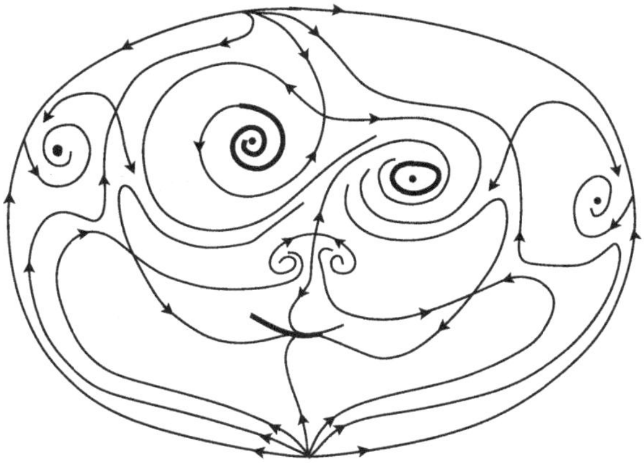

Fig. 7.4 Example of a special phase portrait (a 'phace'). There are four basins (domains of attraction) and six anti-basins (domains of repulsion)

The possible portraits of streamlines in a stationary flow (or, even more pictorially, of hair in a hairdo) are, necessarily, simple. In two dimensions, there are (apart from locally more or less parallel regions) only a few types of 'sinks' and 'sources' (and combinations thereof) possible: the stable (or unstable) node, the stable (or unstable) focus, with the intermediary case of the center, and the saddle point. In addition, there is the attracting (or repelling) limit cycle (Fig. 7.3).

All possible two-dimensional phase portraits can be generated using these four kinds of elements. And for almost all values of the parameters in the underlying rate laws, these elements are mutually separated. Moreover, the attracting and repelling elements are arranged in a simple order, with separating trajectories delineating their domains of attraction (and repulsion, respectively). See Fig. 7.4 for a special 'phase portrait'.

For example, two limit cycles lying on the same torus are in most cases interlaced, that is, cannot be pullet apart. Since all kinds of (locally) two-dimensional manifolds can be embedded in three-dimensional space, all kinds of 'knots' lying on these man-

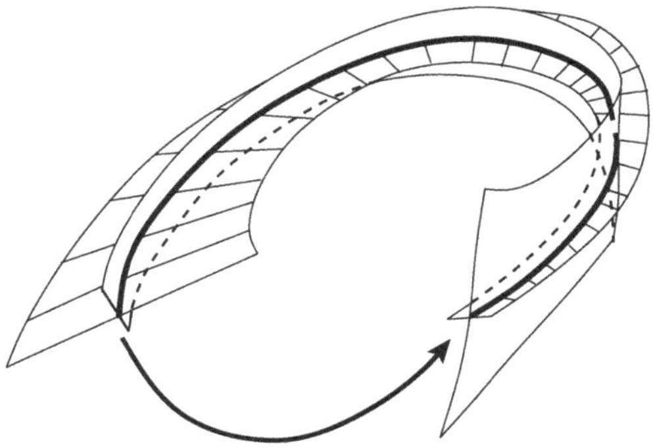

(a) Tangent space to an unstable periodic orbit

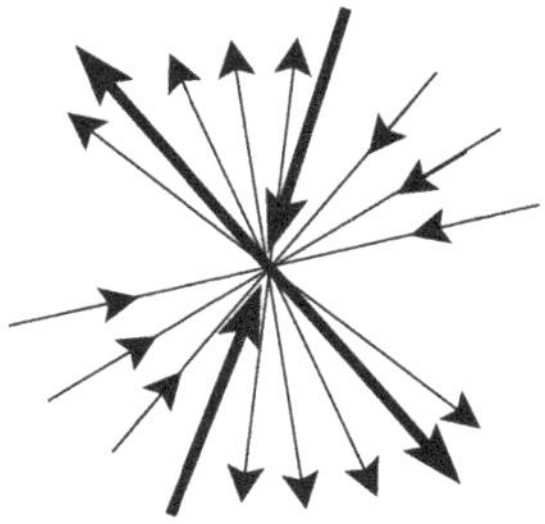

(b) Direction of attraction and repulsion

Fig. 7.5 Behavior of neighboring trajectories close to a both repelling and attracting periodic solution in three-dimensional space (**a**). After the gluing together: formation of an 'average' direction of attraction and repulsion, respectively (**b**)

ifolds can be expected.[4] Still, even fully three-dimensional flows should in principle be easy to follow and understand (when navigating along them in three-dimensional space), because all the paths are rigidly suspended ('frozen') and locally parallel, just as they are in two-dimensional dynamical systems.

'No,' because such stationary flows can already generate infinite complexity,[5] as Poincaré [17] first realized. The magic word is a principle of recurrence or (as stated in the Introduction) 'perichoresis' in the sense of Anaxagoras.

When travelling inside a jungle of stationary streamlines, one may after a while pass close to regions that one has visited already. This means that taking into account a whole neighborhood of sufficient extension and looking at it simultaneously, putting all the local laws together into one describing the fate of all the trajectories simultaneously), one arrives at a more compact description of what is going on. Obviously, the interesting case is that in which one passes through the same cross section ('surface de section') more than once. This is why the term recurrence was chosen by Birkhoff [12].

For example, it is not hard to visualize a flow that never changes very much locally but, after one 'round,' nonetheless arrives with a quite different orientation (Fig. 7.5). Trajectories are approaching a closed trajectory in one direction (vertically), and are repelled from it in another (horizontally), but these local orientations change over the circumference of the closed trajectory. A cross section through the whole bundle then will show only an 'averaged' effect (Fig. 7.5b).

Very close to the saddle (the closed trajectory), where all the directions are infinitesimal, the net law can be described by a smooth two-dimensional local flow. That is to say, one has an ordinary saddle point of the same kind as one would have in

[4]Such a two-dimensional manifold was schemed as the universal template by Robert Ghrist and Philip Holmes [15]. See [16] for an explicit relationships with the Chua circuit.

[5]This should be understood as 'of infinite periodicity'.

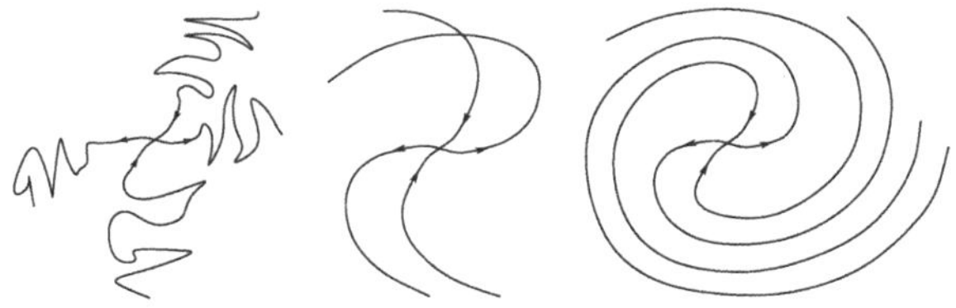
(a) (b) (c)

Fig. 7.6 Three possible 'outside fates' of the saddle's manifolds in a two-dimensional cross section through a three-dimensional flow. See text

a smooth two-dimensional flow (Fig. 7.3d). Somewhat farther away from the saddle, however, things are not so clear-cut. Does the averaged behavior then still translate into smooth 'pseudo trajectories'? At first sight, this is highly unlikely to occur, for the behavior of neighboring trajectories farther away from the closed solution can be subjected to quite differing local influences during their journey, influences whose average need not have anything in common with the average close to the periodic solution.

On the other hand, if one starts out close and closer to the saddle, continuity arguments make a 'prolongability' of the local stable and unstable manifold an inescapable conclusion. Different versions of this so-called stable manifold theorem are to be found in Hartman's [18], and Abraham and Marsden's [19] books. Of the three basic possibilities which offer themselves (Fig. 7.6), thus the medium one has the greatest probability of being true, striking a compromise between an ordinary two-dimensional smooth flow (Fig. 7.6c) and a situation where pseudo-trajectories cease to be definable at all (Fig. 7.6a).

The plausibility of this 'compromise' may be enhanced by the sample flow sketched in Fig. 7.7. In this picture, freely adapted from Fig. 2.5b and Eq. (2.2), there is both a 'horizontal repulsion' away from a periodic trajectory, within a 'sheet' that finally bends over underneath, and a 'vertical attraction' toward the periodic trajectory, within another sheet that finally is being intersected by the former underneath. The purpose of Fig. 7.7 would be served even better by a real three-dimensional set of wires, shaped like those in the Figure, along which one then could move with a co-moving cross section, thereby really convincing oneself that the abstractly postulated 'layers' or 'sheets' indeed can be followed around as suggested. The plausibility would be enhanced even more if these wires could be identified with the 'raisins' within a two-dimensional piece of dough (or taffy), and if the moving frame, following the flow around, was indeed an automatic 'taffy-pulling' arm, rotating about the middle hole of Fig. 7.7a and producing a single folding-over per rotation (rather than 6, as the arm of Fig. 1.2 would).

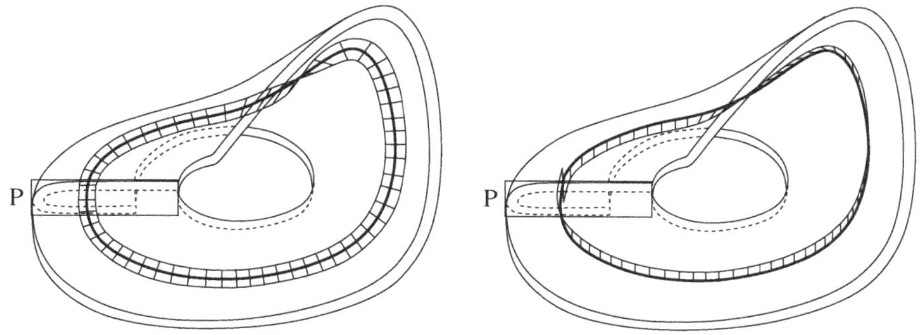

(a) 'Horizontal' sheet (b) 'Vertical' sheet

Fig. 7.7 Possibility of having two intersecting invariant sheets in a flow in three-dimensional space. Compare with Fig. 2.5b

7.3 Homoclinic Point Implies Chaos

Poincaré [17] named the hypothetical intersection point between the stable and the unstable manifold of a saddle point in a cross section (seen at the top of Fig. 7.6b) a homoclinic point, because of its being asymptotic to the saddle in both directions of time. (Homoclinic means: tending toward the same.) Obviously, the crossing of two 'trajectories' in a cross section is only a reflection of some fundamental difference between 'trajectories' in cross sections and trajectories in continuous flows. Rather than trying to follow up and clarify this basic difference, Poincaré [17] simply set out to look at the implications of the single new entity encountered—so as if looking at this part held a promise for understanding the whole.[6]

The first implication of having a trajectory in three-dimensional space that is doubly asymptotic to a closed one, is that this trajectory is bound to stay doubly asymptotic to the periodic trajectory for all (positive and negative) times. In other words, it has to stay on its two leaves ('trajectories' in the cross section) for all times.

This yields the following well-known three pictures: Fig. 7.8. Since the trajectory (homoclinic point) is to approach the saddle more and more closely at subsequent iterates, along the stable manifold, the unstable manifold (which the homoclinic point cannot leave) is bound to intersect the stable one at every subsequent iterate, thereby 'doubling up' on the saddle the mode densely the closer the iterated homoclinic point comes to the saddle (Fig. 7.8a). The same is true for negative t (Fig. 7.8b). Hence also the infinite grid of further homoclinic points near the saddle (Fig. 7.8c). A single doubly asymptotic recurrent trajectory thus implies an infinite number of further such trajectories, each of infinite periodicity by definition.

[6]Poincaré identified homoclinic and heteroclinic points in the third volume of his *Méthodes nouvelles de la mécanique céleste* but "*not even try to draw*" the pictures shown in Fig. 7.8, arguing that "*nothing is more suitable for providing us with an idea of the complex nature of the problem*" [20].

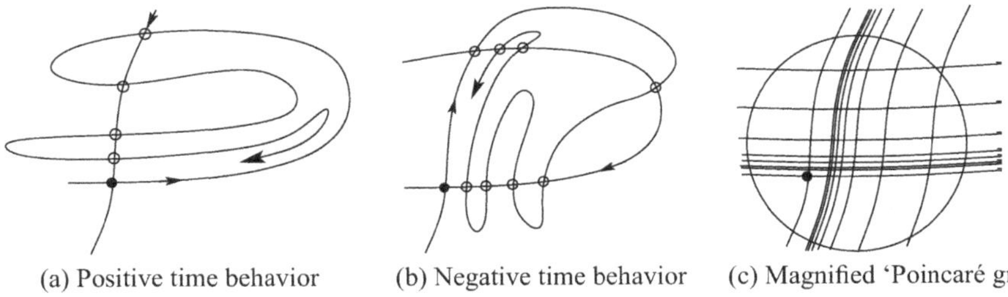

Fig. 7.8 Further homoclinic points (after Poincaré, 1899)

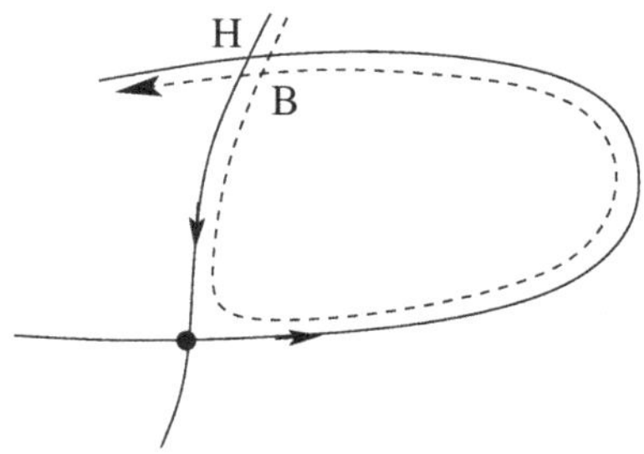

Fig. 7.9 Birkhoff's self-intersection of pseudo-trajectories in a neighborhood of a homoclinic point. H = homoclinic point, B = Birkhoff's point (see text)

Birkhoff [12] later introduced another interesting intersection point, lying close to a homoclinic point (Fig. 7.9). The existence of this point is based on the assumption that a cross section also has a 'phase portrait,' with a 'trajectory' running through every point, and with these 'trajectories' being locally parallel just as those in an ordinary (continuous) phase portrait are. Taking this 'hidden assumption' [7] for granted, Birkhoff [12] showed that, since all these self-intersecting 'trajectories' are punctured most densely near the saddle, they harbor a continuum of fractional 'hitting numbers' inside the loops formed. And although the probability of a single intersection point B harboring a periodic solution is zero, the whole continuum is bound to contain an infinite number of such periodic solutions, each of a differing periodicity. Thus, a single homoclinic point implies also an infinite number of periodic solutions of differing periodicities.[8]

Several years later, Smale [23] showed that the picture of Fig. 7.8a can be redrawn without loss of generality by introducing the assumption that a finite neighborhood to the saddle is described by a linear map, or rather, by two linear pieces of a horseshoe-shaped nonlinear map (Fig. 7.10). One sees that all points to the right of a certain line inside the right-hand linear region (B) will be mapped to the left of the saddle after one iterate (stripe '1' in Fig. 7.10b). Similarly everything in stripe '2' will be mapped into stripe '1' after one iteration, and everything in stripes '3' into stripe '2', and so on. Thus, almost all initial conditions are mapped toward the left of the saddle in the long run. What remains is a Cantor set of (infinitesimally thin) stripes. All

[7] Igor Gumowski, personal communication, 1979.

[8] This was investigated by Leonid Shilnikov [21, 22].

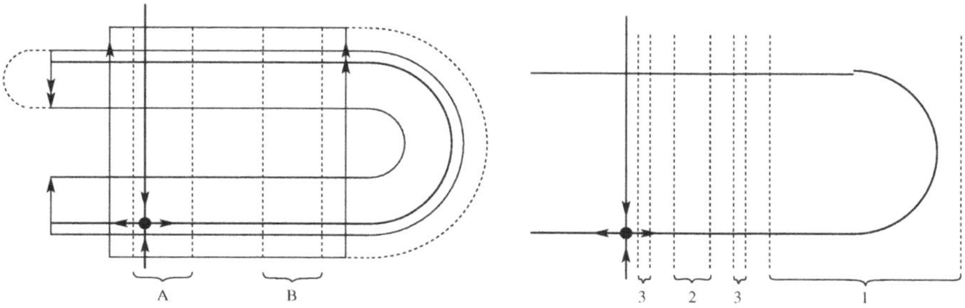

Fig. 7.10 Piecewise linear neighborhood of a homoclinic point. The portions of the two legs inside the box (**a**) are assumed to be linear images of the stripes A and B, respectively. Stripes are being 'thrown out' after 1, 2 and 3 iterates. See text

points inside this Cantor set stay to the right of the saddle for all t. This set contains solutions of all periodicities, including infinite ones.

This is seen most easily by looking only at the horizontal axis of the map of Fig. 7.10a. The values along this axis do not depend only on the last value, as far as the linear region of the map (between the saddle and the beginning of region '1') is concerned. A single-variable description suffices to understand the whole map in the region of interest, because the contraction in vertical directions ensures that there is a fixed point in the whole map for every fixed point of the horizontal submap.

The pertinent one-dimensional map is shown in Fig. 7.11a. Everything outside the box shown is being mapped to the left and beneath the bottom. So is the region 2, after one round, and region 3, after two, and so on. What remains is a Cantor set along the x-axis. To understand better what happens, the second, third and fourth iterate of Smale's map[9] are shown in Fig. 7.11b. One sees that not only more and more 'holes' are formed (of initial conditions having no image within the box), but also that the number of intersection points of the map with the identity line is growing. The first iterate (Fig. 7.11a) has two such intersection points, corresponding to periodic solutions. The second has four, the third eight, the fourth sixteen, and so on. The nth has 2^n. Each corresponds to a periodic solution. Those of the first iterate have period 1, those of the second that are new have period 2 (their total being 2), those of the third that are new have period 3 (there are 4 of them), and those of the nth iterate that are new (namely $2^n - 2^{n-1}$) have period n.

As it is easy to verify (because the slope at the intersection point is larger than 1 or less than -1, respectively), all these periodic solutions are unstable. And their total number, in the limit of n going to infinity, is uncountable. For the cardinality of 2^n in the limit of $n \to \infty$ is that of the continuum [24]. In other words, there are unstable periodic solutions of all periodicities. Since 'all' periods comprise all possible permutations of infinitely many numbers, all real numbers (between 0 and 1, say) are required to 'count' the periodic solutions of infinite periodicity. These solutions are

[9] Smale's map is now commonly called the horseshoe map.

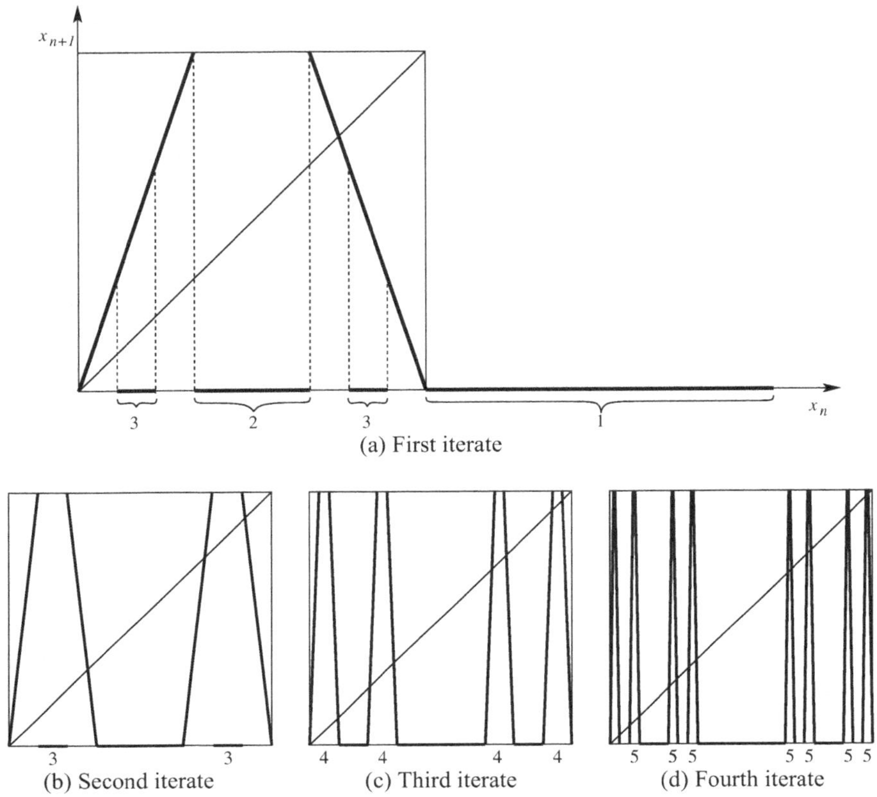

Fig. 7.11 The one-dimensional map underlying Smale's linear submap. Four iterates of this map

sometimes also called 'nonperiodic'. The remaining (nonsingular) solutions in the Cantor set are also nonperiodic.

There are other way to show the presence of solutions of all periods (and nonperiods) in the map of Fig. 7.10. The most common method is that of symbolic dynamics [23]. Thus, a homoclinic point implies unstable solutions of all periods in its neighborhood. The presence of such a set is, however, what Li and Yorke [25] chose for the definition of chaos.[10]

[10]Notice that here is retained the presence of all unstable periodic orbits for ensuring the presence of chaos, and not the too simple "period-3 implies chaos".

7.4 Chaos and Hyperbolic Attractors

Smale [23] has the idea that hyperbolic, that is, internally repelling, regions may nonetheless be attracting to the outside. Smale's example was the 'solenoid,' sketched in Fig. 7.12. There is a solid annulus (made of dough, for example) and its image, which is elongated, shrunk (with volume reduction), and wrapped up once before being put back inside. If the original dough is assumed red, and any new dough, poured into the annular form to fill up after the thinned double loop has been put in (so that the process can be repeated) is assumed white, then it is easy to imagine what happens under iteration: the red core becomes thinner and thinner exponentially, and more and more multiple-stranded exponentially. If the baker allows one to inspect his product at regular intervals by cutting out a slice, there are more and more (and more and more tiny) 'eyes,' a pair inside the position of each former (bigger) eye. In the limit, a single 'red line,' of zero volume, but with 2^n (that is, uncountable many) strands, is formed.

So one has an attractor with a Cantor set structure, formed in the limit of infinite iteration number inside the diffeomorphic map of Fig. 7.12. At the same time, all internal points of the attractor are hyperbolic: any pair of lateral neighbors (two raisins, say) present in the initial annulus are exponentially drifting away from each other during the iteration, just as two lateral neighbors near a saddle do (Fig. 7.3d). The whole limiting object can therefore be called a hyperbolic attractor. Somewhere within the map, there is also a homoclinic point. Assume there is a fixed point in the map of Fig. 7.12 (there always is one, as can be seen by looking at the inverted map), then this fixed point has a two-dimensional stable manifold cutting through the ring, and a one-dimensional unstable manifold running along the ring. The latter is bound to hit the plane after one round, creating a slightly generalized homoclinic point. The latter implies, again, an uncountable set of singular solutions (a 'basic set' in Smale's terminology [23]) and hence, chaos.

The problem of how the two-dimensional stable manifold is to 'double up' on the one-dimensional unstable manifold (under negative t) is, by the way, an interesting spatial puzzle. At any rate, we now have a chaotic regime which at the same time is attracting. The map of Fig. 7.11 is, in this respect, analogous to the map of Fig. 3.2 which produced a Lorenz attractor. (The latter, however, does not possess a homo-

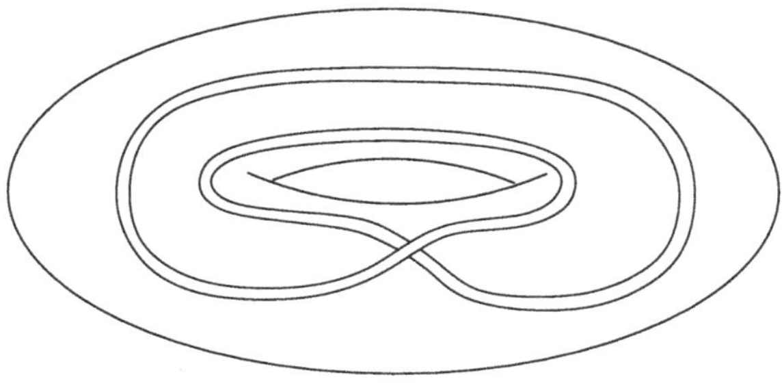

Fig. 7.12 The solenoid [23] as an example of a hyperbolic attractor. Only the zeroth iterate (a solid annulus) and the first iterate (lying inside) are shown

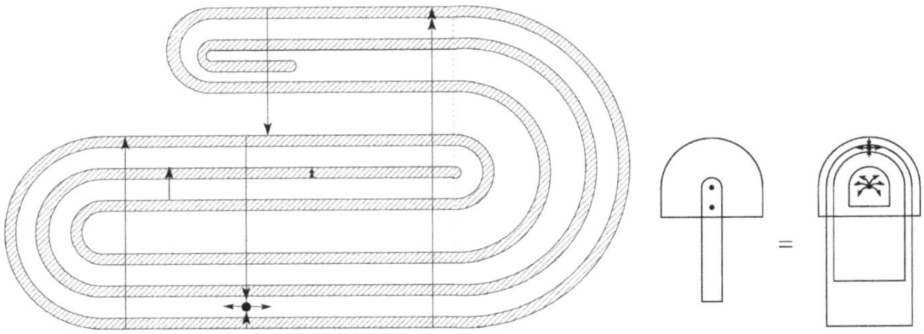

Fig. 7.13 Plykin's diffeomorphism [29], redrawn with some freedom. The two medium pieces are each a (somewhat longwinded and multiply folded) horseshoe map. In addition there are three 'caps maps,' shown in detail at the right. Each has two fixed points (a saddle and an unstable node) and is shaped like a thumbnail for easy remembering. Since the two central portions contain three saddles, there are nine fixed points in all

clinic point *sensu stricto*, since all unstable manifolds are being cut continually; see Sect. 9.5).

Smale [23] called the class of attractors of which he had found the first abstract example, 'axiom $\mathscr{A}$' attractors. This means, essentially, that none of the periodic solutions contained in the diffeomorphism is attracting. Ruelle and Takens [26] later changed that label into 'strange' attractor. The latter term was handy enough to be applied to non-diffeomorphic maps like that of Fig. 3.2 [27]. Williams [28] had used the notion 'expanding' attractor in the context of a non-invertible flow on a branched two-dimensional manifold having the map of Fig. 3.1b as its cross section.

For a while there was the problem whether a two-dimensional diffeomorphism which, like the three-dimensional solenoid-generating map, produces an axiom $\mathscr{A}$ attractor, was at all possible. The first, and so far only, example is Plykin's map [29] as shown in Fig. 7.13. This map consists of two horseshoe maps, each with more than one folding-over, and three 'caps' shaped like a thumbnail each. The three thumbnail maps do the trick: everything that happens to be mapped into one of the curved regions (along the arcs running through them) is going to be expelled from there again, except for the saddle itself. Distances along the arcs are hereby increased. The image, consisting of straight stripes and arcs only, is thus being elongated everywhere—just as in the solenoid. Except that everything is two-dimensional now.

The map is, of course, insensitive to slight variation of its form, as well as to slight variations in the local 'elongation factor' (just as the map of Fig. 7.11 was). On the other hand, the map has to be crafted more or less to shape. For example, the nine fixed points must all retain their properties and approximate relative positions. Also, the upper 'corner' (endowed with a down-pointing marker arrow in Fig. 7.13) is critical as far as the position of its image is concerned. This means that the chances are slim at best that an explicit equation (like Eq. (2.3), but more complicated) meeting all the requirements will be found. A continuous flow possessing this map as a cross section is also unlikely to be indicated soon. Interestingly, an explicit flow or equation for the map of Fig. 7.11 is also still to be found.[11]

The simple maps of Figs. 2.7 and 2.11 lack the provisions, paintstakingly incorporated by Plykin [29] into the map of Fig. 7.13, for making their 'knee' regions

---

[11] Such a flow was proposed by Sergey P. Kuznetsov [30]. According to Plykin's result, hyperbolic attractor on a sphere requires at least four holes which are not belonging to the attractor. Let take these non-visited domains centered around the four points of coordinates $(x, y, z) = (\pm\delta, 0 \pm \delta)$ where $\delta = \frac{1}{\sqrt{2}}$. The flow

$$\left|\begin{array}{l} \left\{\begin{array}{l} \dot{x} = \kappa\left[-\epsilon xy + \zeta(t)(z+\delta)\right] y \\ \dot{y} = \kappa\left[+\epsilon xy - \zeta(t)(z+\delta)\right] x + \gamma y\left(1 - x^2 - y^2 - z^2\right) \quad \text{if } 0 \leq \text{ mod}(t, 2T) < T \\ \dot{z} = 0 \end{array}\right. \\ \left\{\begin{array}{l} \dot{x} = 0 \\ \dot{y} = \kappa\left[+\epsilon yz - \zeta(t)(x+\delta)\right] z + \gamma y\left(1 - x^2 - y^2 - z^2\right) \quad \text{if } T \leq \text{ mod}(t, 2T) < 2T \\ \dot{z} = \kappa\left[-\epsilon yz + \zeta(t)(x+\delta)\right] y \end{array}\right. \end{array}\right.$$

where

$$\zeta(t) = \left|\begin{array}{l} 0 \ \text{ if } 0 \leq \text{ mod}(t, T) < T - \tau \\ 1 \ \text{ if } T - \tau \leq \text{ mod}(t, T) < T \end{array}\right.$$

matches with these requirements. This is not a continuous flow since it is piecewise. The stroboscopic section

$$\mathscr{P}_{\text{stroboscopic}} \equiv \left\{(X_n, Y_n) \in \mathbb{R}^2 \mid t = 2nT\right\}$$

with

$$\left|\begin{array}{l} X_n = \dfrac{z_n - x_n}{x_n + z_n + \sqrt{2}} \\ Y_n = \dfrac{y_n\sqrt{2}}{x_n + z_n + \sqrt{2}} \end{array}\right.$$

of that flow reveals a Pylkin attractor as shown here for parameter values as follows. $\kappa = 1.1$, $\epsilon = 0.1$, $T = 10$, $\tau = 2$, and $\gamma = 0.25$.

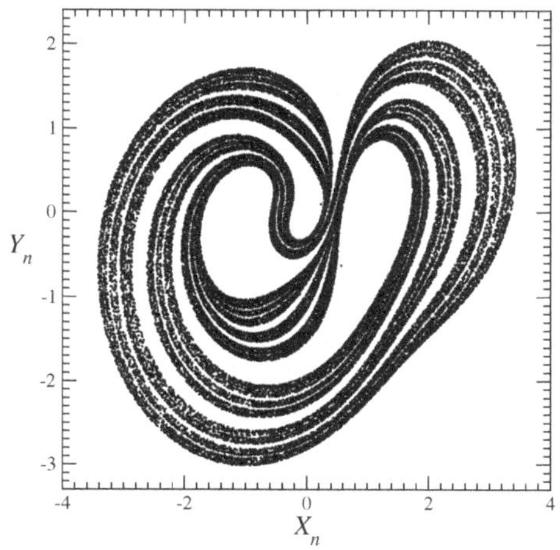

'laterally expelling' throughout. As a consequence, one cannot exclude that somewhere within their knee—or some of its iterates—a 'contracting pocket' forms. On the contrary: the strong 'vertical compression' present in the knee region, which is preserved (and thereby aggravated) over a good many iterates before the 'horizontal elongation' takes over for a while, makes full compensation unlikely in general. Thus, the simple walking-stick map (Fig. 2.7 or Fig. 2.11) will in general contain a contracting region of finite size mapped upon part of itself at some iterate. It implies the presence of a periodic attractor of the same period.

Thus, the chaotic regimes numerically simulated or calculated in any of the previously presented systems, except the 'Lorenzian' systems of Sects. 3.1–3.3, cannot be said to possess a strange attractor (and hence to produce chaotic solutions for all times). No explicit example of axiom $\mathscr{A}$ flow is presently known. In other words, there is no difference in principle between Smale's original horseshoe map (Fig. 7.11, where a point sink (periodic attractor of period 1) was assumed to be present to the left of the saddle, and the map of Fig. 2.7 (or Fig. 2.11) where a periodic sink of high unknown period is likewise present 'in general.' The corresponding (high) iterate simply represents a multiply folded nonlinear horseshoe map with point sink.

On the other hand, the 'region of dominating influence' of the point sink (where neighboring trajectories have the same periodicity already as their final attractor) normally is very small. Even there was no external noise (truncation errors in the calculation), the 'chaotic monoflop' of Fig. 2.7 possessed unpredictability long wandering times. This fact of 'manifest chaos' over long (and unpredictable) times provides a practical justification for applying the word chaos not only to systems showing chaotic solutions for all times for almost all initial conditions—that is, to systems with a strange attractor—, but also to systems which merely possess a 'chaotic attractor.' The latter is an attracting regime containing an uncountable set of singular solutions and, perhaps, also some associated attracting periodic solutions. A strange attractor is a chaotic attractor, but not *vice versa*.[12] A non-practical, purely

[12]At that time, the word chaos designated aperiodic solutions that could not be described more accurately. The terms "strange" or "chaotic" were used equivalentely at least up 1984, that is, up to the paper entitled "strange attractors that are not chaotic" by Celso Grebogi, Edward Ott, Steve Pelikan and James Yorke [31]. These scientists choose to use these two words as follows:

> *Chaotic* refers to the dynamics on the attractor, while *strange* refers to the geometrical structure of the attractor. [...]
>
> *Definition.* A chaotic attractor is one for which typical orbits on the attractor have a positive Lyapunov exponent.

Otto Rössler therefore already understood that there was a difference to make between strange and chaotic. Today, chaotic means "sensitive to initial conditions" and strange is associated with the property announced by Ruelle and Takens [26], that is, an "attractor which is locally the product of a Cantor set and a piece of two-dimensional manifold." Thus, although Guckenheimer and coworkers used strange, the formal definition can be taken for chaotic. It was [27]:

> By a "[chaotic] attractor", for a map $f(\cdot)$ we mean an infinite set Λ with the following properties:
>
> 1. Λ is invariant under $f(\cdot)$, i.e. $f(\Lambda) = \Lambda$.

mathematical reason for this terminological choice can be given also (see Sect. 7.2). Presently, we may return to the problem of hyperbolic attractors.

Besides axiom $\mathscr{A}$ systems with a basic set, there are also axiom $\mathscr{A}$ without a basic set. That is, there exist hyperbolic attractors which are not chaotic. They were recently discovered by Sato [33] and Dankwart [34]. The (non-explicit) diffeomorphisms presented by these authors are five-dimensional, however, so that a simple picture catching the gaze of their argument is somewhat hard to present. So it may suffice to state that hyperbolic attractors within which periodic solutions are not dense, exists too. This means that just as not all chaotic attractors are hyperbolic attractors, so not all hyperbolic attractors are chaotic attractors.

References

1. * A.J. Wesselink, Stellar variability and relaxation oscillations. Astrophys. J. **89**, 659–668 (1939)
2. * J.K. Cannizzo, D.A. Goodings, Astrophys. J. ii, **334**(1), L31–L34 (1988)
3. B. van der Pol, On "relaxation-oscillations". Philos. Mag. **7**(2), 978–992 (1926)
4. * B. van der Pol, *Norsk Riks-Kring Kastings Forelesninger* (1934), p. 234
5. * P. le Corbeiller, *Les systèmes auto-entretenus et les oscillations de relaxation* (Hermann (Paris), 1931)
6. * A. Eddington, *The Internal Constitution of the Stars* (Cambridge, 1926), p. 198
7. O.E. Rössler, Continuous chaos, in *Synergetics. A Workshop*, ed. by H. Haken (Springer, Berlin, 1977), pp. 174–183
8. J.D. Farmer, J.P. Crutchfield, H. Frœling, N.H. Packard, R.S. Shaw, Power spectra and mixing properties of strange attractors. Ann. New York Acad. Sci. **357**, 453–472 (1980)
9. R. Shaw, *The Dripping Faucet as a Model Chaotic System* (Aerial Press Edition, 1984)
10. J.P. Crutchfield, J.D. Farmer, N.H. Packard, R.S. Shaw, Chaos. Sci. Am. **254**(12), 46–57 (1986)
11. * W. Duddell, On rapid variations in the current through the direct-current arc. J. Inst. Electr. Eng. **30**(148), 232–267 (1900)
12. G.D. Birkhoff, On the periodic motions of dynamical systems. Acta Math. **50**, 359–379 (1927)
13. N. Bhatia, G.P. Szegö, *Dynamic Systems: Stability and Applications* (Springer, New York, 1967)

---

2. Λ has an orbit which is dense in Λ.
3. Λ has a neighborhood a consisting of points whose orbits tend asymptotically to Λ : $\lim_{t\to\infty} f^{(t)}(a) \subset \Lambda$.

The requirement that Λ be infinite guarantees that Λ consists of more than a single periodic orbits.

The central—and startling—fact about chaotic attractors is that orbits on or near them may behave in an essentially chaotic and unpredictable fashion. Thus, despite the fact that the model is completely deterministic, the dynamical behavior of trajectories can only be predicted statistically!

It is interesting to note that the properties of determinism was only implicit in Grebogi *al.* definition, a lack of explicitness which induced a huge amount of papers only proving with a positive Lyapunov exponent the chaotic nature of the behavior which was investigated. But as remarked by Glass in the late eighties, "prior to asserting that a behavior is chaotic, there should be a clear evidence that deterministic equations govern its dynamics" [32].

14. Th. Bröcker & K. Jänich, *Einführung in die Differentialtopologie*, Springer-Verlag (Berlin), 1973
15. * R.W. Ghrist, P.J. Holmes, An ODE whose solutions contain all knots and links. Int. J. Bifurc. Chaos **6**(5), 779–800 (1996)
16. * C. Letellier, R. Gilmore, The universal template is a subtemplate of the double-scroll template. J. Phys. A **46**, 065102 (2013)
17. H. Poincaré, Sur le problème des trois corps et les équations de la dynamique. Acta Math. **13**, 1–271 (1890)
18. P. Hartman, *Ordinary Differential Equations* (Wiley, New York, 1964)
19. R. Abraham, J.E. Marsden, *Foundations of Mechanics*, 2 enlarged edn. (Benjamin/Cummings, Reading, 1978)
20. H. Poincaré, *Les méthodes nouvelles de la mécanique céleste*, vol. iii (Gauthier-Vilars, 1899)
21. * L.P. Shilnikov, A case of the existence of a denumerable set of periodic motions. Soviet Math., Doklady **6**, 163–166 (1965)
22. * L.P. Shilnikov, The existence of a denumerable set of periodic motions in four-dimensional space in an extended neighborhood of a saddle-focus. Soviet Math., Doklady **8**(1), 54–58 (1967)
23. S. Smale, Differentiable dynamical system. Bull. Am. Math. Soc. **73**, 747–817 (1967)
24. G. Cantor, Über unendliche, lineare Punktmannigfaltigkeiten v (On infinite linear point-manifolds). Math. Ann. **21**, 359–379 (1883). (Also published separately as Grundlagen einer allgemeinen Mannigfaltigkeitslehre, Leipzig, 1883)
25. T.-Y. Li, J.A. Yorke, Period-3 implies chaos. Am. Math. Mon. **82**, 985–992 (1975)
26. D. Ruelle, F. Takens, On the nature of turbulence. Commun. Math. Phys. **20**, 167–192 (1971)
27. J. Guckenheimer, G.F. Oster, A. Ipaktchi, Periodic solutions of a logistic difference equation. J. Math. Biol. **4**, 101–147 (1976)
28. R.F. Williams, Expanding attractors. Publications Mathématiques de l'Institut des Hautes Etudes Scientifiques **43**, 169–203 (1974)
29. R.V. Plykin, Sources and sinks of $\mathscr{A}$-diffeomorphisms of surfaces. Math. USSR-Sbornik **23**(2), 233–264 (1974)
30. * S.P. Kuznetsov, Plykin type attractor in electronic device simulated in MULTISIM. Chaos **21**, 043105 (2011)
31. * C. Grebogi, E. Ott, S. Pelikan, J.A. Yorke, Strange attractors that are not chaotic. Phys. D **13**(1–2), 261–268 (1984)
32. * L. Glass, Chaos and heart rate variability. J. Cardiovasc. Electrophys. **10**, 1358 (1999)
33. K. Sato (1979)
34. P.V. Dankwerts, Continuous flow systems: distribution of residence times. Chem. Eng. Sci. **2**(1), 1–13 (1953)

Chapter 8
Maps

8.1 Chaos and Structural Stability

Poincaré's discovery [12] that two-dimensional phase portraits do not change 'qualitatively' for a while if someone turns the knobs (that is, changes the parameters of the underlying equation for a sufficiently small finite amount), was recognized as a principle by Andronov and Pontryagin [2]. Systems with this 'coarse' property are nowadays called 'structurally stable' [18]. For a while, there was a lively debate going on whether this property is a 'generic' property of dynamical systems in general (that is, essentially, whether 'almost all' systems of nonlinear differential equations possess this property). A first counter-example in terms of an abstract four-dimensional diffeomorphism was given by Smale [16]. Since then, numerous ways to 'amend' the definition of structural stability have been proposed and abandoned, all with the aim to find a property of dynamical systems of arbitrary dimensionality that indeed is insensitive to sufficiently small finite parametric changes. The argument was that natural systems, in order to be observable at all, must be structurally stable in a certain sense [18]. See Chillingworth [8] for a detailed account of these attempts.

The original definition of structural stability in two dimensions focused on separatrices (see Sect. 7.2). Whenever a basin (domain of attraction of an attractor) disappeared, or a new one appeared, one said that the structure of the system had changed qualitatively. The set of parameters at which this change occurred was called a bifurcation set, and the system was said to be 'structurally unstable' along the point and boundaries in parameter space made up by this set. One hoped that in parameter space, such points were 'rare' (so that an arbitrarily picked parameter value would almost never belong to such a set—unless the system was in the sense 'constrained,' for example, symmetrical).

An easy way to recognize changes in the basin structure is to focus on the separatrices (stable and unstable manifolds) of saddles. As long as these do not merge with an approaching critical point, or an approaching limit cycle ('approaching' under sufficiently small finite changes of the parameters), the basin structure remains unchanged in two-dimensional systems. Changes of stability of a former attractor (as in the Hopf

O. E. Rössler and C. Letellier, *Chaos*, Understanding Complex Systems,
https://doi.org/10.1007/978-3-030-44305-4_8

bifurcation, where a formerly stable focus becomes unstable) do also fit into this picture. Since the state space remains more or less unchanged outside, automatically the formation of a new separatrix (an infinitesimally small stable limit cycle surrounding the newly hatched unstable focus) is implied [9]. A thorough description of the two-dimensional possibilities—with many pictures—is to found in Andronov et al. [3]. Thom [18] showed that gradient systems always behave in essentially the same manner, irrespective of their dimensionality, because such systems by definition always have a simple basin structure.

Moving over to three dimensions, we are confronted with Poincaré's finding [13] that the generalized stable and unstable manifolds which are now possible (imposing on a cross section as 'trajectories') are no longer separatrices in many cases, because they can intersect in 'points' which are not singular (limiting) points. See Fig. 7.6b and the next following Figures. This embarrassing finding meant that points lying to the right of a 'separatrix' (namely, the stable manifold of a saddle in the cross section) no longer necessarily stay there. In terms of Smale's picture (Fig. 7.10), one sees that the sink to the left of the saddle has a basin that is infinitely 'splintered': there is an (uncountable) infinitude of 'stripes' to the right of the saddle which nonetheless, after finite or infinite times, are going to be attracted by the sink. A basin in a three-dimensional space thus may consist of an uncountable number of disconnected regions.

An even clearer picture of the new situation is obtained if one looks at the trajectories in Birkhoff's picture [7] (Fig. 7.8), but over a somewhat longer extension: see Fig. 8.1. Assuming, as Birkhoff did [7], that a neighboring non-saddle 'trajectory' (dashed manifold in Fig. 8.1) keeps a certain distance to the (undashed) stable and unstable manifolds of the saddle to which it is 'parallel', we observe that such 'trajectory' is bound to be crossed by those doubled-up segments of the unstable that come closer to the saddle than it does itself.

On the other hand, the dashed line by the same assumption keeps running along these doubled-up segments (as a kind of 'envelope'). This means that it must cross itself again and again, not only along its 'vertical' portion that accompanies the stable

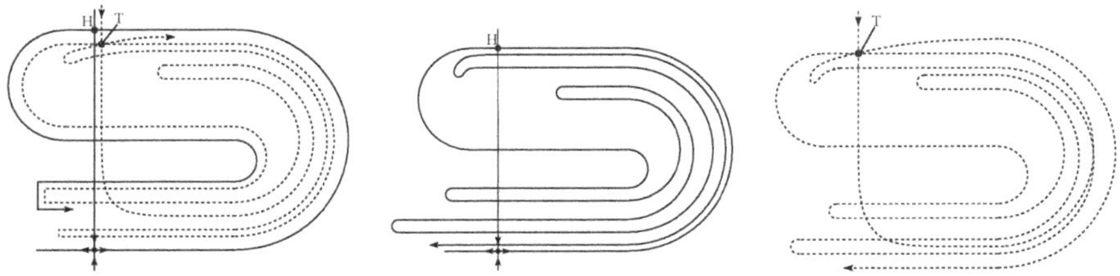

(a) Whole picture
(b) Detail of the saddle's manifolds ('doubling up structure')
(c) Detail of a non-saddle manifold

Fig. 8.1 Birkhoff's picture (prolongation of Fig. 7.8). H = homoclinic point, T = point of triple self-intersections (see text)

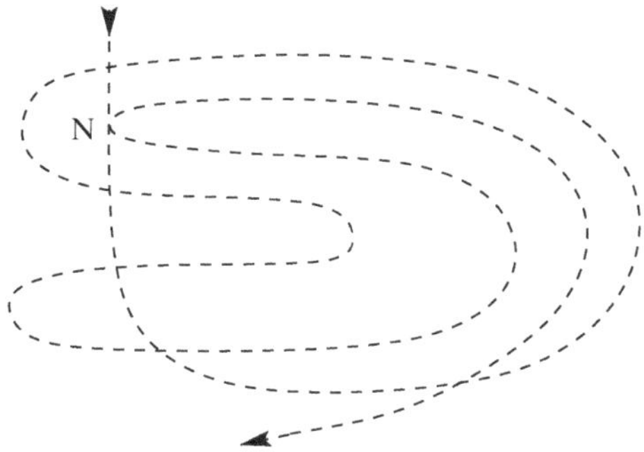

Fig. 8.2 Another singular implication of Fig. 7.8: a non-transversal point (N). Compare with Fig. 8.1c

manifold, but also beyond the region of the saddle. As a consequence, triple crossings (as shown) become in principle possible.

Such triple crossings are, of course, 'infinitely rare'. Nonetheless, since there is a whole family of (roughly) parallel such Birkhoff manifolds, there must a countable infinite number of triple crossings be present within the family. The points T of Fig. 8.1c indeed bear a close relation to the points B of Fig. 7.8: triple self-intersection points are periodic points whenever there is an odd number of crossings in between (as in the case shown in Fig. 8.1c), as it is easy to verify.

The main reason why Fig. 8.1 was presented here is, however, because it implies a second 'singular' situation (apart from that of Fig. 8.1c), shown in Fig. 8.2: there is also a countable infinite number of cases present within the same family of manifolds in which a tangential (non-transversal) self-intersection of a 'trajectory' occurs. This case is just as likely and unlikely as the former.

Returning to the two-dimensional flow of Fig. 7.4 (or any other two-dimensional phase portrait) and the conditions for its structural stability, we recall that these flows were structurally stable whenever everything to the right of each separatrix (defined for one or the other direction of time) stayed to the right of it, especially the sink. The single way this could be violated was if the separatrix merged with the attractor. In case the separatrix contained a singular point (saddle), this saddle point had to merge with the attractor.

Since an attractor (point or limit cycle) to the right of a saddle always also attracts the right-hand unstable manifold of the saddle, the above structurally unstable situation can be characterized by the condition: tangency between a stable and an unstable manifold.

Thus, a 'non-transversal' situation analogous to that found in Fig. 8.2 defines a structurally unstable situation in two-dimensional flows (see [1]). The same violation-of-transversality condition has been applied also to cross sections: whenever a stable and an unstable manifold of a saddle intersect non-transversability in a diffeomorphism, is the underlying flow called structurally unstable [16].

This saddle case does not occur in general in the simple horseshoe map. The latter system therefore is considered to be structurally stable [17]. An exception is the case of two horseshoe maps, combined in such a way that each is the sink of the other (see

[15], for a continuous example); this case imples a 'cycle' in the sense of Newhouse and Palis [11].

If non-saddle manifolds indeed exist and behave as if they did not know they are not saddle-manifolds, then the simplest chaos-generating diffeomorphism (the horseshoe map) is already structurally unstable (Fig. 8.2). A future qualitative theory of 'phase portraits' (in quotes, that is, in two-dimensional cross sections) will provide the final answer.

In a more practical sense, the horseshoe map is structurally unstable already according to the classical definition (tangency of a saddle-manifold)—although not so with respect to infinitesimal perturbations. The doubling-up structure of Fig. 8.1b contains an infinite number of 'in between' situations, each like that depicted in Fig. 8.1b (there is one 'nose' that has already crossed the stable manifold, and there is a 'next' nose that has not yet done so; or *vice versa*). Obviously, turning knobs for an arbitrarily little (although mathematically finite) amount suffices in general for obtaining a different qualitative situation. Thus, even if there is not structural instability toward infinitesimal perturbations (as Fig. 8.2 is suggesting), there is at least 'pratical' structural instability.

Williams showed that the Lorenz attractor is structurally unstable [19], and this in a structurally stable way, that is, for a dense set in parameter space. If Fig. 8.2 is admissible, the same holds true for each horseshose map, and hence for all of the (folded-cover) chaotic flows discussed so far.

8.2 The Baker's Transformation

One of the first chaos-producing maps, as we now know, is Eberhard Hopf's example of a 'mixing transformation' proposed in 1934 [10]. It is sketched in Fig. 8.3. The picture is self-explanatory: a slab of dough is subjected to a number of professional manipulations. The map (Fig. 8.3b) underneath describes the mathematical definition. An equation generating this map is, for example,

$$\begin{cases} p_{n+1} = 2\, p_n - H(p_n - 0.5) \\ q_{n+1} = 0.5\, q_n + 0.5\, H(p_n - 0.5) \end{cases} \tag{8.1}$$

where H is the Heaviside function $H(\zeta) = 0$ if $\zeta \leq 0$ and $H(\zeta) = 1$ if $\zeta > 0$.

The inverse of the baker's transformation is identical to the baker's transformation itself, as Fig. 8.4 shows. A corresponding equation is

$$\begin{cases} p_{n+1} = 0.5\, p_n + 0.5\, H(q_n - 0.5) \\ q_{n+1} = 2\, q_n - H(q_n - 0.5)\,, \end{cases} \tag{8.2}$$

being transformed into Eq. (8.1) by an exchange of p and q.

It is now possible to put the equation into a computer and look at its behavior. One can also determine directly what happens. The second iterate looks like the right-

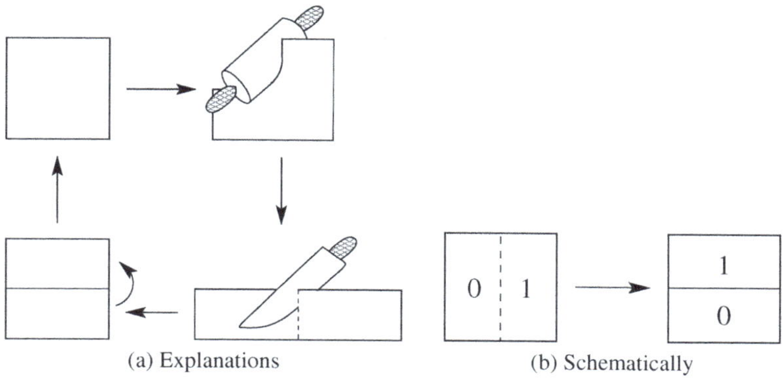

Fig. 8.3 Baker's transformation

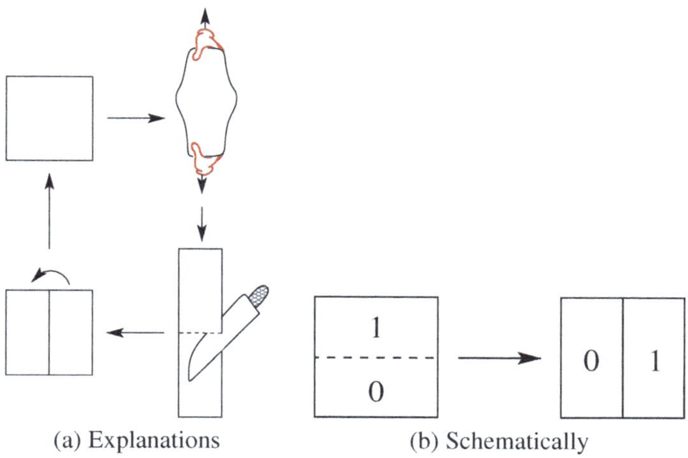

Fig. 8.4 Inverse baker's transformation

hand side of Fig. 8.3b, but with 4 rather than 2 parallel horizontal stripes stacked upon each other in a 'mixing' manner: see Fig. 8.5a. The third iterate analogously 'horizontalizes' 8 vertical stripes, again with a simple law appearing in their binary position numbers within the stack. Subsequent iterates are easy to derive by analogy. (The right-hand sides of Fig. 8.5 incidentally also show the positions of the fixed points of the corresponding iterate. They can be verified, for example, by putting a rectangular grid on the respective left-hand stripes.)

Thus, if the baker keeps rolling out, cutting, and then putting layer upon layer, in the limit there will be 2^n, that is, uncountable many horizontal stripes of infinitesimal width ('lines') which together just fill the original square.

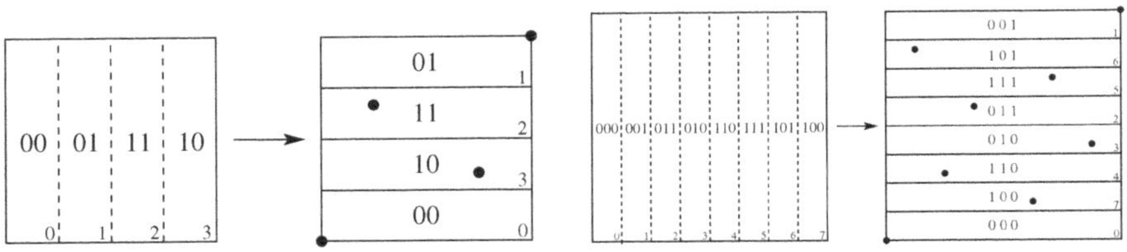

(a) Second iterate (b) Third iterate

Fig. 8.5 Second and third iterate of the baker's transformation. Stripes are labelled successively (on the left) by binary as well as ordinary numbers. The dots (on the right of each sub-figure) mark the periodic points

Hopf introduced the baker's map as an illustration to certain measure theoretic principles [10]. Prigogine looked at the same map under the aspect of defining an observable (a Lyapunov functional) which keeps decreasing under iteration even though the area of the map is preserved (meaning that the map itself is 'nondissipative') [14]. In the present context, the interest in the map again has a different reason.

It is the problem of transfinite invertibility: the baker's map of Fig. 8.3 is still invertible even after an infinite number of iterations? Of course, one cannot expect that in going back through the infinite sequence, all points of the original square will fall back in place. All those points lying on the stable manifold of a saddle will have become indistinguishable after infinite time. On the other hand, such points are necessarily of measure zero within the original domain (since the set of periodic points is of measure zero along the p-axis).

8.3 A Toroidal Analogue

The map shown in Fig. 8.6 is closely related to that of Fig. 8.3. The only differences are that it (i) is stretched more strongly between one round and the next (making four labels necessary rather than two), (ii) takes place on a torus (meaning that the upper and lower, and left and right, edges are pairwise identical), and (iii) is a diffeomorphism (there is no cutting involved anywhere).

The last assertion is startling. Checking it out carefully (by matching corresponding edges on the left and the right of the arrow) is therefore necessary to believe it. An equation for the map of Fig. 8.6 is

$$\begin{cases} p_{n+1} = p_n + q_n \,, & p \bmod 1 \\ q_{n+1} = p_n + 2\,q_n \,, & q \bmod 1 \,. \end{cases} \tag{8.3}$$

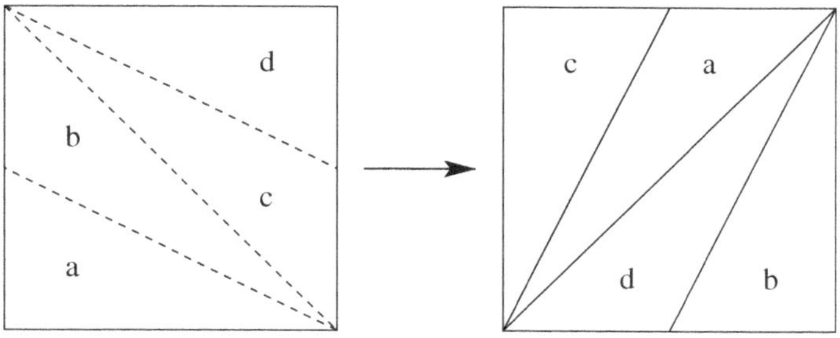

Fig. 8.6 An Anosov diffeormorphism. Compare with the baker's map shown in Fig. 8.3 and text

Equation (8.3) was studied extensively by Dmitri Anosov [5]. The same equation was invented by René Thom in the late 1950s (see [17][1]). While Eq. (8.3) is the simplest of its kinds, it is not the single one. The next-simplest is

$$\begin{cases} p_{n+1} = p_n + q_n \,, & p \bmod 1 \\ q_{n+1} = 2\,p_n + 3\,q_n \,, & q \bmod 1 . \end{cases} \tag{8.4}$$

It looks essentially like the second iterate of the former one. On the other hand, all nontrivial area-preserving linear diffeomorphisms of the torus—having a nonsingular matrix of integers whose determinant is unity—behave in the same manner [4].

An alternative way to draw the map of Fig. 8.6 is presented in Fig. 8.7. One here does not as easily recognize the diffeomorphic nature of the map. On the other hand, one now sees more clearly what happens under iteration: the square gets more and more elongated, and thinner and thinner, while still being wrapped around the torus. Thus, the map of Fig. 8.6 imposes as a diffeomorphic analogue to Hopf's map (Fig. 8.3). Conversely, Hopf's map is a non-diffeormorphic (and non-toroidal, and more robust) analogue to the map of Fig. 8.6.

Iterating the 'toroidal baker's transformation' yields 'multiple stripes' again (see Fig. 8.8). They are segments of triangles now rather than rectangles. Note the similarity between Fig. 8.8 (second iterate of the toroidal baker's transformation) and Fig. 8.5b (third iterate of the ordinary baker's transformation). The left and the right hand side of Fig. 8.8 possess an identical 'tiling pattern', except for a rotation by 90°, just as the map of Fig. 8.6 (and those of Figs. 8.3 and 8.5) did.

[1]In 1967 Smale wrote: "The two dimensional toroidal example was first communicated to me by Thom to show that there was an open set in Diff(T^2) of diffeomorphisms with no contracting periodic points, therefore implying that diffeomorphisms satisfying (2.2) were not dense. After adding some geometry to the example, I showed it to Anosov when I spoke in the Soviet Union in 1961. By 1962 Anosov announced his theorem on structural stability in the context of what is called here Anosov diffeomorphisms. Proofs have now appeared [5].".

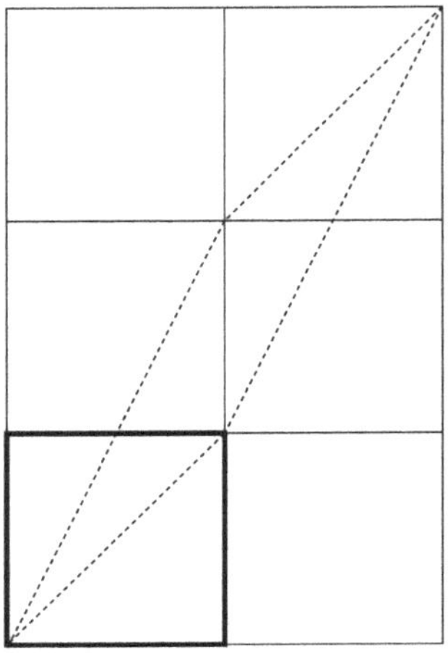

Fig. 8.7 Another way to draw the map of Fig. 8.6. Bold square: original; dashed rhombus: image. All six squares represent the same torus. (Adapted from [6])

The present diffeomorphic version of the baker's transformation has the asset that its asymptotic properties are somewhat better known. Proceeding with the iteration process, again an uncountable set of 'lines' that happen to be square-filling is obtained in the limit. Most of these 'lines' moreover 'contain' a segment of either branch of the unstable manifold of the saddle (one branch lying wholly within the stripe marked a_1 on the right-hand side of Fig. 8.8b, the other lying wholly within d_1). Of the stripes seen on the left-hand side of Fig. 8.8c, only c_4 and d_1 (and b_4 and a_1, respectively) are spared one of the two branches, as is readily verified.

Concentrating on only one of the two branches of the expanding manifold, it will—in the limit—be wrapped densely around the torus. Since the same thing happens—under negative iteration numbers—with the stable manifold (one branch of which lies within and follows b_1 on the left-hand side of Fig. 8.8b), one obtains a dense intersection grid of stable and unstable manifolds. That is to say, in every neighborhood of every point on the square there is a homoclinic point [6].

Of course, there is also an 'explosion' of periodic points under the iteration, letting their number grow like 2^n. (The right-hand side of Fig. 8.8b displays those applying for $n = 2$.) And there is also Smale's Cantor set of singular solutions, almost all of them nonperiodic. The demonstration of this set is less easy in the present case than in that of the horseshoe map (Fig. 7.10), however. The similarity of the present map with the solenoid (Fig. 7.12) suggests to look again only at a reduced one-dimensional map (along the unstable manifold), in order to verify this assertion in a non-quantitative sense at least.

We may now return to the problem of transfinite invertibility. If the ordinary baker's transformation (see Sect. 8.2) is transfinitely invertible, so is the present map and *vice versa*. But: *Is the present map indeed transfinitely invertible?*

Fig. 8.8 First, second and 32nd iterate of the Anosov map shown in Fig. 8.6 and Eq. (8.3). The correspondence with the labeling used in Fig. 8.6 is such as: black = a, red = b, green = c and blue = d. There are five disjoint periodic points now. Note the similarity to the map shown in Fig. 8.7

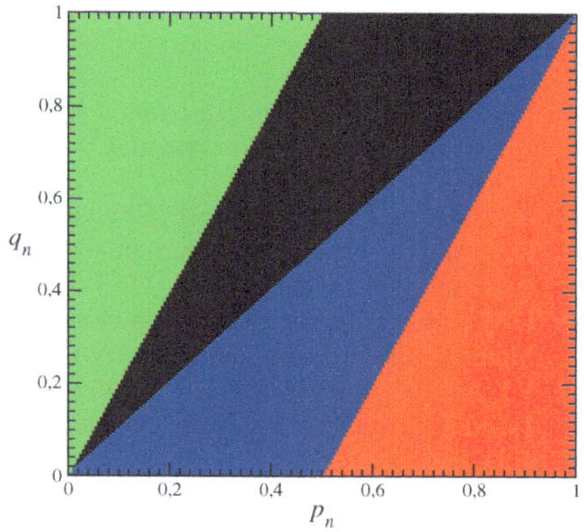

(a) First iterate

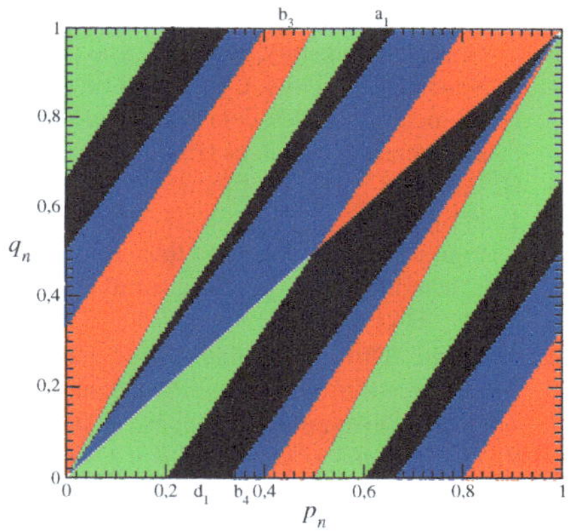

(b) Second iterate

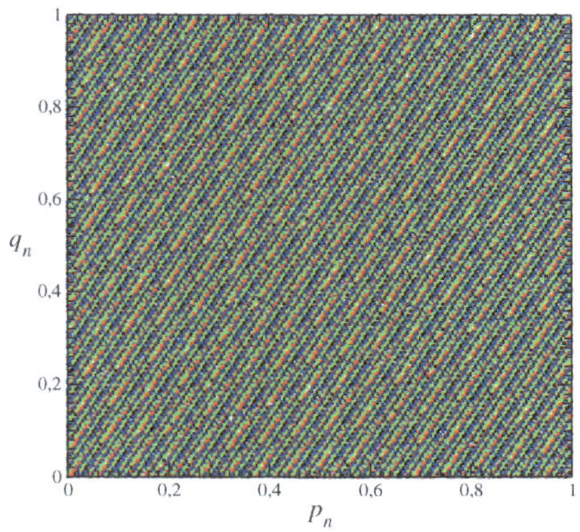

(c) 32nd iterate

References

1. R. Abraham, J. Robbin, *Transversal Mappings and Flows* (W. A. Benjamin, New York, 1967)
2. A.A. Andronov, L.S. Pontryagin, Coarse systems. Dokl. Akad. Nauk SSSR **14**(5), 247–250 (1937)
3. A.A. Andronov, A.A. Vitt, S.E. Chaikin, *Theory of Oscillators* (Pergamon Press, Oxford, 1966)
4. D.V. Anosov, Ergodic properties of geodesic flows on closed Riemannian manifolds of negative curvature. Sov. Math. Dokl. **4**, 1153–1156 (1963)
5. D.V. Anosov, Geodesic flows on closed Riemannian manifolds of negative curvature. Trudy Matematicheskogo Instituta imeni VA Steklova **90**, 3–210 (1967)
6. V.I. Arnold, A. Avez, *Ergodic Problems of Classical Mechanics* (Benjamin, New York, 1970); 1st edn. in French, *Théorie Ergodique des systèmes dynamiques* (Gauthier-Vilars, Paris, 1967)
7. G.D. Birkhoff, On the periodic motions of dynamical systems. Acta Math. **50**, 359–379 (1927)
8. D.R.J. Chillingworth, *Differential Topology with a View to Applications*. Research Notes in Mathematics, vol. 9 (Pitman, London, 1976)
9. O. Gurel, Partial peeling, in *Dynamical systems*, vol. 2, ed. by L. Cesari, et al. (Academic, New York, 1976), p. 25
10. E. Hopf, On causality, statistics, and probability. J. Math. Phys. **13**, 51–102 (1934)
11. S. Newhouse, J. Palis, Cycles and bifurcation theory. Astérisque **31**, 44–140 (1976)
12. H. Poincaré, Sur les courbes définies par une équation différentielle. Comptes Rendus de l'Académie des Sciences (Paris) **90**, 673–675 (1880)
13. H. Poincaré, Sur le problème des trois corps et les équations de la dynamique. Acta Math. **13**, 1–271 (1890)
14. I. Prigogine, *From Being to Becoming* (Freeman, San Francisco, 1980)
15. O.E. Rössler, Quasi-periodic oscillation in an abstract reaction system (abstract). Biophys. J. A **17**, 281 (1977)
16. S. Smale, Diffeomorphisms with many periodic points, in *Differential and Combinatorial Topology*, ed. by S. Cairns (Princeton University Press, Princeton, 1965), pp. 63–80
17. S. Smale, Differentiable dynamical system. Bull. Amer. Math. Soc. **73**, 747–817 (1967)
18. R. Thom, *Stabilité structurelle et Morphogénèse* (Interéditions, Paris, 1972)
19. R.F. Williams, The structure of Lorenz attractors. Publications Mathématiques de l'Institut des Hautes Études Scientifiques **50**(1), 73–99 (1979)

Chapter 9
Non-sink Attractors

9.1 The Anaxagoras Conjecture

Anaxagoras (see Introduction) inaugurated the philosophical hypothesis that a perfect mixture, existing since eons,[1] could be unmixed. By adding the word 'again', we arrive at a mathematical version of the same conjecture. The deterministic maps of the preceding two sections are candidates examples. Even in its mathematical version, the conjecture has a strong philosophical appeal. And indeed, at first sight it seems as if the answer to it were largely a matter of definitions.

For example, take a simple linear area-preserving expanding map, like

$$\begin{cases} p_{n+1} = 2\, p_n \\ q_{n+1} = 0.5\, q_n \,. \end{cases} \tag{9.1}$$

It transforms an initial square $(-0.5 \ldots 0.5, -0.5 \ldots 0.5)$ into an oblong rectangle, and this into an even longer and thinner object, and so on. In the limit there is an infinitely long, infinitely thin 'line'. This line contains the unstable manifold of the (saddle) periodic point at the origin $(0, 0)$ of the map. Now comes the question: *Should one say that this unstable manifold attracts all transients within the map (starting from finite initial conditions)?*

Putting a half-sphere above the origin, and projecting the map and its images radially onto the surface of the half sphere (Poincaré's projection), one clearly sees that in the limit only three attracting points remain: the saddle, and the two points at infinity making the end points of the longer and longer 'line'. The connecting line may be included in a definition of the attracting set, but almost all of its 'mass' is of course concentrated in the two end points.

On the other hand, one sees just as clearly that **without** the Poincaré transformation, the limiting rectangle has the same width and 'density' everywhere. So if there

[1] In planetology, the history of earth is described by a geologic time scale relating stratigraphy to time. Earth's history is divided in four eons, namely Hadean, Archean, Proterozoic, and Phanerozoic.

O. E. Rössler and C. Letellier, *Chaos*, Understanding Complex Systems,
https://doi.org/10.1007/978-3-030-44305-4_9

is an ‘attractor’, it must be the whole ‘line’. This whole ‘line’ cannot be an attractor, however, since it possesses the same volume (measure) as the original square. It also cannot contain an attracting ordinary line of the same length (the saddle’s unstable manifold, for example), since any such line will possess zero volume. Thus, if one adheres to the non-transformed situation, it is possible to conclude that the above unit square is transfinitely invertible under the map of Eq. (9.1). Only the stable manifold of the saddle periodic point at the origin, connecting the two initial points (0, 0.5), (0, 0.5), is being mapped irreversibly onto the origin.

Both the map of Fig. 8.3 (ordinary baker’s transformation) and the map of Fig. 8.6 (diffeomorphic baker’s transformation) are transfinitely invertible under the same convention. Thus, Anaxagoras’ conception of the mind as a force fine enough to conceive of (and initiate) the inversion of a perfect mixture is, perhaps, justified also from a mathematical point of view.

Anaxagora’s original formulation is reproduced here in Fig. 9.1 and is translated as follows [3].

> Whereas all other things contain a portion of everything, the Mind (Nous) is unlimited and autonomous, mixing with nothing but being alone by itself. For it were not self-contained but mixed with anything else, it would partake with everything, since in everything there is a part of everything as I said before. Whatever were intermingled with it would prevent it from having power over anything, of the kind it has now being alone by itself. It is the finest of all things and the purest having every knowledge about everything and exerting the greatest power. Whatever, great or small, exists (has psyche), is controlled by the Mind. The whole recurrence was started. Then more got involved in the about-moving (perichoretic) process, and even more will be involved in it. What is mixed together, and what is separate and distinguished, was all known to the Mind. All that was to come into being—all that has been but no longer is, all that is now, and all that there will be—got neatly led out by the Mind. Even the recurrent motion that governs the stars and the sun and the moon and the air and the æther, separate as they are to date, was included. The recurrence in turn generated the separation process, from the thin there got separated the dense, from the cold there warm, from the dark the glowing, from the moist the dry. Many parts are there of the many things. For all of them it holds true that nothing gets separated off and distinguished, the one from the other, except through the Mind. The Mind as a whole is self-similar no matter whether it refers to the large or the small. Of the other things, however, none is similar to any other: always that which is the most strongly represented in each thin determines its properties.

9.2 An Ideal Chaotic Attractor ...

The preceding baker’s transformation were all area-preserving, just like the ‘taffy-pulling map’ of Fig. 1.2c was. How about assuming a little loss of volume (due to fluid loss, say) while the doug is under the rolling pin? The resulting ‘contracting baker’s transformation’ is sketched in Fig. 9.2.

An appropriate equation is (compare with Eq. (8.1):

$$\begin{cases} p_{n+1} = 2\,p_n - H(p_n - 0.5) \\ q_{n+1} = (0.5 - \epsilon)\,(q_n + 0.5\,\epsilon) + 0.5\,H(p_n - 0.5)\,. \end{cases} \tag{9.2}$$

τὰ μὲν ἄλλα παντὸς μοῖραν μετέχει, νοῦς δέ ἐστιν ἄπειρον καὶ αὐτοκρατὲς καὶ μέμεικται οὐδενὶ χρήματι, ἀλλὰ μόνος αὐτὸς ἐφ' ἑαυτοῦ ἐστιν. εἰ μὴ γὰρ ἐφ' ἑαυτοῦ ἦν, ἀλλά τεωι ἐμέμεικτό ἄλλωι, μετεῖχεν ἂν ἁπάντων χρημάτων, εἰ ἐμέμεικτό τεωι· ἐν παντὶ γὰρ παντὸς μοῖρα ἔνεστιν, ὥσπερ ἐν τοῖς πρόσθεν μοι λέλεκται· καὶ ἂν ἐκώλυεν αὐτὸν τὰ συμμεμειγμένα, ὥστε μηδενὸς χρήματος κρατεῖν ὁμοίως ὡς καὶ μόνον ἐόντα ἐφ' ἑαυτοῦ. ἔστι γὰρ λεπτότατόν τε πάντων χρημάτων καὶ καθαρώτατον, καὶ γνώμην γε περὶ παντὸς πᾶσαν ἴσχει καὶ ἰσχύει μέγιστον· καὶ ὅσα γε ψυχὴν ἔχει καὶ τὰ μείζω καὶ τὰ ἐλάσσω, πάντων νοῦς κρατεῖ. καὶ τῆς περιχωρήσιος τῆς συμπάσης νοῦς ἐκράτησεν, ὥστε περιχωρῆσαι τὴν ἀρχήν. καὶ πρῶτον ἀπό του σμικροῦ ἤρξατο περιχωρεῖν, ἐπὶ δὲ πλέον περιχωρεῖ, καὶ περιχωρήσει ἐπι πλέον. καὶ τὰ συμμισγόμενά τε καὶ ἀποκρινόμενα καὶ διακρινόμενα πάντα ἔγνω νοῦς. καὶ ὁποῖα ἔμελλεν ἔσεσθαι καὶ ὁποῖα ἦν, ἄσσα νῦν μὴ .ἔστι, καὶ ὅσα νῦν ἔστι καὶ ὁποῖα ἔσται, πάντα διεκόσμησε νοῦς, καὶ τὴν περιχώρησιν ταύτην, ἣν νῦν περιχωρεῖ τά τε ἄστρα καὶ ὁ ἥλιος καὶ ἡ σελήνη καὶ ὁ ἀὴρ καὶ ὁ αἰθὴρ οἱ ἀποκρινόμενοι. ἡ δὲ περιχώρησις αὕτη ἐποίησεν ἀποκρίνεσθαι. καὶ ἀποκρίνεται ἀπό τε τοῦ ἀραιοῦ τὸ πυκνὸν καὶ ἀπὸ τοῦ ψυχροῦ τὸ θερμὸν καὶ ἀπὸ τοῦ ζοφεροῦ τὸ λαμπρὸν καὶ ἀπὸ τοῦ διεροῦ τὸ ξηρόν. μοῖραι δὲ πολλαὶ πολλῶν εἰσι. παντάπασι δὲ οὐδὲν ἀποκρίνεται οὐδὲ διακρίνεται ἕτερον ἀπὸ τοῦ ἑτέρου πλὴν νοῦ. νοῦς δὲ πᾶς ὅμοιός ἐστι καὶ ὁ μείζων καὶ ὁ ἐλάττων. ἕτερον δὲ οὐδέν ἐστιν ὅμοιον οὐδενί, ἀλλ' ὅτων πλεῖστα ἔνι, ταῦτα ἐνδηλότατα ἓν ἕκαστόν ἐστι καὶ ἦν.

Fig. 9.1 Fragment 12 of Anaxagoras reproduced from [3]

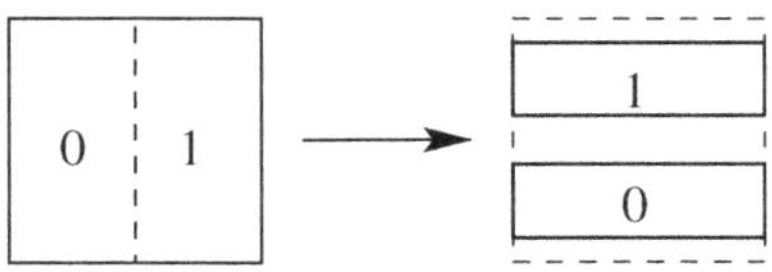

Fig. 9.2 Contracting baker's transformation. (The non-contracting case was shown in Fig. 8.3)

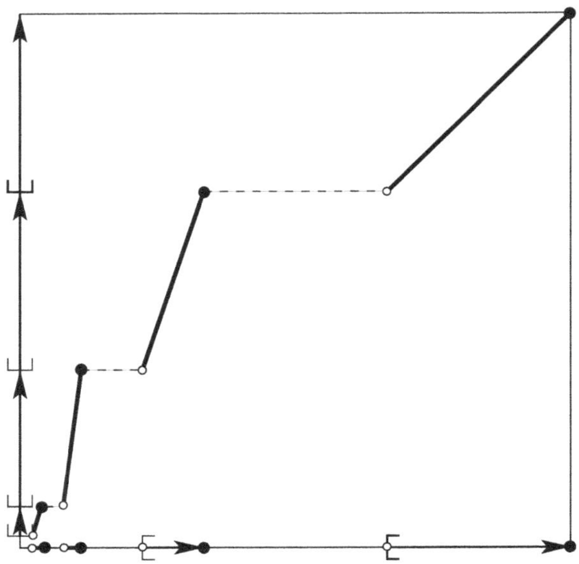

Fig. 9.3 Demonstration of the possibility of a bijection between a real line (ordinate) and a Cantor set (abscissa). The corresponding intervals (disconnected at the abscissa, connected at the ordinate) are indicated. A filled-out circle means that the respective end point belongs to the intervals, an empty circle that it does not

In the limit $\epsilon \mapsto 0$, Eq. (8.1) is reobtained. The new map is, despite the area-contraction, closely related to the ordinary baker's transformation (Fig. 8.3). Again, the map is a bijection at each step—if care is taken in the definition of the pieces (such as to avoid nonuniqueness in the definition of the cutting edges in accordance with physical reality; see below).

Apart from this bijective nature of the map itself, there exists also an abstract (artificial) bijection, applying between the map of Fig. 9.2 on the one hand and the map of Fig. 8.3 on the other. This abstract bijection involves (i) a bijection between the two original sets (squares), (ii) a bijection between the two image sets (two blocks in either case), and (iii)—by consequence—a bijection between the two arrows.

Such bijections between whole maps are called 'bijective functors' in category theory [5]. Such a bijective functor exists not only for the first iterate, but for all finite iterates of the two maps, as is readily verified by inspection. Moreover, the functor even still exists in the limit of infinite iteration number.

The last assertion is nontrivial. It follows from the fact that in the limit, there is a Cantor set of stripes on the right-hand side of either map. One of the two Cantor sets (formed in the map of Fig. 8.3) is area-filling, the other (formed in the map of Fig. 9.2) is not. The 'gaps' which have zero width in the former have finite width in the latter.

Cantor showed that a bijection between a Cantor set and a real line is possible if care is taken that non-invertibility is avoided at the cutting edges. Figure 9.3 is a modification of his Fig. 3 [2]. For simplicity, the continuing fractionation is displayed only for one interval at a time. A similar argument was presented by Meschkowski [6].

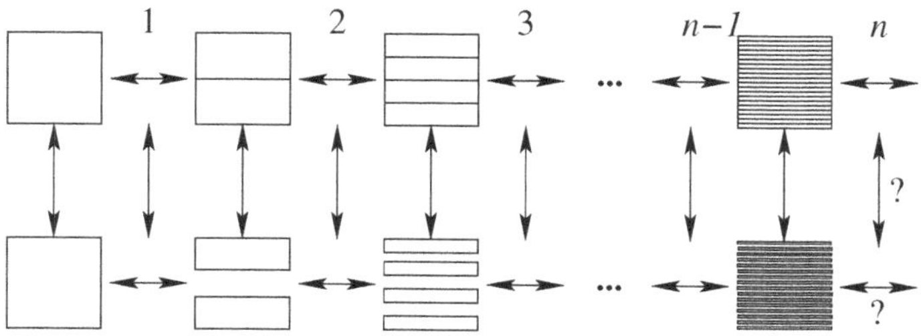

Fig. 9.4 Bijective relationships in and between the two maps of Figs. 8.3 and 9.3 if $n \to \infty$. Each double-arrow corresponds to a bijection. See text for the two arrows with question marks

Thus, we have the diagram of relationships shown in Fig. 9.4. If indeed all arrows except those with a question mark are bijections, then the whole diagram commutes. Thus, all arrows are bijections. This means that if the ordinary baker's transformation is transfinitely invertible (except for a set of measure zero), then so is the contracting baker's transformation. In the light of the preceding section, we may thus conclude that the contracting baker's transformation of Fig. 9.2 is transfinitely invertible in the same sense as the ordinary baker's transformation, and the map of Eq. (9.1), is.

This conclusion seems natural in the present context. On the other hand it is highly unexpected from the viewpoint of the theory of attractors. This is because the Cantor set formed in the contracting baker's transformation in the limit of infinite iteration number (symbolically represented in the lower right corner of Fig. 9.4) is an attractor. Specifically, it is an attractor closely related to the Lorenz attractor (see Sect. 8.3).

Up till now, the two notions of an 'attractor' and a 'sink' were taken to be equivalent [4]. Any attractor which like the present one is transfinitely invertible with its domain of attraction (except for a set of measure zero) is not a sink, however. It rather...

9.3 ... Is a Non-sink Attractor

Two examples of 'ordinary' attractors which at the same time are sinks are presented in Fig. 9.5. The first, formed in a uniformly contracting diffeomorphism, is an attracting closed line. (The structural instability of this attractor—see Sect. 6.2—is of no concern presently.) Interpreting the two maps as cross sections through a three-dimensional flow (which, according to Smale [7], is always possible), the point attractor corresponds to an attracting limit cycle in a three-dimensional flow, and the closed line attractor to an attracting torus. An example of such a flow was given above in Fig. 6.14.

The 'sink' property of an attractor is something very intuitive. It means that transients having reached the attractor (which occurs in the limit of infinite time and/or infinite iteration number) are lost. All transients that initially were a finite distance way already from the sources (an assumption necessary because of the

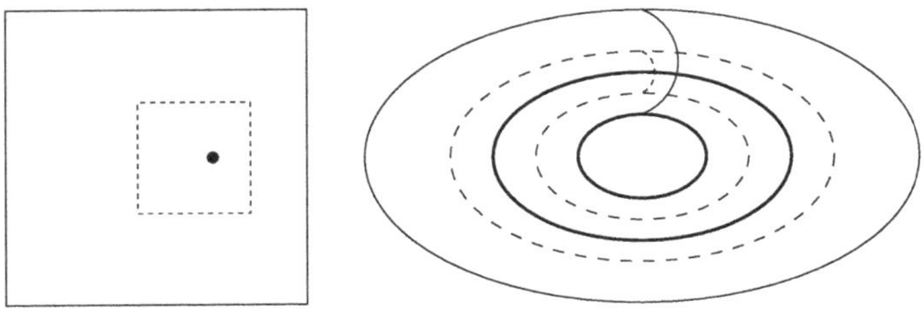

(a) Continuous square (b) Continuous ring

Fig. 9.5 Two area-contracting diffeomorphisms. (**a**) Continuous square = initial, dashed square = first iterate, and point = limiting iterate. (**b**) Continuous ring = initial, dashed ring = first iterate, closed line = limiting iterate. The two arcs mark the boundaries of initial and first iterate, respectively

time-inversion symmetry of sources and sinks) are irreversibly 'sucked in' by the sink after infinite iteration number.

The same property can be stated more precisely as follows (after La Salle [4]: a sink formed in a map under iteration is a set that (i) is invariant under the map (meaning that points originally in the stay in the set under iteration) for iteration numbers (including $n \to \infty$); (ii) is of measure zero within the domain of the map; (iii) attracts a domain (which is in the simplest case is connected and without exceptional points); and (iv) is invariant under the map also for all negative iteration numbers (including $n \to \infty$).

The attractors formed in the maps of Fig. 9.5 certainly possess all four properties. In contrast, the attractor formed in the contracting baker's transformation (Fig. 9.2) fulfills only properties (i) through (iii), but not (iv). Property (iv) is met only by a set of measure zero within the attractor (the set of periodic points and their unstable manifolds). This both positively and negatively invariant subset of the domain of attraction of the whole attractor which consists of the stable manifolds of the periodic points. This external subset may be called the anti-skeleton. Just as the skeleton is of measure zero within the attractor, so is the anti-skeleton of measure zero within the domain of attraction.

Thus, in contrast to the classical opinion, not all attractors are sinks. Some attractors may only contain a (skeleton) sink, but themselves—with almost all points—be transfinitely invertible with respect to their domain of attraction. An example is the attractor formed in the contracting baker's transformation.

9.4 A Philosophical Implication

Nonsink attractors, if they indeed exist, have some irritating properties that cannot be extracted in principle in finite times. Even if it was possible to let the attractor run backwards with absolute accuracy, an infinite amount of time was needed to extract the initial configuration.

Of course, one usually thinks of dynamical systems as possessing a single state only (corresponding to a point in a cross section). But as soon as there are two chaotic systems running simultaneously, one already has a two-point system, and so on. That is to say, a single state of a larger system may correspond to many (even infinite many) states of subsystems. Thus it makes sense to model all those infinite many states into a single chaotic machine in which they are all present simultaneously (as in the taffy puller of Fig. 1.2).

Such a 'big machine', if transfinitely invertible, has the unsettling property that with a bit of fore-knowledge (see Fig. 9.1), a 'Mind' could have manipulated it in such a way as to bring into existence a particular constellation at a particular instant of time. This constellation may contain an infinite amount of information, that is, be infinitely improbable from the point of view of those observers having seen the machine run for a finite amount of time.

Thus, 'bombs of hidden knowledge' can in principle exist in a deterministic machine. These bombs (and any hints as to their existence) are inextractable in principle even if the whole machine is known in all its details with absolute accuracy. Equation (2.1) is a possible example for such a machine.

Lovecraft, the mysteries' author, used to say that there exist truths that are so terrible that once having learned of them one forever wishes himself back toward the innocent state before the news are broken. While this is not quite the case with the present tentative result, it has a ring of the same quality—whether bad or good.

Of course, physics is not deterministic. Only the word of the wave functions is. So one will have to wait some time until a quantum-mechanical analogue to the present class of systems will have been formulated. The preceding remarks were not quite serious, of course. They served to provide some more motivation to find the hidden flaw in the argument that attractors that are not sinks be possible.

9.5 The Lorenz Attractor as a Non-sink Attractor

The contracting baker's transformation of Fig. 9.2 is unnecessarily idealized. Some of its 'straight' properties can be relaxed. Three examples are presented in Fig. 9.6. For these modified maps it is somewhat harder to show that they generate a potential non-sink attractor in the limit. However, by replacing the 'ordinate' used as the basis for comparison in Fig. 9.4 by a (more or less vertical) stable manifold in each of the present maps, a 'third row' may be added to Fig. 9.4. In it, each of the present ideal maps may be juxtaposed, piece by piece, with the sequence of maps of the lower row. The main problem hereby is that the lengths of the horizontal pieces are no longer identical in the two rows. Nonetheless, the gist of the argument—that the limiting attracting set consists solely of 'lines' which correspond to a square-filling set of more or less vertical 'lines' of the original—can be maintained.

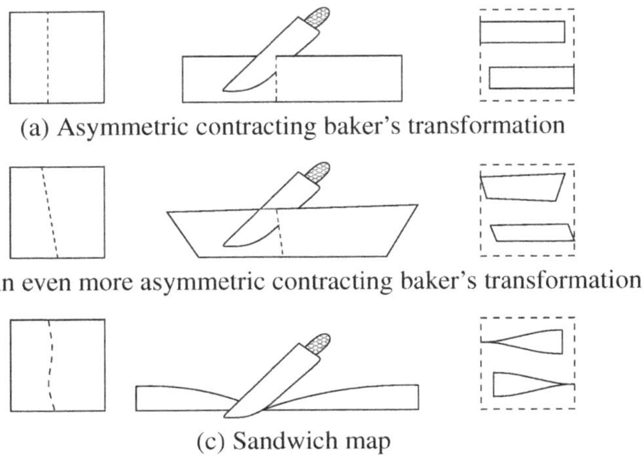

(a) Asymmetric contracting baker's transformation

(b) An even more asymmetric contracting baker's transformation

(c) Sandwich map

Fig. 9.6 Modified contracting baker's transformations (compare with Fig. 9.4)

In the present context, the most interesting, because the most realistic, case is that of Fig. 9.6c.[2] There a Lorenz attractor is being formed in the limit (see Sect. 2.7). This map contrasts with the other two insofar as there is one line in the original (or rather two lines, with an 'empty gap' in between, of one is again as careful with definitions as in Fig. 9.3) which is contracted to a point in the image (or rather, two points). Such a line is, of course, non-invertible. This means that in contrast to the situation of Fig. 9.4, where all individual maps were fully invertible, we now have an analogous sequence of maps in which every individual submap is invertible only in the 'weak' sense meaning that there are exceptions of measure zero. As a consequence, the overall map also can be invertible in the weak sense at best. Indeed, the limiting map will contain a countable infinitude of lines that are non-invertible. However, as we saw in Sects. 7.2–7.4, the limiting map in the map of Fig. 9.2 also is weakly invertible only. Thus, nothing has been changed in a qualitative sense.

This means that if the ideal contracting baker's transformation (Fig. 9.2) generates a non-sink attractor, so does the sandwich map (Fig. 9.6c). The same conclusion can be stated in a slightly different way: if non-sink attractors exist at all, the Lorenz attractor is an example.

[2]As well suggested by Fig. 9.4, there is a tearing or a cutting mechanism necessarily involved in the Lorenz dynamics. This is an important property that distinguish the Lorenz attractor from the Rössler one. See [1] for more details.

References

1. * G. Byrne, R. Gilmore, C. Letellier, Distinguishing between folding and tearing mechanisms in strange attractors. Phys. Rev. E **70**, 056214 (2004)
2. G. Cantor, Über unendliche, lineare Punktmannigfaltigkeiten v (On infinite linear point-manifolds). Mathematische Annalen **21**, 359—379 (1883) (Also published separately as Grundlagen einer allgemeinen Mannigfaltigkeitslehre, Leipzig, 1883)
3. * H. Degn, A.V. Holden, L.F. Olsen (eds.), Chaos in biological systems. NATO ASI Ser. A: Life Sci. **138**, x (1987)
4. J. LaSalle, *The Stability of Dynamical Systems* (SIAM, Philadelphia, 1976)
5. S. Mac Lane, *Categories for the Working Mathematician*, vol. 5, Graduate Texts in Mathematics (Springer, Berlin, 1971)
6. H. Meschkowski, *Hundert Jahre Mengenlehre* (Die Tachenbuch-Verl, Münschen, 1973)
7. S. Smale, Differentiable dynamical system. Bull. Am. Math. Soc. **73**, 747–817 (1967)

Chapter 10
Chaos and Turbulence

10.1 Three Higher-Order Baker's Transformations

Figure 10.1 shows three straightforward generalizations of the baker's transformation of Fig. 8.3. The rolling pin and the knife have been omitted for simplicity. What one sees is easy to interpret: the first case is trivial, the second row shows how to make pancake, the third, noodles. Under iteration, only the second procedure produces fluffy pastry. The first ought to produce something similar, but is shunned by bakers since it generates 'lumps' in one direction. The third generates an interesting modification of fluffy pastry which bakers, apparently, have never tried to make.

The procedures of Fig. 10.1b and 10.1c do both generate a bijection across dimensions, just as the ordinary baker's transformation (Fig. 8.3) does. Only that Fig. 10.1b generates a bijection between a cube and an uncountable stack of planes, while Fig. 10.1c generates a bijection between a cube and an uncountable bundle of lines. Thus, the map of Fig. 10.1b generates a bijection between three- and two-dimensional objects while the map of Fig. 10.1c generates a bijection between three- and one-dimensional objects.

This is astonishing insofar as the two maps just mentioned are mutually identical under time reversal. Thus, Cantor's contention [6] that it be just as easy to generate a bijection across two dimension numbers as across one has found a direct illustration.

Obviously also, both nontrivial three-dimensional baker's transformations generate a non-sink attractor (if again a slight volume contraction is admitted at each step): see Fig. 10.2. If again the initial cube (slab of dough) is assumed red while new white dough is being used for making up for the lost volume at each step, the two limiting objects (attractors) formed will impose as a measure-zero stack of red sheets and a measure-zero bundle of red lines, respectively.

The object formed in a contracting version of the first map of Fig. 10.1 is less easy to define because an infinitesimal perturbation of the original map suffices to destroy its behavior: if area-preservation along the third dimension (depth) is not perfect, there will be either an attracting red line (of the same type as formed in Fig. 9.2) or a repelling red line of the same type. Moreover, the red line will meander through the

O. E. Rössler and C. Letellier, *Chaos*, Understanding Complex Systems,
https://doi.org/10.1007/978-3-030-44305-4_10

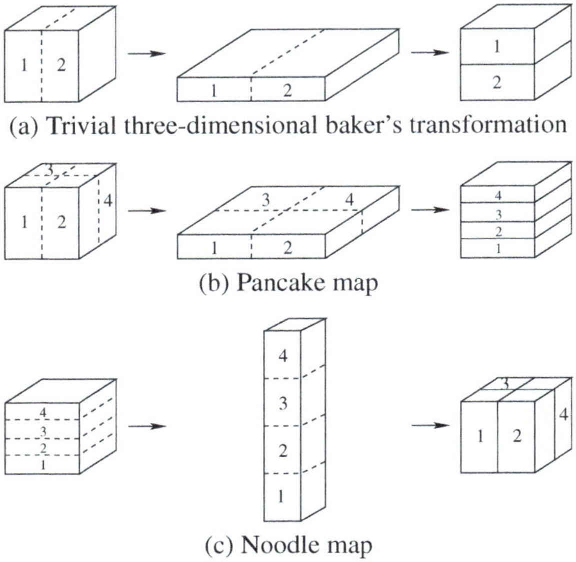

Fig. 10.1 Three three-dimensional baker's transformations. See text

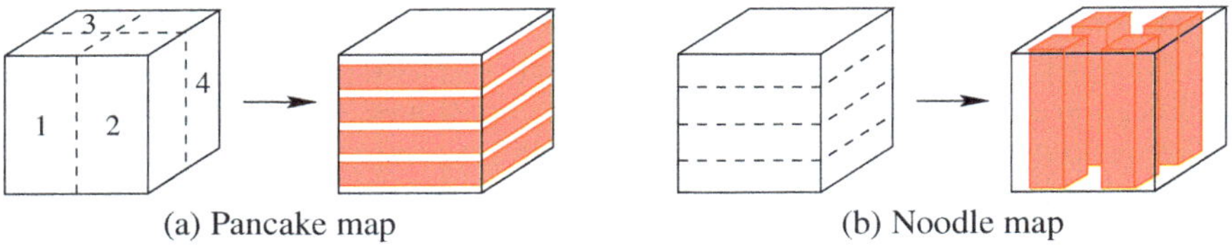

Fig. 10.2 Contracting maps. Compare with Figs. 10.1 and 9.2

whole slab (that is, will be non-differentiable along the depth dimension) if there is the slightest coupling from one of the two main variables toward the depth variable. This explains also the nonhomogeneous structure (and uneven taste) of such bakery products.

Coming back to the notion of chaos, it is clear that all three maps of Fig. 10.1 generate chaos (if thought as cross sections through flows, for example). To make the maps more realistic one may modify them along the lines of Fig. 9.6, obtaining higher-order analogues to the (Lorenz-attractor generating) 'sandwich map' of Fig. 9.6c, so-called 'big maps'.

Coming back to the notion of higher chaos, it is equally clear that only the map of Fig. 10.1b (and its contracting analogue of Fig. 10.2a) generates it. Does the series of maps of Fig. 10.1 allow for a deeper understanding of the nature of hyperchaos?

10.2 Space-Filling, Big and Small

All three maps of Fig. 10.1 generate a space-filling lower-dimensional object in the limit. The two maps of Fig. 10.1b, c generate a non-trivial limiting object each. Either object has a zero-volume analogue (see Fig. 10.2) that still is bijectively related to its domain of attraction. One of them is locally two-dimensional, the other is locally one-dimensional, but both are non-sink attractor.

Thus, there exist both 'big' and 'non-big' space-filling objects and non-sink attractors, respectively. The big ones (formed in Figs. 10.1b and 10.2a) are of a 'neighboring' dimensionality while the non-big ones (Figs. 10.1c and 10.2b) have an even lower dimensionality ($d_{\mathrm{loc}} n - 1$, if n is the dimensionality of their pre-image).

'Big' space-filling objects (and their contracted analogues) do the space-filling in an especially laborious way, so to speak: for every single lower dimension (up to $n - 1$), they are being folded over an uncountable number of times.

This is a bit of paradoxical. One would have expected that the lower-dimensional a space-filling object (with respect to the space that it is to fill), the harder would be its 'filling job'. Now it turns out that on the contrary there is virtually no difference between the filling of a two-dimensional space by lines (see Fig. 8.5) for some of the steps) and the filling of a three-dimensional space by lines (see Fig. 10.1c). Of course, the 'weaving rules' are different in the two cases; but there are also many alternative rules admissible within each class already.

It would be nice if it were possible to formalize this difference, so that one would come up with a single number measuring the 'bigness' of an attractor. Unfortunately, this is not easy to do. There does exist a number that is appropriate to characterize the 'dimensionality' of an object that 'would' possess a higher dimensionality than it locally has—the Hausdorff measure [11, 26].

For example, it is not hard to compute that the Hausdorff measure (or 'fractal dimensionality', [26]) of the attractor formed in the ordinary contracting baker's transformation (Fig. 9.2) is

$$D = \frac{\log 2}{\log 3} + 1 ,$$

if it is assumed that the ordinate is an ordinary (triadic) Cantor set, since the Hausdorff measure of a triadic Cantor set is $\frac{\log 2}{\log 3}$ [26]. This ratio means that if one linearly magnifies the object by a factor of 3 (see the number in the denominator), one gets only 2 (rather than 3) copies of the original object while scanning it at the higher magnification (see the number in the numerator). The formula $N = S^2$ (number of objects formed = magnification factor raised to the power of the dimension applying) yields a 'fractal' D in the present case.

Nonetheless, this measure is not quite the 'bigness indicator' we are looking for. This is because the same method, applied to the two objects of Fig. 10.2, yields a fractal dimensionality between 2 and 3 in **both** cases ($D = \frac{\log 2}{\log 3} + 1$ and $D = \frac{\log 2}{\log 3} + 2$, respectively, if a volume loss of $\frac{4}{9}$ is assumed per step). Thus, a more complicated indicator of the bigness (or 'maximality') of an attractor is needed.

Fig. 10.3 The conjecture of maximality (almost n-dimensionality of a set)

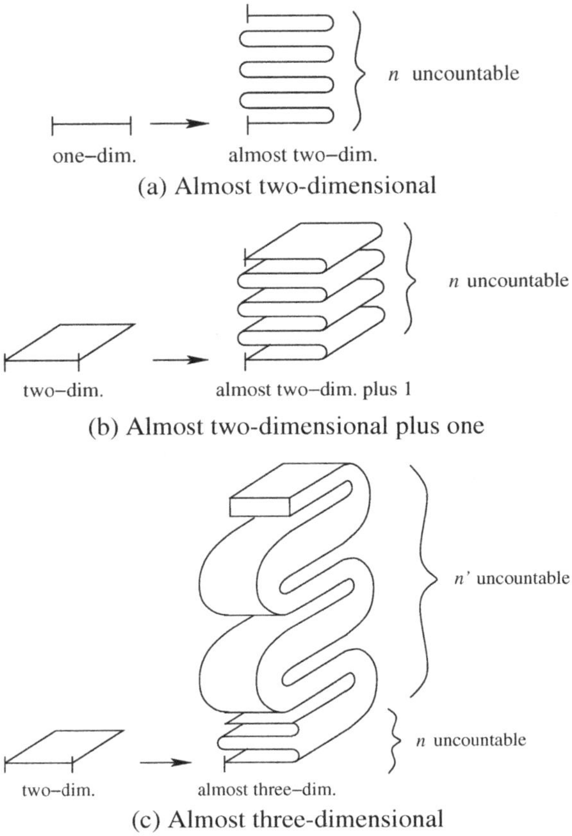

How the eventual solution to this problem will look like is hard to say. One straightforward possibility will be to introduce an independent 'magnification factor' S_i $(i = 1, \ldots, n)$ for each dimension of the original object. The resulting number will then consist of n summands, with up to $n-1$ of them non integers.

10.3 'Maximal Chaos'

The unsolved problem of the preceding section is summarized in Fig. 10.3: *Is it possible to define sets of 'maximal thickness'?*

Without having arrived at a satisfying definition, we can nonetheless already operate with the notion of 'maximality' as a conjecture. In terms of this conjecture, we can state that a chaotic system shows 'maximal chaos' if its limiting regime (in the simplest case: chaotic attractor) has maximal thickness.

Being able to state that a chaotic system has maximal chaos is important because in that case there is no way to simplify the system by decomposing it. And of course, questions as to the type of chaos (sandwich map chaos, meaning Lorenzian- or, if one of the two portions of the map is inverted, pseudo-Lorenzian-behavior, or walking-stick map chaos, which is the second basic type to expect in two-dimensional maps) are worth posing only after the question of the dimensionality of the chaotic regime has been settled. This can even be done empirically (using the method of characteristic exponents), as will be explained below in Sect. 12.3.

10.4 Turbulence in Its Own Right

The phenomenon of turbulence is, in a sense, too beautiful to be explained.[1] It not only has infinite complexity[2]—which is something we are accustomed to by now—but is non-repetitive in a much more creative way: it 'draws' a whole new flow pattern all the time.

Nice simple examples are the weather—especially as it imposes when looking at a satellite map as it changes from day to day—and the pattern of vapor that blurs the transparency of part of the upper portion of an open tea pot made of glass (and containing near boiling water). Even if the environment of the tea pot is completely perturbation-free, the 'weather' inside changes all the time in a fascinating manner.

More familiar—and less large-scale—phenomena are the sudden change in macroscopic properties of a fluid sent through a pipe at increasing speed (laminar/turbulent transition[3]), and the well-known series of 'morphogenetic' changes observable if a fluid is put between two rotating cylinders [42] or if a fluid is heated from below [4]. Beautiful pictures of the latter two phenomena may be found in Katchalsky [14] and Haken [10]. Especially the Taylor experiment has been scrutinized in recent years (See Swinney et al. [41]). When increasing the rotation speed of the outer cylinder, with the inner one stationary, the fluid first becomes organized into 'scrolls' of opposite rotation directions (the Taylor vortices), then these rings become wavy, then wavering, and then they disappear. Actually, this is not quite true, as Swinney et al. found [41]. Over a long range of rotation speeds, the waviness just contains some irregular little 'fishes'. (The water contains littles leaves of aluminium for better observability; so everything is silvery and slippery to look at.) Even if there are fishes than regularity, locally, still the original stripes (vortices) are observable. Thus, the vorticity is the same in two completely different realms (two different fluids, so to speak), one laminar and oily, the other turbulent and frisky.

---

[1]Classical texts describing turbulence were provided by Joseph Boussinesq [3], Ludwig Prandtl [31] or George Batchelor [2].

[2]With an infinite periodicity.

[3]A review about this transition can be found in [1] and an historical account on this transition by intermittencies in [23].

The interesting observations of Swinney et al. [41], obtained with laser doppler interferometry, of two frequencies of 'wavering', then (with increasing angular speed) a 'broadening' of one toward a 'band', are only mentioned here. Crutchfield et al. observed a similar phenomenon in Eq. (2.2), by the way [5].

How about a qualitative understanding of turbulence? The single most important insight to keep in mind is, perhaps, that turbulence is not something unexpected and hard to explain but, on the contrary, one of the most straightforward phenomena to expect. (Thus, a similar attitude is required as with respect to cancer: the problem is not how to explain the occurrence of the phenomenon in organism all composed of potentially autonomous subunits, but its relative rarity.)

Turbulence is to be expected as a gross phenomenon whenever one has a network of 'fluidic amplifiers' (each containing a strong 'main stream' that can be switched between two or more alternate directions by means of a weak 'side stream'), coupled in a random fashion. And of course, any fluid (or gas) in motion is a 'continuous approximation' to such a random network.

Thus, the 'dancing flames' of a fire are just as expected as the irregular trace obtained when pouring honey from a certain height. The latter system is perhaps especially appropriate for an experimental investigation of fluid turbulence under simplified conditions. From a theoretical point of view, the possibility of having a huge randomly coupled network is reminiscent of cellular metabolism, for one thing, or the brain, for another, or evolution as a whole. Why is there no highly evolved dynamical off-spring of billions of years of terrestrial weather, and why are there, to our knowledge, no highly evolved dynamical patterns present in the interior of the sun?

Or, to put it the other way: what is the distinguishing feature of that particular turbulent reaction network of which biology (with the cetacean brain) is an outgrowth? A certain 'rigidity', allowing for the differential conservation (and recursive accumulation) of certain 'self-maintaining' subprocesses, is a characteristic of evolutionary chemical networks [37]. On the other hand, one cannot exclude that even apparently rigidity-free turbulent systems are full of long-term after-effects.[4] So there might, again, be more to a simple turbulent process than meets the eye.

10.5 Turbulence and Coupled Oscillators

Since genuine turbulence is too complicated (and beautiful) to be modeled directly, one is bound to introduce simplifications. One of the most natural things to do is to

1. omit convection,
2. make the locally possible behaviors the same,
3. reduce the number of dimensions,
4. choose a finite number of cells.

---

[4]Crispin Gardiner, personal communication during a visit in Tübingen, 1980.

With these simplifications, one arrives at a finite set of coupled ordinary differential equations. In the simplest case, they involve the same two variables with the same local laws. The local behavior can be either oscillatory (damped; undamped; excitable) or non-oscillatory but stable (globally stable node, bistability, tristability...). In general, all possible phase portraits of two-variable systems are admitted. The coupling is, in the simplest case, of diffusion type, that is, linear.

Historically, such a system was apparently first conceived of, in the context of nonlinear systems with nontrivial behavior, by Fermi et al. [7]. The just-invented tool of digital computers was considered an ideal prerequisite for its study. Unfortunately, not much of interest (or at least nothing easily interpretable) was found in that first study.

Next came Ruelle and Takens' broad conjecture that four and more infinitely weakly coupled nonlinear oscillators should generate a near-quasi-periodic Axiom $\mathscr{A}$ attractor (see Sect. 6.2). This idea was an outgrowth of the concept due to Landau [21] and Hopf [13], that turbulence may be equivalent to an infinity of infinitely weakly coupled oscillators.[5]

Experimenting with nonlinear coupled oscillators during the early fourties, Rolf Landauer[6] observed highly irregular and 'apparently not quasi-periodic' behavior. Gollub et al. reproduced these unpublished experimental observations [9]; two nonlinear oscillators turned out to be sufficient for apparently chaotic behavior.

In 1976, Kuramoto and Yamada considered the problem of diffusion-coupled chemical oscillators [18–20]. In computer experiments backed by analytical calculations, they found evidence for 'chaotic' behavior (like an exponentially decaying autocorrelation function). In the same year, chaos was found if only two identical (not necessarily undamped) chemical oscillators were coupled by diffusion [36]. The type of diffusion coupling assumed was the same in both cases (cross-inhibitory 'morphogenetic', rather than cross-activating 'triggering', type).

Other set-ups to generate chaos in diffusion coupled identical two-variable systems were found soon. Following a suggestion by Art Winfree,[7] an irregular 'meandering' of the core of a self-reproducing rotating spiral pattern was found in computer experiments of a two-variable excitable medium [35, 38]. Kuramoto found a Lorenz attractor in his system [16], just beyond a certain critical coupling value. Soon thereafter, he also found three-variable chaos of the (simpler) type of Fig. 2.3 under similar

[5] Hopf was aware that the lack of coupling between the oscillators was a limitation of such a theory: viscous coupling was considered by Kolmogorov [15], Carl von Weizsäcker [44], and Werner Heisenberg [12]. The assumption for nearly non-coupled oscillators results from a remark made by Prandtl who stated that the resistance to the flow in turbulent behavior was not due to viscosity but to a collective effect [32].

[6] Rolf W. Landauer (1927–1999), personal communication during a visit of Otto and Reimara Rössler at their house near Yorktown Heights at IBM where Rolf worked, 1977. Landauer was a German-American physicist who contributed to the thermodynamics of information processing, condensed matter physics, and the conductivity in disordered media. He established that, in any irreversible operation manipulating information, such as, for instance, erasing a bit of memory, an associated amount of energy is dissipated as heat, thus inducing an increase of the corresponding entropy [22].

[7] A. Winfree, personal communication, 1975. See [24] for some details about this suggestion.

conditions (see [17]). Then, he discovered a 'licking wavefront', namely a wavefront whose shape is unstable so small symmetry-breaking perturbations [17]. Thus, a wealth of different mechanisms leading to turbulent behavior in linearly coupled nonlinear oscillators (continuum or cellular approximation) have already turned up so far.

Ulam's early conjecture about the significance of the vibrating nonlinear string [7] was thereby vindicated. Early chemical experiments of Marek et al. also found a potential new interpretation [25]. See also Neu for an alternative mathematical interpretation of these experiments [29].

Ortoleva et al. [39] took up Hopf's problem of diffusion coupled three-variable oscillators [13], and Pismen studied quasi-periodic and perhaps chaotic standing (and moving) patterns with analytical means [30]. Mimura and Yamaguti also considered the possibility, observed numerically by Meinhardt [27], of having standing patterns of arbitrary complexity ('frozen chaos', to use a term suggested by Kuramoto[8]) [28].

In the theory of flames, interest in 'macroscopically turbulent', that is, chaotic, phenomena emerged also [40]. In models of hot plasma, which are of interest in the context of the containment problem for fusion reactors, chaotic motions were also found [45].

10.6 Coupled Oscillators and Boiling

So far, we saw that the idea to replace the turbulence problem by a coupled oscillators problem is valid in the sense of producing a lot of numerical and analytical observations. The main question—whether indeed the full complexity of turbulence can be captured this way—has yet to be addressed, however. *Is there a class of coupled identical oscillators which (i) mimic a natural 'turbulent' system, (ii) can be analyzed analytically, and (iii) produce more than just 'ordinary' chaos?*

One particular turbulent phenomenon which lends itself rather naturally to such an analysis is the phenomenon of 'boiling-type turbulence.' The best example is spinach sauce heater in a saucepan from below. What occurs is easy to describe. Somewhere, a steam bubble is forming near the bottom of the sauce, absorbing heat during its growth (thereby preventing neighboring bubbles from growing). After having grown big enough, the bubble leaves the bottom (or directly hits the surface by virtue of its growth, respectively), with the consequence of bursting. The sauce collapses locally and the 'wound' heals. A neighboring bubble takes the lead in growing. And so forth.

The result is the typical 'boiling pattern' observable not only in spinach sauce, but also (quite well) in mashed potatoes and, at a much faster pace and on a much smaller scale, in ordinary ('non-thick') fluids heated from below. If the bubbles are reabsorbed after take-off (upon reaching higher and cooler layers of the fluid), one sometimes speaks of 'sizzling.' Sizzling is interesting because there goes a characteristic noise

[8] Y. Kuramoto, personal communication, 1978. Otto met for the first time Yoshiki Kuramoto in 1976 when he was visiting Hermann Haken in Stuttgart for half a year.

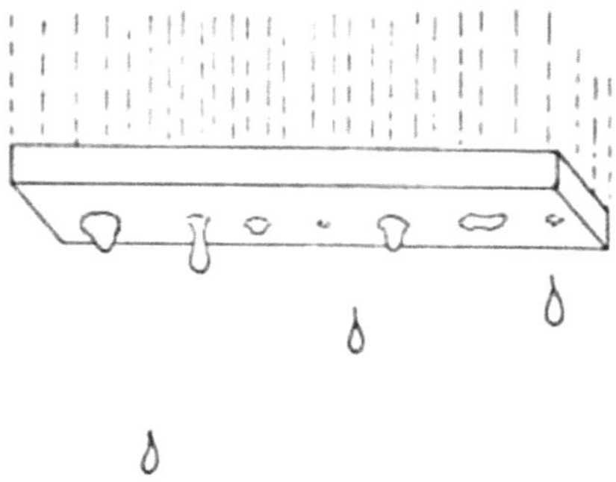

Fig. 10.4 The dripping hand-rail. See text

with it. The same 'noise' should be produced in bubbling spinach sauce—only slowed down by three orders of magnitude.

Still, such a slowed-down sizzler is not yet simple enough to allow for a complete qualitative understanding of what happens. Therefore, the notion of a 'one-dimensional saucepan' may be introduced. It contains the same material as before and is heated in the same way as before. Its width is not zero, but a constant (and just sufficient to harbor a full-grown bubble, before it breaks or takes off, respectively). The one-dimensional saucepan, in turn, has an analogue which shows the same behavior but at the same time is even more intuitive and easy to understand: the dripping hand rail standing in the rain. It is depicted in Fig. 10.4.

Again, there is (i) a local relaxation oscillation going on (the falling drop corresponding to the off-taking bubble of the former example), and (ii) a local cross-inhibition involving the slowly growing variable. The local formation and water-draining growth of a 'water sac' corresponds to the local formation and heat-draining growth of a 'steam bubble'. Both systems thus behave like a 'morphogenetic process' as first described by Rashevsky [33] and Turing [43] in the context of explaining the spontaneous onset of 'polarity' (differentiation) in a homogeneous cell or tissue. There, the local elements are also assumed cross-inhibitory [34]. In the present case, the cross-inhibitory elements are made up by the slow variables of a set of relaxation oscillators. The results is a 'growing morphogenesis' (resembling a growing skyline). This growing pattern then is subjected to a constraint: Whenever locally a skyscraper is suddenly turned into a minimum (is displaced toward zero). Thereafter, another growing pattern is formed, and so forth. Thus, what we have could be called a 'modulo morphogenesis' for short. A possible equation is

$$\dot{x}_i = 1 - a\,(x_{i-1} - x_{i+1}) + b\,x_i\,, \ \ \text{mod } 1 \qquad (10.1)$$

where $i = 1, \ldots, n$ and (in the simplest case of a ring) $x_0 = x_n$. Here n is the number of cells in the string (ring).

This equation is an extended version of the simplest relaxation oscillator

$$\dot{x} = 1\,, \ \ \text{mod } 1\,. \qquad (10.2)$$

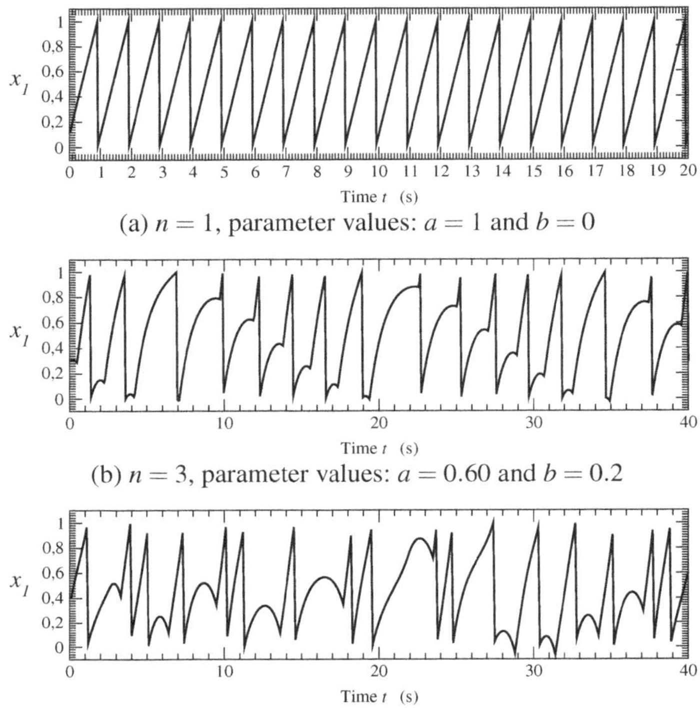

(a) $n = 1$, parameter values: $a = 1$ and $b = 0$

(b) $n = 3$, parameter values: $a = 0.60$ and $b = 0.2$

(c) $n = 5$, parameter values: $a = 0.75$ and $b = 0.2$

Fig. 10.5 Time behavior of an idealized dripping handrail (see Fig. 10.4). Numerical simulation of Eq. (10.1) with varying n. When $n = 1$ (case a), it corresponds to Eq. (10.2). Initial conditions: $x_1 = 0.1$ and $x_i = 0$ $(1 > i > n)$

which produces a sawtooth oscillation (because as soon as x in its constant growth hits the value of unity, the 'mileage counter' measuring x is reset to zero, see Fig. 10.5a). The main difference between Eqs. (10.1) and (10.2) is that there is a whole battery of such oscillations in Eq. (10.1), and that they are coupled in a 'cross-inhibitory' fashion: the higher one variable, the stronger the decay of its neighbors.

Simulations of the time behaviors of Eq. (10.1), with 1, 3 and 5 cells, are shown in Fig. 10.5. The interactions between neighboring elements giving rise to this overall behavior can be followed in details (see below).

If the saucepan is replaced by one of larger diameter (a frying pan), the effects of heating spinach sauce are slightly different. The reason is that the formerly laterally homogeneous situation during the buildup of the vapor layer now has undergone a morphogenetic (spatial symmetry breaking) bifurcation: there is the typical picture of a landscape with more or less sinusoidal hills and valleys as described by Turing [43] for reaction diffusion systems. If one were able to halt the heat flow through the saucepan at the right moment, this profile would be stable. With the heat transfer going on, however, it is clear what happens: the biggest 'hill' that has formed is suddenly being punctured by a hole through which much steam escapes while the hill collapses. After the hole is 'healed,' the landscape rearranges itself, with the consequence that a different spot now takes the lead in the overall growth; and so forth. The resulting nonrepetitive pattern of waxing and waning may be termed boiling-type turbulence.

There are, of course, many natural phenomena obeying the same pattern. A horizontal handrail standing in the rain, with fine droplets falling on its top, develops a chain of budding large drops on its bottom side. Again, a very short piece of handrail shows a periodic relaxation oscillation, due to a single drop forming and falling off at a time; and a somewhat longer piece shows a symmetry breaking (there is never a horizontal bar of water forming); and so forth.

10.7 An 'Ideal' Example

The simplest equation likely to produce the 'boiling-type' phenomenon is

$$\begin{cases} \dot{x} = 2.1 - y - z\,, & x \bmod 1 \\ \dot{y} = 2.1 - z - x\,, & y \bmod 1 \\ \dot{z} = 2.1 - x - y\,, & z \bmod 1\,. \end{cases} \tag{10.3}$$

Here, three relaxation oscillators of the type of Eq. (10.2) are coupled pairwise in a cross-inhibitory fashion. If only the linear parts are considered (omitting the mod 1 constraints), the cross-inhibition present between the three components is responsible for the formation of an unstable symmetrical steady state of saddle type in three-dimensional space, located at $x = y = z = 1.05$. All symmetrical initial conditions tend towards this point. The saddle has three unstable eigendirections, each leading to an increase in one variable and a simultaneous decrease in the two others. The flow in the lower portion of the positive orthant of state space therefore looks like the 'dust threads' of certain flowers, being grouped around the flower's axis and veering away from the latter the more strongly the farther out they start.

Adding the 'modulo' constraints in Eq. (10.3) means that there is a maximum height which can be reached in any of the three directions of space. Thus, the flower has been imbedded into a cube, stretching within it from one corner (the origin) across the center toward the opposite corner (1, 1, 1). If one of the three upper sides of the cube is being hit by a motion that originally came from one of the three lower sides, the system's state suddenly is displaced toward the opposite lower side. From there, it then starts moving up again, and so forth.

Nonetheless, there is a geometric argument showing that the flow described thus far cannot produce a higher form of chaos. Due to the fact that the sides of the cube are pair-wise identified, and the flow inside also invertibly connects those sides, at the specific parameters assumed, the whole flow necessarily takes place on a 3-torus. Its cross section therefore is a two-dimensional invertible map (homeomorphism or diffeomorphism) rather than a noninvertible map (endomorphism). Two-dimensional maps must be nonuniquely invertible, however, if they are to generate higher chaos.

This drawback is a consequence of the fact that Eq. (10.3), as written, is too much idealized. There are several ways to make it more realistic while leaving it simple. Straightforward change consists in reducing the constant 2.1 to 1.1. Since the saddle has now been pulled into the cube, trajectories no longer only enter the three lower sides. They nonetheless hit their mod 1 threshold, only now outside the cube. If they have not been veering away too strongly, they land still close enough to the cube in order not to leave it (or at least not to leave it equally far) next time, if the parameters are right. As a consequence, the lids of the cube now become endomorphisms rather than diffeomorphisms. The resulting flow now indeed can show hyperchaotic behavior—but, unfortunately, only as a transient. At the parameter values tested so far (around 1.1), the at first hyperchaotic appearing flow either escapes after a while toward an external limit cycle (because some trajectories are veering away too strongly after having left the cube), or finally settles for an asymptotic toroidal attractor inside. Thus, purely quantitative (from a qualitative point of view accidental) reasons prevent Eq. (10.3) from showing the desired behavior.

There are several possibilities to amend this situation. For example, one might slant the upper sides of the cube. Or, very realistically, one might round off the sharp edges that apply between the three upper sides of the cube. This can be done by inserting a slanted flat 'rim' at the edges. As a consequence, there is no longer complete independence between the thresholds of the three sub-oscillators throughout the whole range. The second solution has been tested and functions nicely. Both solutions nonetheless completely destroy the appealing simple appearance of Eq. (10.3). Remains the possibility to take into account the fact that each relaxation oscillator in reality always involves two variables, a fact which implies both 'rounding off' and 'delay' effects. Such an explicit example follows below.

An alternative way of proceeding would be to stick to Eq. (10.3) as written above and simply increase the number of cells. Possibly, again a hierarchy of more and more complicated motions, beginning at a somewhat higher cell number, will be found when studying such (piecewise linear, C^0 type) flows on n-tori.

10.8 A Smooth Example

The six-variable equation

$$\begin{cases} \dot{x} = 20 - y - z - w \\ \dot{y} = 20 - z - x - v \\ \dot{z} = 20 - x - y - u \\ \dot{w} = 0.01 + w(x - 10) \\ \dot{v} = 0.01 + v(y - 10) \\ \dot{u} = 0.01 + u(z - 10) \end{cases} \tag{10.4}$$

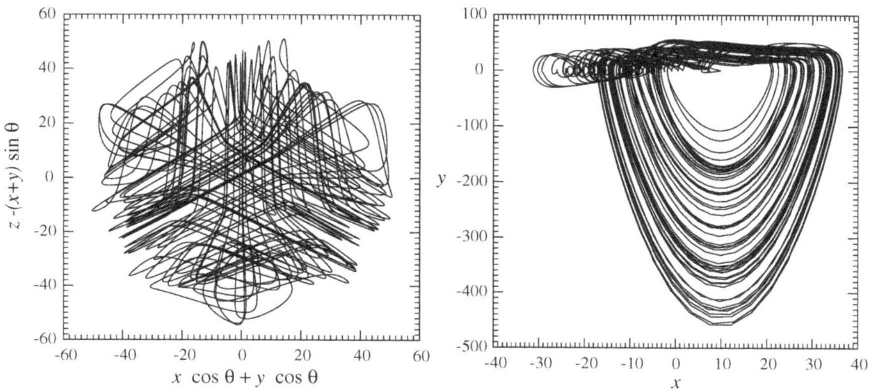

Fig. 10.6 Trajectorial behavior of three single-threshold relaxation oscillators whose slow variables are coupled in a cross-inhibitory fashion, projected into the slow sub-space. Numerical simulation of Eq. (10.4). Initial conditions: $x_0 = 0.05$, $y_0 = 0.05$, $z_0 = 0.07$, $w_0 = 0.05$, $v_0 = 0$, and $u_0 = 0$

is a 'smooth' realization of Eq. (10.3): There are once more three 'morphogenetically' coupled linear variables (x, y, and z) and three 'switches' (w, v and u), one for each linear variable. That is, Eq. (10.4) can again be said to consist of three single-threshold relaxation oscillators coupled in a cross-inhibitory manner. The fact that each sub-oscillator has a stable focus (which is a consequence of the extreme simplicity of the sub-systems; a second quadratic term would be needed to make the individual relaxation oscillators undamped) has no consequences farther away from the unstable symmetrical state. The behavior of the system's trajectories in the three-dimensional 'slow' sub-space (x, y, z) is displayed in Fig. 10.6.

The behavior in this state sub-space is very similar to that displayed by Eq. (10.4). The main difference is that the upper portion of the 'cube' no longer consists of three orthogonal pieces separated by sharp edges. The 'cap' now has a rounded-off, soft shape. There is a bunch of trajectories diverging from the middle of the left-hand picture (corresponding to the origin of the 'cube'). And there is a reinjection 'back downstairs' once the three upper sides of the cube are hit. By tracing back trajectories from one of the three upper sides, one verifies that there is non-uniqueness of trajectorial origin at some places (implying overlap of the underlying map), and so along different directions. See, for example, the first bundle to the right of the z-axis (Fig. 10.7).

Figure 10.8 shows the time behavior of the six variables. Focusing only on the 'switches' (displayed in the three lower rows), one is tempted to interpret the spikes as falling drops (in terms of the above handrail example). One sees that nearly always two sub-oscillators are engaged in a mutual oscillation of 'push pull' type, with the third being idle. These sub-regimes last for differing periods and alternate irregularly.

Whereas Eq. (10.4) is too complicated to allow an analytical proof that its cross section[9] has the form of a towel that is folded in a more or less orderly way (Fig. 10.7),

[9]The Poincaré section was defined as

$$\mathscr{P} \equiv \left\{ (y_n, z_n, u_n, v_n, w_n) \in \mathbb{R}^5 \mid x_n = 10, \dot{x}_n < 0 \right\} .$$

The first-return map is in fact constructed using variable $\dot{x}_n$.

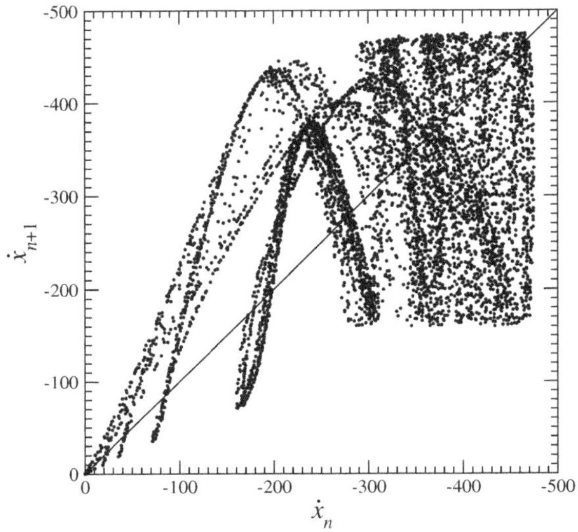

Fig. 10.7 First-return map to a cross section of the attractor produced by one of the sub-oscillators. Same numerical simulation as in Fig. 10.6

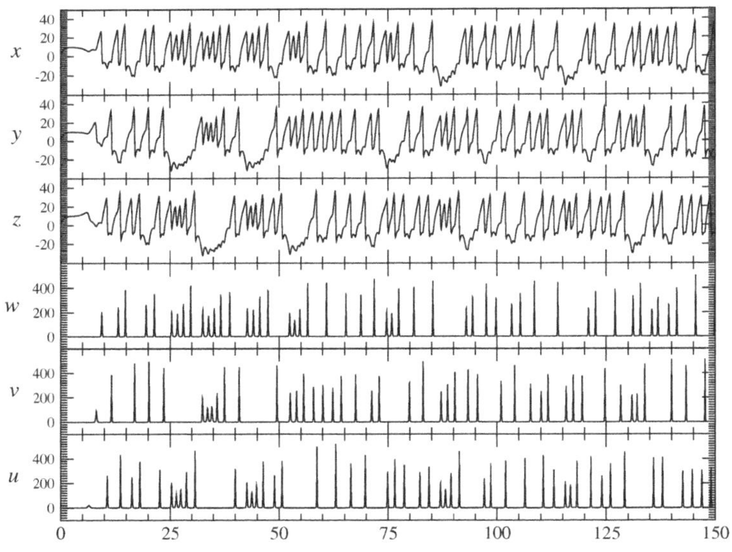

Fig. 10.8 Time series produced by Eq. (10.4). Numerical simulation as in Fig. 10.6

the argument which led to Eq. (10.4), via Eq. (10.2) as a vehicle, suggests that there exists an ideal version of Eq. (10.4), with infinitely fast relaxation (that is, with a structure similar to Eq. (10.2)), for which the presence of an endomorphism of folded towel shape can be proven analytically. Assuming this to be the case with the present ring (or plane) of three cells, it is natural to expect that the principle will carry through (with shape-dependent modifications) to the case of four and more cells. The trajectories, hitting now the sides of hyper cubes in order to be reinjected toward the opposite side, should again do so in a not uniquely invertible manner as far as the slow sub-space is concerned, and with more and more directions opening up at the same time in the underlying, higher-dimensional maps. ('Folded-in cloud map,' etc.)

10.9 A Hierarchy in Boiling-Type Turbulence

The model of boiling proposed in the preceding section is very crude. For example, only one space dimension was assumed.[10] Even more incisively, a discretization was imposed: instead of a continuum, only a small finite number of local elements was taken into consideration. It is possible to relax these simplifications. A smooth equation possessing the same structure as Eq. (10.4) is, for example,

$$\begin{cases} \dot{x}_1 = a + bx_1 - c(x_2 + w) \\ \dot{x}_2 = dx_l - ex_2 + D(y_2 + z_2 - 2x_2) \end{cases} \tag{10.5}$$

where x_1 and x_2 replacing the former x, and so forth. The resulting nine-variable analogue to Eq. (10.4) can be expected to show the same behavior as Eq. (10.4) at appropriate parameter values. The new system has the disadvantage that each individual relaxation oscillator involves three variables. Its asset is that the whole system can now in the limit be written as a simple parabolic partial differential equation. The system can be turned into an explicit three-variable reaction-diffusion system [8] by multiplying the 'non-chemical' terms, $-c(x_2 + w)$, etc., in the first lines each with a Michaelis–Mentens term, $\frac{x_1}{x_1+\kappa}$, etc. whereby κ may approach zero [43, p. 42]. Three-variable reaction-diffusion systems are unnecessarily complicated as far as the generation of ordinary chaos is concerned, since two variables are sufficient. However, two-variable reaction-diffusion systems are probably no less prone to produce higher chaos. The above three-variable equation therefore serves only a heuristic (and, perhaps, didactic) purpose.

A simulation of Eq. (10.4) with $n = 3$ is presented in Fig. 10.6. One sees that neighboring initial conditions (in the x-z plane, for example) diverge not only in one

[10]This is actually the case since only a ring of oscillators is considered.

direction, but in two. Indeed, one can show that an analogous 'folding over' as it occurs in the neighborhood of saddle fixed point in a two-dimensional state space occurs in two directions in the flow of Fig. 10.6. 'And so forth.' The last three words mean that with four cells in a ring, one has a folding-over of a three-dimensional 'side' of a hyper cube, and so in three independent directions; and with five cells the 'side' is four-dimensional, with four directions of folding over; and so on. Thus, we observe the presence of a chaotic hierarchy in Eq. (10.4): for n sub-oscillators, the resulting chaos is of the hyper^{n-2} kind, to use the terminology of Sect. 2.8.

So far it is not known whether this result of a hierarchy of higher and higher chaos (proportional to n) in boiling type turbulence is confined to discrete models like Eq. (10.4), or whether an analogous result (with n replaced by L, the length of the spatial interval) holds true for non-discretized equations. It seems, however, a reasonable hypothesis to presume that boiling and sizzling in general correspond to high or very high forms of higher chaos. Whether or not other kinds of empirical turbulence follow the same pattern is, of course, even harder to say. Nonetheless, an analogous hypothesis as that just made about 'boiling type turbulence' seems to be admissible in general.

References

1. * D. Barkley, Theoretical perspective on the route to turbulence in a pipe. J. Fluid Mech. **803**, P1 (2016)
2. * G. Batchelor, *An Introduction to Fluid Dynamics* (Cambridge University Press, Cambridge, 2012)
3. * J.J. Boussinesq, Essai sur la théorie des eaux courantes. Comptes-Rendus l'Académie Sci. **23**, 1–680 (1877)
4. H. Bénard, Etude expérimentale des courants de convection dans une nappe liquide. — Régime permanent; tourbillons cellulaires. J. Phys. III **9**, 513–525 (1900)
5. * J.P. Crutchfield, N.H. Packard, Symbolic dynamics of one-dimensional maps: entropies, finite precision, and noise. Int. J. Theor. Phys. **21**(6–7), 433–466 (1982)
6. G. Cantor, Über unendliche, lineare Punktmannigfaltigkeiten v (On infinite linear point-manifolds). Math. Ann. **21**, 359–379 (1883). Also published separately as Grundlagen einer allgemeinen Mannigfaltigkeitslehre, Leipzig (1883)
7. E. Fermi, J. Pasta, S. Ulam, *Studies of Nonlinear Problems*, Document Los Alamos 1940, Reprinted in *Collected Papers of Enrico Fermi* (University of Chicago Press, Chicago, 1965)
8. P.C. Fife, Propagating waves and target patterns in chemical systems, in *Dynamics of Synergetics Systems*, ed. by H. Haken (Springer, Berlin, 1980), pp. 97–106
9. J.P. Gollub, S.V. Benson, Chaotic response to periodic perturbation of a convecting fluid. Phys. Rev. Lett. **41**(14), 948–951 (1978)
10. H. Haken, *Synergetics – An Introduction* (Springer, Berlin, 1976)
11. F. Hausdorff, Zur Theorie der linearen metrischen Räume. J. Reine Angew. Math. **167**, 294–311 (1932)
12. * W. Heisenberg, Zur Statischen Theorie der Turbulenz. Z. Phys. **124**, 628–657 (1948)
13. E. Hopf, A mathematical example displaying features of turbulence. Commun. Pure Appl. Math. **1**, 303–322 (1948)
14. A. Katchalsky, P.F. Curran, *Nonequilibrium Thermodynamics in Biophysics*, 3rd edn. (Harvard University Press, Cambridge, 1974)

15. * A.N. Kolmogorov, The local structure of turbulence in incompressible viscous fluid for very large Reynolds numbers. Dokl. Akad. Nauk SSSR **30**(4), 299–303 (1941). Translated by V. Levin, Proc. R. Soc. Lond. A **493**(9–13), (1991)
16. Y. Kuramoto, Diffusion-induced chaos in reaction systems. Prog. Theor. Phys. (Supplement) **64**, 346–367 (1978)
17. Y. Kuramoto, Instability and turbulence of wavefronts in reaction-diffusion systems. Prog. Theor. Phys. **63**(6), 1885–1903 (1980)
18. Y. Kuramoto, T. Yamada, A reduced model showing chemical turbulence. Prog. Theor. Phys. **56**(2), 681–683 (1976)
19. Y. Kuramoto, T. Yamada, Pattern formation in oscillatory chemical reactions. Prog. Theor. Phys. **56**(3), 724–740 (1976)
20. Y. Kuramoto, T. Yamada, Turbulent state in chemical reactions. Prog. Theor. Phys. **56**(2), 679–681 (1976)
21. L.D. Landau, On the problem of turbulence. Dokl. Akad. Nauk SSSR **44**, 339–342 (1944)
22. * R.W. Landauer, Irreversibility and heat generation in the computing process. IBM J. Res. Dev. **5**, 183–191 (1961)
23. * C. Letellier, Intermittency as a transition to turbulence in pipes: a long tradition from Reynolds to the 21st century. Comptes Rendus de Méc. **345**(9), 642–659 (2017)
24. * C. Letellier, L.A. Aguirre, On the interplay among synchronization, observability and dynamics. Phys. Rev. E **82**, 016204 (2010)
25. M. Marek, E. Svobodova, Nonlinear phenomena in oscillatory systems of homogeneous reactions – Experimental observations. Biophys. Chem. **3**, 263–273 (1975)
26. B. Mandelbrot, *Fractals — Form, Chance and Dimension* (Freeman, San Francisco, 1977)
27. H. Meinhardt, Morphogenesis of lines and nets. Differentiation **6**, 117–123 (1976)
28. M. Mimura, N.M. Yamaguti, Some diffusive prey-predator systems and their bifurcation problems. Am. N. Y. Acad. Sci. **316**, 490–510 (1979)
29. J.C. Neu, Interacting nonlinear chemical oscillators. SIAM J. Appl. Math. **35**, 536–547 (1978)
30. L.M. Pismen, Asymmetric steady states revisited. Chem. Eng. Sci. **34**(4), 563–570 (1979)
31. * L. Prandtl, Bericht uber Untersuchungen zur ausgebildeten Turbulenz. Z. Angew. Math. Mech. 136–139 (1925)
32. * L. Prandtl, Ergebnisse, Neuere, der Turbulenzforschung. Z. Ver. Dtsch. Ing. **7**(5), 105–114 (1933). English translation: Recent results of turbulence research. N. A. C. A. Technical Memorandum **720**, (1933)
33. N. Rashevsky, An approach to the mathematical biophysics of biological self-organization and of cell polarity. Bull. Math. Biophys. **2**, 15–25, 65–67, 109–121 (1967)
34. R. Rosen, *Dynamical System Theory in Biology* (Wiley, New York, 1970)
35. O.E. Rössler, Chaos in abstract kinetics. Two prototypes. Bull. Math. Biol. **39**, 275–289 (1979)
36. O.E. Rössler, Chemical turbulence: Chaos in a simple reaction-diffusion system. Z. Naturforschung A **31**, 1168–1172 (1976)
37. O.E. Rössler, Ein systemtheoretisches Modell zur Biogenese. Z. Naturforschung B **26**, 741–746 (1971)
38. O.E. Rössler, C. Kahlert, Winfree meandering in a 2-dimensional 2-variable excitable medium. Z. Naturforschung A **34**, 565–570 (1979)
39. O.E. Rössler, P.J. Ortoleva, Strange attractors in 3-variable reaction systems. Lecture Notes in Biomathematics, vol. 21 (1978), pp. 67–73
40. G.I. Sivashinsky, Hydrodynamics theory of flame propagation in an enclosed volume. Acta Astronaut. **6**, 631–645 (1979)
41. H.L. Swinney, P.R. Fenstermacher, J.P. Gollub, Transition to turbulence in a fluid flow, in *Synergetics, a Workshop*, ed. by H. Haken (Springer, Berlin, 1977), pp. 60–69
42. G.I. Taylor, Experiments on the motion of solid bodies in rotating fluids. Philos. Trans. R. Soc. A **223**, 289–343 (1923)
43. A.M. Turing, The chemical basis of morphogenesis. Philos. R. Soc. Lond. B **237**, 37–72 (1952)
44. * C. von Weizsäcker, Das Spektrum der Turbulenz bei groβen Reynoldsschen Zahlen. Z. Phys. **124**, 614–627 (1948)
45. P.K.C. Wang, Nonperiodic oscillations of Langmuir waves. J. Math. Phys. **21**, 398–407 (1980)

Chapter 11
When to Expect Chaos

11.1 Suspecting Chaos

Lefschetz once remarked that Poincaré, after having discovered the possibility of homoclinic points, was well aware that from now on, the majority of simple dynamical systems could be expected to show the same complexity of behavior [8].

In light of this remark, and in light of the simple structure of Eqs. (2.2), (4.1), and (6.17)—to mention only examples containing a single quadratic nonlinearity—, one gains the impression that indeed of a majority multiple-loop (and even single-loop) feedback systems of nonlinear type may be capable of chaotic behavior in a finite portion of their parameter spaces.

11.2 Two Exceptional Classes

Notwithstanding the general probability argument of the preceding Section, there are at least two classes of highly nonlinear, highly complex systems which do **not** show chaos. One is the class of 'dynamical automata', as they have been called [14]. An example is the class of electrically realized automata, the so-called digital computers. These systems are properly described by at least $2n$ coupled ordinary differential equations, if n is the number of flip-flops involved, but at the same time they do admit a 'short-hand description' in terms of the language of finite-state automata. A special subclass is that of chemical automata [14].

The second big class of nonlinear systems which do not show a chaotic limiting regime is the class of evolutionary chemical systems. These systems consist, mathematically speaking, of a very large set of coupled ordinary differential equations (if well-stirredness is assumed for simplicity): about 10^{10^6} or so nonlinear equations.

O. E. Rössler and C. Letellier, *Chaos*, Understanding Complex Systems,
https://doi.org/10.1007/978-3-030-44305-4_11

Such numbers are called a universal library,[1] because the set of all possible books of a given length has about the same size. The 'booklet' in the present case are the many different kinds of molecules that can be formed spontaneously out of a small initially given set of 'pooled', that is, exogenously maintained and resupplied (if consumed) substances, if at least one of them is 'energy rich' (photons could play this role) and at least one is capable of forming strings (like a –C–C–C– or –B–N–B–N– 'backbone').

The huge, randomly wired nonlinear system that is defined by the set of initially non-zero pool substances is not quite an ordinary dynamical system, however. This is because molecules come only in discrete units. And no more than (say) 10^{45} molecules fit into an ocean. So the system consists of a huge number of threshold type variables: apart from the usual (quadratic) nonlinearities of reaction kinetics, the right-hand sides include each in addition a 'switching type' nonlinearity of singular type (so as if the z-variable in Eq. (2.1), with both ϵ and δ tending to zero, had been added to the respective variable), allowing it to become nonzero only for sufficiently large finite influx. A more realistic description would involve Monte-Carlo processes instead, for instance.

Inspite of this complication, it is possible to derive some features of the qualitative behavior of such a network [13]. It turns out that, due to the opening of more and more non-zero 'channels' in the mathematically but not physically existing network, again and again a new 'functionally isolated' subsystem ('species') of unstable growth behavior is triggered into physical existence, in order to quantitatively dominate the picture for a while. Thus, while many variables at each moment in the development of the system may be engaged in a chaotic motion, there is all the time also an 'emancipation' from this overall dynamics going on, with the consequence that the overall behavior is pushed into the background for a while. Such systems may be called 'ultra-unstable', due to their analogy to Ashby's 'ultrastable' homeostat [1].

Thus, there is a class of highly nonlinear, physically 'open' dynamical systems which apparently do not approach a chaotic limiting regime over very long finite times at least. This class of systems thus is fundamentally different from ordinary turbulent systems: it combines turbulence with rigidity, so to speak (see Sect. 9.5).

11.3 Mass-Action Type Chaos

When trying to establish a probability (like: 'What is the frequency of chaos?'), one needs a well-defined universe of discourse. The class of nonlinear systems in general is too abstract for this purpose. A much better defined class is that of mass-action type reaction systems, forexample. Feinberg and Horn [3] developed a promising approach how to 'narrow down' what to expect dynamically from a given open reaction system of 'mass action type.' (Here the substances react pairwise, with

[1] Otto discovered the universal library, which was introduced by Gustav Theodor Fechner, in a book entitled *Die Universalbibliothek* written in 1904 by Kurd Lasswitz.

quadratic rates, and there is quantitative conservation of constituents.) They found a way how to identify all (or at least a good proportion of all) systems with 'dull' behavior, that is, with a single globally attracting steady state.

Willamowski [16] and Feinberg [2] recently showed that the relative frequency of such dull reaction schemes decreases very rapidly with the number of complexes (that is, reactions) involved. What remains as a vast majority is systems producing either multiple steady-state or oscillatory behavior (periodic or not) for some ranges of their parameters.

Most realistic chemical oscillators that are presently known (see Noyes et al. [4], for a review) involve more than five—usually about fifteen—variables, and about twice as many quadratic terms on the right-hand side of the rate equation. In general, neither the actual number of variables is known precisely (there may be 'fast' intermediates which have escaped measurements), nor is the precise form of the all functions on the right-hand side (with a small finite number of alternatives being open), nor—of course—are the precise (or approximate) values of all parameters known.

Nonetheless, if one assumes (as necessary) a 'most probable' reaction scheme, with the 'most probable' rate constants for a given temperature, one usually finds that there are several alternative ways open for 'reducing' this actual scheme to a more manageable, lower-dimensional ideal scheme. (This is usually done by assuming certain variables to 'relax' at an infinite rate.) Thus, one knows that in general more than one 'basis oscillator' of simple type is hidden in a given realistic oscillator.

This fact can be taken as evidence that these multiple-variable, multiple-loop systems contain enough structure to be capable not only to limit cycle oscillations, but also of chaotic oscillations [15]. It may even be worth looking for more than one hierarchy-level of chaos in a given chemical oscillator. So, far, it is known what is the relative frequency of potentially oscillatory reaction schemes in relation to merely multiple-steady state (but non-oscillatory) ones. Multi-stability is known to need fewer variables (and loops) than a limit cycle oscillation: In the former case, two variables of mass-action are required, in the latter, three [7].

Since the number of possibilities for loops increases combinatorially with the number of reactants, it seems that the oscillatory (and with them the chaotic) systems are the more probable the larger the reaction system. The curious fact mentioned that (i) two variables are necessary for multiple-steady state behavior in mass action systems (while one variable, like $\dot{x} = -x^3$, suffices in ordinary nonlinear systems) and (ii) three variables are necessary for limit cycle behavior (instead of two), let it appear doubtful at first whether three mass action variables indeed suffice for chaos as had been conjectured. An example is the following three-variable mass action system shown in Fig. 11.1.

With the third variable (Z) removed and the second-order outflux from X also omitted (that is, only the first line retained), the system of Fig. 11.1 is identical with Lotka's first abstract chemical oscillator [10]. This oscillator possesses only a 'conservative' type of oscillation (a continuum of closed trajectories around a center), but no limit cycle; see Rosen [11] for a derivation of the system's Hamiltonian function. With the second variable (Y) removed, in contrast, the system of Fig. 11.1

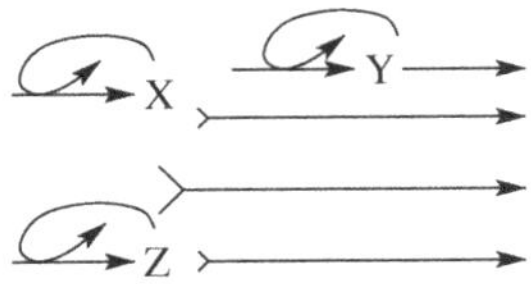

Fig. 11.1 A three-variable mass action system with chaotic behavior. Constant pool (substrates and products) have been omitted from the scheme as usual

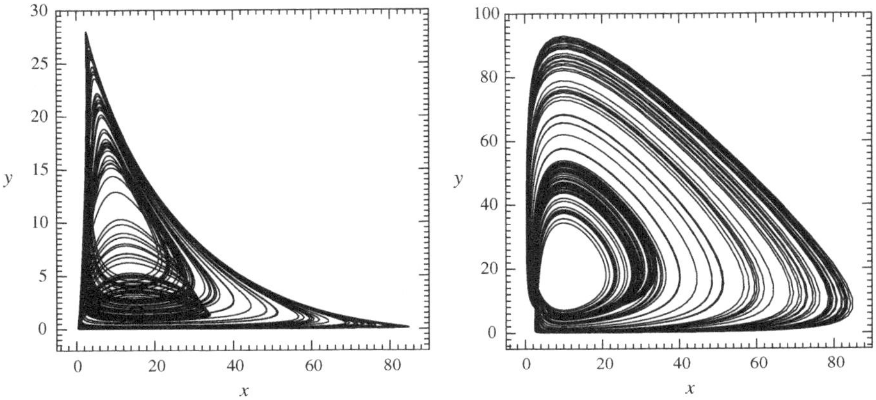

(a) x-y plane projection (b) x-z plane projection

Fig. 11.2 Numerical simulations of Eq. (11.1). Parameter values: $a = 30, b = 0.25, c = 1, d = 1, e = 10^{-3}, f = 0.01, g = 10, h = 0.01, k = 16.5$, and $l = 0.5$. Initial conditions: $x_0 = 10, y_0 = 80$, and $z_0 = 0.1$

becomes identical to the Gause switch of theoretical ecology [6], a bistable system which was later discussed also in the context of mass action kinetics [5]: Either one of the two auto-catalytic variables (X or Z) is bound to prevail, depending on the initial conditions. Put together, the two 'classical' systems of abstract reaction kinetics and abstract ecology (which lacks mass conservation constraints) illustrate well the heuristic principle (Sect. 1.3) that an 'oscillator' and a 'switch' combined are likely to produce chaos.

The system of Fig. 11.1 is, under the usual conditions of well-stirredness, isothermy, an appropriate concentration range, and with all reactions assumed reversible, described by the following set of rate equations (x = concentration of X, etc.) [17]:

$$\begin{cases} \dot{x} = x(a - bx - cy - dz) + ey^2 + f \\ \dot{y} = y(cx - ey - g) + h \\ \dot{z} = z(k - dx - lz) + f\,. \end{cases} \tag{11.1}$$

A numerical simulation of Eq. (11.1) is presented in Fig. 11.2.

When the system of Fig. 11.1 (irreversible version, that is, the constants e, f, and h being equal to zero) was found empirically, the range of parameters yielding chaotic behavior was very small [12]. Further combined numerical and analytical studies [17] showed, however, that the range of parameters with chaotic behavior is not small and, moreover, can be indicated approximately with analytical means. For a differential-topological investigation of a similar system, see Newton [9].

To come back to the main context: the system of Fig. 11.1 is certainly 'rare' among the set of all three-variable reaction systems with up to five complexes (reaction arrows). Moreover, not all abstract reaction systems have the same degree of chemical 'realisticness'; even a single autocatalytic reaction is rather unrealistic already, for example. So one is lead to guess that among the set of 'realistic' four-variable systems (a notion which is hard to define abstractly), oscillations (and especially chaotic oscillations) will still be relatively rare. With five variables, the statistics may already reach its turning point, however. Fortunately, this is a question which can be answered in principle, and probably will be in the not too distant future. For a well-defined class of systems, one will then know how 'likely' or unlikely chaos really is.

References

1. W.R. Ashby, Homeostasis, in *Cybernetics: Transactions of the Ninth Conference* (Josiah Macy Foundation, 1952), pp. 73–108
2. M. Feinberg, Complex balancing in general kinetic systems. Arch. Ration. Mech. Anal. **49**(3), 187–194 (1972)
3. M. Feinberg, F. Horn, Dynamics of open chemical systems and the algebraic structure of the underlying reaction network. Arch. Ration. Mech. Anal. **49**, 172–186 (1972)
4. R.J. Fields, R.C. Noyes, Oscillations in chemical systems iv, limit cycle behavior in a model of a chemical reaction. J. Chem. Phys. **60**, 1877–1884 (1974)
5. F.C. Frank, On spontaneous asymmetric synthesis. Biochim. Biophys. Acta **11**, 459–463 (1953)
6. G.I. Gause, *The Struggle for Existence* (Williams & Wilkins, Baltimore, 1934)
7. P. Hanusse, De l'existence d'un cycle limite dans l'évolution des systèmes chimiques ouverts. Comptes-Rendus de l'Académie des Sciences **274**, 1245–1247 (1972)
8. S. Lefshetz, Geometric differential equations: recent past and proximate future, in *Proceeding of the International Symposium on Differential Equations and Dynamical Systems* (Mayaguez, Porto Rico, 1965) (Academic Press, NewYork, 1967), pp. 1–14
9. R.B. Leipnik, T.A. Newton, Double strange attractors in rigid body motion with linear feedback control. Phys. Lett. A **86**, 63–87 (1981)
10. * A.J. Lotka, Analytical note on certain rhythmic relations in organic systems. Proc. Natl. Acad. Sci. (USA) **6**(7), 410–415 (1920)
11. R. Rosen, *Dynamical System Theory in Biology* (Wiley, New York, 1970)
12. O.E. Rössler, Chaos and strange attractors in chemical kinetics, in *Synergetics — Far from Equilibrium*, ed. by A. Pacault, C. Vidal (Springer, Berlin, 1978), pp. 107–113
13. O.E. Rössler, Ein systemtheoretisches Modell zur Biogenese. Zeitschrift für Naturforschung B **26**, 741–746 (1971)
14. O.E. Rössler, Chemical automata in homogeneous and reaction-diffusion kinetics. Lect. Notes Biomath. **4**, 399–418 (1974)
15. O.E. Rössler, Chaotic behavior in simple reaction systems. Zeitschrift für Naturforschung A **31**, 239–264 (1976)

16. K.-D. Williamowski, O.E. Rössler, Irregular oscillations in a realistic abstract quadratic mass action system. Zeitschrift für Naturforschung A **35**, 317–318 (1979)
17. R.F. Williams, The structure of Lorenz attractors. Publications Mathématiques de l'Institut des Hautes Études Scientifiques **50**(1), 73–99 (1979)

Chapter 12
How to Prove Chaos

12.1 Looking at Maps

If the presumable chaotic regime is low-dimensional and simple, the easiest thing to do is to prepare a next-amplitude map or some other one-dimensional projection of a cross section.[1] The meaning of the next-amplitude (= next extremum) map is explained in Fig. 12.1.

All simply folded-over flows admit the same type of maps as 'approximate cross section'. The two basic types involved differ because in the one type the whole cross section (in one direction) is involved, in the other only a '2π-segment' of it (between one turn of a spiral or screw and the next). Map "5" in Fig. 12.1 belongs to this second class.

The same two types of approximative cross sections are again encountered in funnel-type chaos (Fig. 12.2). Note that the two types of maps can be generated from a single measured (or computed) variable in each case. Obviously, it is not necessary to draw the whole flow (or to measure all variables): all that could be gained in this way are slanted cross sections (which are, of course, no better).[2]

Not all chaotic flows do, of course, belong to the preceding two simple types or to the next, only slightly more complicated 'composed' types (like the torus flow, or like a Bonhoeffer-van der Pol flow). Moreover, the cross section of some three-dimensional chaotic flows are not 'locally almost one-dimensional' due to their low intrinsic damping (divergence).

From a practical point of view, the above-described technique nonetheless constitues the most important method to prove the presence of chaos in simple realistic

[1] We would like to remind the reader that by the early 1980s, it was not yet clear how to prove chaos. Even today, this remains a great challenge for many systems. Thanks to the Li-Yorke theorem, "period-three implies chaos", Otto used the first-return map to a Poincaré section to have a "proof" for a chaotic attractor. Such an approach was used for instance by Lars Olsen and Hens Degn [7].

[2] By using a three-dimensional space reconstructed with a derivative or delay coordinates [9], it is possible to define a cross section and, consequently, to have a first-return map better defined.

O. E. Rössler and C. Letellier, *Chaos*, Understanding Complex Systems,
https://doi.org/10.1007/978-3-030-44305-4_12

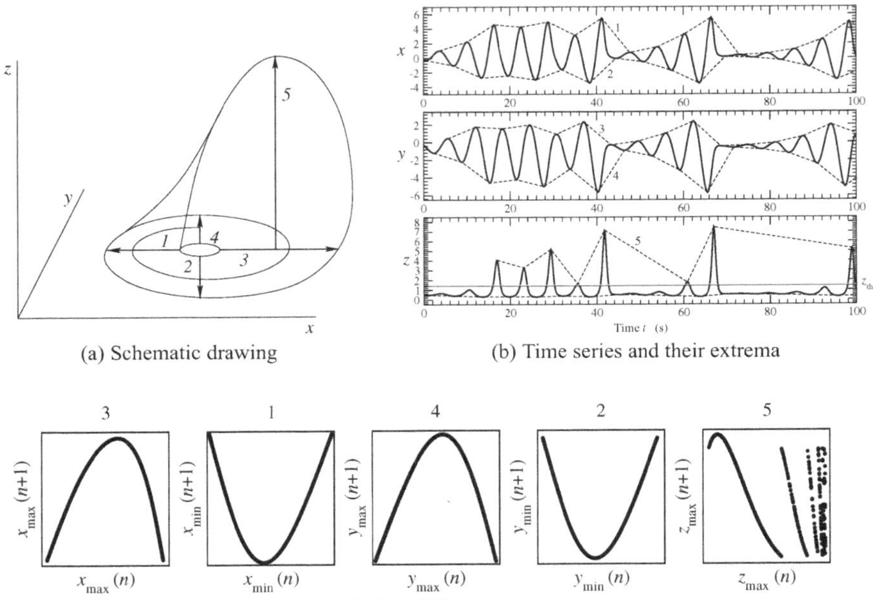

(a) Schematic drawing

(b) Time series and their extrema

(c) The five different maps

Fig. 12.1 The five possible next-extremum maps in the flow of Fig. 2.5b. The 'attracting boxes' labelled from "1" to "5" have been scaled for having the same size. "1" = next-maximum map of variable x, "2" = next-minimum map of variable x, "3" = next-maximum map of variable y, "4" = next-minimum map of variable y, "5" = next-maximum map of variable z. In map "5", a threshold range $z_{th} = 1.5$ has been excluded (see text). Parameter values: $a = 0.43295$, $b = 2$ and $c = 4$

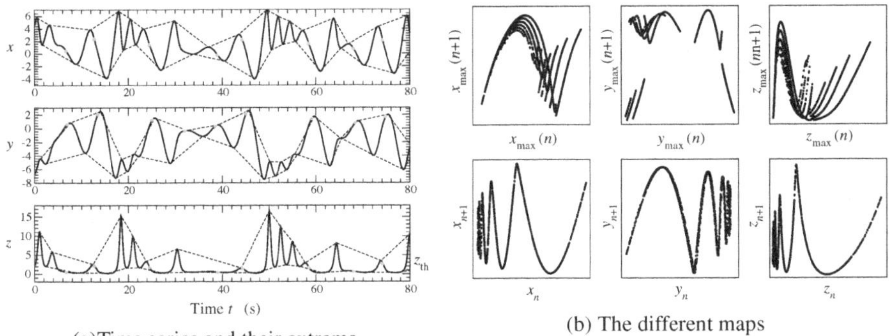

(a)Time series and their extrema

(b) The different maps

Fig. 12.2 Typical approximate one-dimensional cross sections through a simple funnel-type chaotic flow produced by Eq. (2.2) with parameter values: $a = 0.556$, $b = 2$ and $c = 4$

systems.[3] This is because most realistic systems are strongly damped. If one is not interested in showing that a given system produces a higher form of chaos (which is less easy—see the next Section), one can moreover always play with the parameters in such a way that the chaotic behavior has just appeared out of a limit cycle oscillation. Appearing or disappearing chaos induced in many cases seems to belong to a simple three-variable type, even if the system itself contains many variables (see [3], for an infinite-dimensional example).

12.2 Looking at More Complicated Maps

If the chaotic flow is more complicated, a single recorded variable is usually no longer sufficient for obtaining a representative projection of its cross section. Reading off two variables simultaneously in order to get a cross section is, of course, much less convenient.[4] Moreover, there is the problem of which successive instances in time to choose. Here, the rule to select the maxima or minima of the variable with the sharpest extrema can be helpful.[5] If one is lucky,[6] one obtains a folded-towel (or folded handkerchief) like picture. These maps look, under iteration, like a multiply-folded transparent veil: there are regions where the number of layers is higher. This 'pattern of diamonds' becomes the more pronounced the longer the simulation (observation), and it moreover repeats visibly on a finer and finer scale. This is because the 'folding edges' of higher and higher iterates become visible. An example of such a two-dimensional endomorphism is presented in Fig. 12.3.

Such a pattern is indirect proof of higher chaos. Or rather, it allows one to have a look into an indefinite stretching and folding process along two directions. Any such process, however, is bound to generate a set of singular solutions which is a product of a two Cantor sets.

Such a discontinuous point set, while possessing topological dimension zero, possesses a fractal dimensionality greater than unity (if not too large a fraction is cut out at each step). Assuming that the median third is cut out (triadic Cantor set), the fractal dimensionality is

---

[3]This is a very important statement here which is still so often forgotten, even in the most recent literature.

[4]When this book was written (1980), the technique for reconstructing the state space using delay or delay coordinates was not yet published. This technique was suggested by David Ruelle and first published by Packard *et al.* [9] and then proved by Ricardo Mañé [6] and Floris Takens [11], the latter having worked with Ruelle in the 1970s.

[5]This trick to build a first-return map was heuristically proposed by Edward Lorenz in his 1963 paper [4].

[6]There is no luck for having a simple or (over) complicated return map. The examples shown in Figs. 12.1 and 12.2 evidence that not all cross sections are equivalent for constructing a first-return map. Typically, the domain where there is a folding should be avoided. These figures also show that using minima or maxima is not always the optimal way for constructing a first-return map.

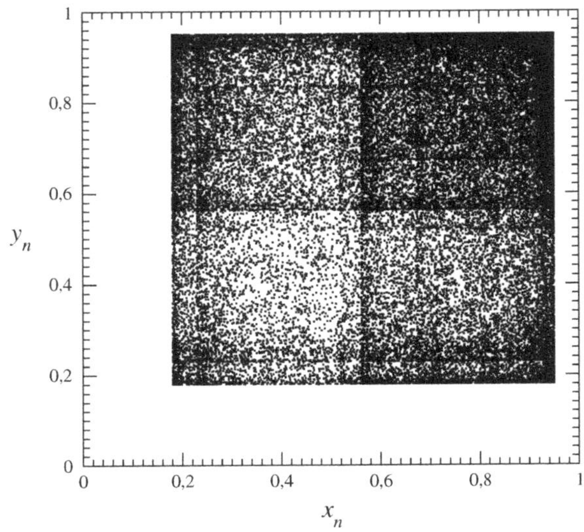

Fig. 12.3 A 'multiply-folded veil' pattern. Numerical simulation of two orthogonal logistic maps, $x_{n+1} = 3.8\,x_n(1 - x_n)$. Calculation of 50 000 sucessive iterates of the initial point (0.50, 0.51)

$$D = \frac{\log 4}{\log 3} > 1$$

(see Sect. 9.3). This means that the chaotic regime contains a set of periodic solutions (including infinite periodicities) that is 'near two-dimensional' in the map. Ordinary chaos, in contrast, is only a 'near one-dimensional' set of singular solutions, namely an ordinary Cantor set.

However, this is about as far as geometric methods can carry. The next higher-dimensional case (a cloud folded over in three directions) already yields three-dimensional diamonds which are much harder to identify through inspection (unless cuts, possessing the same structure as Fig. 2.6, are to be looked at). Also, the simulation or observation times become very long.

This leads to the question whether there are other, more direct methods to prove the presence of chaos and higher chaos.

12.3 Lyapunov Characteristic Exponents

Lyapunov had the insight that every trajectory in a dynamical system can be chosen as the point of reference for the behavior of all other trajectories [5]. Every trajectory can be reduced to a point, with its neighbors still forming a neighborhood, and so forth farther out.

Formally, this is very easy to do: one just looks at the deviations of all other trajectories $\{x\}$ from the given one (called $\{y\}$):

$$\frac{d}{dt}(x - y) = f(x) - f(y). \tag{12.1}$$

If $y(t)$ is known, so is the differential equation of the 'perturbed system' (12.1). One then just has to look at the stability properties of the steady state of Eq. (12.1) in order to know everything about the behavior of $y(t)$'s neighbors. The linear stability analysis of the steady state yields the usual results (exponential growth/decay in n eigendirections), that is, the picture of a node, a saddle, or a focus in every plane spanned by two eigendirections.

The only problem is that $y(t)$, which needs to be known for all t, is in fact unknown. If it was known, in the case of a chaotic system, one could easily check whether in the region of interest (an attracting box about the attractor) indeed all trajectories have an unstable eigendirection (corresponding to a pure chaotic attractor), or more than one, or even $n - 1$ unstable directions (pure hyperchaotic attractor or maximal attractor, respectively).

One could also check, in case not all trajectories in the region have the same behavior, whether nonetheless 'most' trajectories have such (chaotic or higher) behavior; and whether the rest are (along their basins of attraction) intimately 'buried' within the chaotic convolute (case of an ordinary chaotic or higher attractor); and whether their relative volume is small enough for them to be non-detectable in the presence of an error or noise of a certain maximal amplitude (10^{-10^6}, or 10^{-2}). One could then speak of having proven the presence of 'manifest chaos' (ordinary, or higher, or maximal) down to that error limit (including zero).

All this is only wishful thinking, as we know. Nonetheless, the idea is good enough to be exploited—not mathematically, but from a practical, approximation point of view. Lyapunov [5] solved the problem for the case of analytically known periodic solutions. Here one can calculate the 'transition matrix' of the local linear system; the eigenvalues of the (constant) exponent-matrix of this time-dependent solution-matrix then are the desired stability indicators of the periodic solution (see, for example, [12]).

While this can be done analytically in principle, in praxis it has to be done by computer even if the time-dependent periodic solution is known explicitly. Moreover, the same algorithms (and programs) can be used to calculate the characteristic exponents of any solution, not just analytically given ones — but of course, only over finite times. Such 'finite-time Lyapunov characteristic exponents' were calculated for numerically obtained chaotic solutions of Eqs. (2.2) and (4.1) by Packard et al. [9]. An appropriate program (used for evaluating the stability of analytically approximated periodic solutions) has also been provided by Seelig [10].

A close connection between the theory of Lyapunov characteristic exponents (cf. also [8]) and the notion of Kolmogorov entropy, which also measures trajectorial dilatation [2], has been pointed out by Benettin et al. [1]. A simple numerical procedure to calculate the latter was indicated and its 'equivalence' to the former method shown. The way to proceed is as follows: Calculate not only one solution, but two, the second with a given finite initial distance. Then re-scale the deviation after a fixed (or floating) time-interval and multiply the rescaling factor with the last one, and so forth. This yields an (averaged) coefficient of exponential deviation after a while, called a 'maximal Lyapunov characteristic number' (and 'Kolmogorov entropy') by the authors.

This procedure can be generalized, to proving 'numerically' not only the presence of manifest chaos, but also the presence of 'manifest higher chaos', if pertinent: calculate at least two (and in general $n - 2$, if n is the system dimensionality) neighboring solutions along with the main one, and rescale the deviations in the same linear manner as before. A more accurate analogous procedure calculates a polygonal circle ($n - 3$ sphere) about the trajectory and rescales toward a circle, with the multiplication factor of its volume being recorded. If the volume grows exponentially, the system shows maximal chaos (up to that time).

References

1. G. Benettin, L. Galgani, J.-M. Strelcyn, Kolmogorov entropy and numerical experiments. Phys. Rev. A **14**, 2338–2345 (1976)
2. A.N. Kolmorogov, A refinement of previous hypotheses concerning the local structure of turbulence in a viscous incompressible fluid at high Reynolds number. J. Fluid Mech. **13**, 82–85 (1962)
3. Y. Kuramoto, Instability and turbulence of wavefronts in reaction-diffusion systems. Prog. Theor. Phys. **63**(6), 1885–1903 (1980)
4. E.N. Lorenz, Deterministic nonperiodic flow. J. Atmos. Sci. **20**, 130–141 (1963)
5. A.M. Lyapunov, Problème général de la stabilité du mouvement (1907), French translation of 1892 Russian memoires, reprinted by Princeton University Press (1949)
6. * R. Mañé, On the dimension of the compact invariant sets of certain nonlinear maps. Lect. Notes Math. **898**, 230–242 (1981)
7. * L.F. Olsen, H. Degn, Chaos in an enzyme reaction. Nature **267**, 177–178 (1977)
8. V.I. Oseledec, A multiplicative ergodic theorem. Lyapunov characteristic numbers for dynamical systems. Trans. Moscow Math. Soc. **19**, 197–231 (1968)
9. N.H. Packard, J.P. Crutchfield, J.D. Farmer, R.S. Shaw, Geometry from a time series. Phys. Rev. Lett. **45**(9), 712–716 (1980)
10. F.F. Seelig, Nonthermal instabilities of optimized chemical reactors – hysteresis jumps in binary reactions with homogeneous catalysis near the state of maximal yield even under strictly isothermal conditions. Ann. N.Y. Acad. Sci. **316**, 338–349 (1979)
11. * F. Takens, Detecting strange attractors in turbulence. Lect. Notes Math. **898**, 366–381 (1981)
12. R.F. Williams, Expanding attractors. Publications Mathématiques de l'Institut des Hautes Etudes Scientifiques **43**, 169–203 (1974)

Chapter 13
Some Open Problems

13.1 Spectral Properties of Chaos

Chaotic systems produce a characteristic noise, as described in Sect. 10.6. *Does it suffice to diagnose chaos and, perhaps, also its higher forms?*

Farmer et al. calculated the power spectrum produced by chaotic flow of Eq. (2.2) (spiral-type chaos) and found, to their atonishment, that there is not only a 'broad component' in the spectrum, but also a sharp peak [5]. This phenomenon, which resembles the coexistence of a broad component and a sharp peak in the measurements on Taylor vortex turbulence by Swinney et al. [17], is not easy to explain. A corresponding, apparently non-decaying oscillatory tail in the autocorrelation function is also present [6].

Equation (2.2) probably is still close enough to a toroidal chaos-producing system based on a periodic forcing (see Sect. 6.1). Such systems should behave in the same way. Whether spiral-type chaos is frequently—or only exceptionally—close to a periodically forced system is presently unknown.

The spectral features of turbulence were re-examined by Mandelbrot under the notion of '$\frac{1}{f}$ noise' [12]. It is an attractive open question whether a hyperbola-shaped $\frac{1}{f}$ spectrum is a characteristic of high-order chaotic systems in general. If so, the qualitative interpretation of turbulence as high-order chaos would gain credit.

13.2 Chaos and Linear Systems

At first sight, the mentioning of linear systems in the context of chaos appears absurd. On the second glance, one remembers the 'mixing behavior' described by Montroll and Mazur for the infinite harmonic chain with one heavy element [13]. Thus, an infinite number of coupled harmonic oscillators, all identical except one (which is slower) can generate something that is close to chaos.

O. E. Rössler and C. Letellier, *Chaos*, Understanding Complex Systems,
https://doi.org/10.1007/978-3-030-44305-4_13

Kozak et al. provided an analytical calculation of the interaction of an atom with a field (Wigner-Weisskopf problem) [3]. They found solutions which look chaotic, even though the problem again is linear.[1] Even a finite chain of linear oscillators behaves in the 'same' way over long finite times [10]. Thus, the problem of the relationship between quasi-periodic and chaotic systems poses itself in a new light. Another problem in the same context is that of chaos in linear partial differential equations. A one-dimensional analogue to the Montroll-Mazur case can be analyzed [10].

A two-dimensional problem of the same type is the chaotic billard ball table (if one wall is slanted, for example, see Degn et al. [4]), with a Hamiltonian fluid poured onto it. A Hamiltonian medium like

$$u_{tt} = +u_{xx} , \tag{13.1}$$

the wave equation, subjected to impermeable (reflecting) boundaries in the shape of a chaos-generating billard table is bound to possess wave solutions the normals of which follow the path of a chaotic billard ball [10]. Thus, Kac's problem [9] whether one can 'hear the shape of a drum' acquires new facet.

13.3 Chaos and Finite State Machines

All numerically computed chaotic solutions are bound to be periodic.[2] A similar fact is stated by Zeeman's C^0-approximation theorem [18]: there is a C^0 neighborhood to every horseshoe in the sense of Smale [16] in which there is only a finite number of singular (periodic) solutions. On the other hand, there is no doubt that digital computers are very good (arbitrarily good in a finite sense) at producing chaos. *Is it possible to say something mathematical about the relationship?*

A good point to start with is probably the class of tolerant systems [2, 15, 19]. Zeeman took up (unknowingly) an idea of Poincaré to subject metric objects (like the objects in space) to the constraint that features below a certain discrimination threshold (tolerance) cannot be distinguished [14].

Dal Cin showed that automata with a tolerance can possess 'attractors with a tolerance' [2]. Zeeman's result that continuous system with tolerance and discrete systems with the same tolerance are 'equivalent' [19] can perhaps be used to carry over much of the results of continuous mathematics into the realm of discrete systems.

---

[1] Kozak, personal communication.

[2] This assertion anticipated René Lozi's results who found that indeed numerical chaotic solutions are in fact giga-periodic solutions [11].

13.4 Chaos in Non-point Systems

Dynamical systems are usually considered under the point of view of a (moving) point. This is because human beings are, to a finite approximation, points moving in space and time. It can nonetheless be rewarding to consider the orbits not only of points but also of lines or even of planes, and so on. The theory of invariant manifolds may profit from such an approach.

A nice experimental tool to study the behavior of continuous sets under iteration is the 'short-circuited' home T.V. (with camera). If one points the camera on the screen one obtains a realization of an (approximately) linear diffeomorphism. Abraham suggested to generate general diffeormorphisms in the same manner by inserting a distorted glass [1]. For some pictures of the linear situation, see [8]. This is because the different points in the picture now can have different initial times. Thus, what we have is a 'highly parallel diffeomorphism'—like that of Fig. 1.2, but time-distributed.

13.5 A Speculation

Perhaps, the class of highly parallel maps is the most interesting in its near-linear range already, yet with a finite tolerance imposed. It is virtually certain that chaos cannot arise in this case. But thinking about these maps leads to the impression (yet to be verified) that they generate some other mathematical objects, which to know may be helpful for a theory of consistent thought about chaos.

It seems that near identity diffeomorphisms with a tolerance, subjected to a highly parallel mode of action, generate a finite number of 'fixed patterns' as a function of all possible initial patterns. Moreover, these patterns differ in a characteristic manner from their 'originals': they are much more regular. Thus, the present class of systems resembles that of equations acting morphogenetically in a synergetic fashion [7]. Specifically, highly symmetric and periodic patterns are a performed state: there is the perfect circle, the straight line, the two parallels, the right angle, the point, and so forth.

In a sense, these systems generate the opposite of chaos. On the other hand, it may be that precisely such 'automatic idealizations', performed in the human brain, are necessary in order to be able to follow such a non-physical entity as 'deterministic chaos' with one's intuition. For nowhere in nature exists a system that was capable of following a deterministic chaotic flow over more than a few iterates. When thinking about chaos, on the other hand, the greatest fun lies clearly in trying to visualize its 'limiting properties'—precisely those which have no physical meaning. This does not mean that understanding these ultrafine properties may have no macroscopic significance—on the contrary.

I wanted to close with Anaxagoras, but this is now no longer necessary.

References

1. R. Abraham, Dynasim: Exploratory research in bifurcations using interactive computer graphics. Ann. N.Y. Acad. Sci. **316**, 676–684 (1979)
2. M. Dal Cin, Fault tolerance and stability of fuzzy-state automata. Comput. Sci. **2**, 36–44 (1973)
3. R. Davidson, J.J. Kozak, On the relaxation to quantum-statistical equilibrium of the Wigner-Weisskopf atom in a one-dimensional radiation Field. VIII. Emission in an infinite system in the presence of an extra photon. J. Math. Phys. **19**, 1074–1086 (1978)
4. H. Degn, L.F. Olsen, J.W. Perram, Bistability, oscillations, and chaos in an enzyme reaction. Ann. N.Y. Acad. Sci. **316**, 623–637 (1979)
5. J.D. Farmer, J.P. Crutchfield, H. Frœling, N.H. Packard, R.S. Shaw, Power spectra and mixing properties of strange attractors. Ann. N.Y. Acad. Sci. **357**, 453–472 (1980)
6. C.W. Gardiner, O.E. Rössler, *in preparation* (1980)
7. H. Haken, *Synergetics - An introduction* (Springer, Berlin, 1976)
8. D.R. Hofstäder, *Gödel, Escher, Bach — An Eternal Golden Braid* (Basic Book, New York, 1979)
9. M. Kac, S.M. Ulam, *Mathematics and Logic* (Praeger, New York, 1968)
10. J.J. Kozak, O.E. Rössler, Weak mixing in a quantum system. Zeitschrift für Naturforschung A **37**, 33–38 (1982)
11. * R. Lozi, Giga-periodic orbits for weakly coupled tent and logistic discretized maps, in *Modern Mathematical Models, Methods and Algorithms for Real World Systems*, ed. by A.H. Siddiqi, I.S. Duff, O. Christensen (Anamaya Publishers, New Delhi, 2006), pp. 80–141
12. B. Mandelbrot, *Fractals — Form, Chance and Dimension* (Freeman, San Francisco, 1977)
13. E.W. Montroll, P. Mazur, Poincaré cycles, ergodicity, and irreversibility in assemblies of coupled harmonic oscillators. J. Math. Phys. **1**, 70–84 (1960)
14. H. Poincaré, *Science et Méthode* (Flamarion, 1905)
15. T. Poston, I. Stewart, Taylor expansion and catastrophes, *Research Notes in Mathematics*, vol. 7 (Pitman, London, 1976)
16. S. Smale, Differentiable Dynamical System. Bull. Am. Math. Soc. **73**, 747–817 (1967)
17. H.L. Swinney, P.R. Fenstermacher, J.P. Gollub, Transition to turbulence in a fluid flow, in *Synergetics, a Workshop*, ed. by H. Haken (Springer, Berlin, 1977), pp. 60–69
18. E.C. Zeeman, *Catastrophe Theory. Selected Papers 1972–1977* (Addison-Wesley, Reading, 1977)
19. E.C. Zeeman, Twisting spun knots. Trans. Am. Math. Soc. **115**, 471–495 (1965)

Appendix A
A Psychological Astrologically Oriented Portrait

by Christophe Letellier and Valérie Messager

To briefly sketch the personality of Otto E. Rössler, we used his astrological chart as guidelines and combine what we were able to extract with Otto's own recollections. Let us start with a brief introduction to this ancient art.

A.1 Astrology: A Brief Introduction

In order to draw a portrait of Otto E. Rössler, we will use his astrological chart. Astrology is a field of knowledge describing the effects that the celestial bodies (the planets and the moon) could have on humans. The sky is divided in twelve regions known as the signs of the Zodiac (Aries, Taurus, Gemini, and so on).

The astrological sign of a person is associated with the position of the sun at the day of birth: it is strongly dependent on the longitude in the ecliptic. Due to its mass, the sun is the most influent celestial body on terrestrial physics as well as on humans. Astronomers had to switch from a geocentric system to a heliocentric one due to that: the center of mass of the solar system is very close to the sun and the description of the planetary motions is more simple than in other reference frame.

On the earth, gravitation is the main component of the influences of the sun, even if its light is the most obvious effect on the daily life with the circadian rhythm [8]. Indeed, there is an approach to cure some patients by considering that, due to the mass of the earth, gravitation is an important element in the general state of a human: this is the *structural integration* or Rolfing [35]. A third influence of the sun on terrestrial physics is the solar wind, a stream of charged particles emanating from the sun, directly related to the level of solar activity. If the earth magnetic field—as suggested by Walter Elsasser (1904–1991) in the 1940s [10, 11, 12]—deflects most of the solar wind as well as some cosmic rays, it is strongly distorted by it. A zone of the earth magnetosphere, the Van Allen radiation belt, may trap electrons and protons emanating from the sun [52]. Some scientists investigated the relationships

O. E. Rössler and C. Letellier, *Chaos*, Understanding Complex Systems,
https://doi.org/10.1007/978-3-030-44305-4

between the solar activity and the evolution of the weather [1, 20, 55]. It is therefore not surprising that there is a field of knowledge which investigate the relationships between the sun and the humans, although it is still characterized by a strong lack of scientific methodology.

Astrology is a very old field of knowledge, and one of the relevant contributions was Claudius Ptolemy's *Tetrabiblos* written in the second century [34]. Astrology implicitly results from statistics made on humans' psychological potential and their realizations during the lifetime. There is a few studies that show interesting correlations between certain occupations of people and the positions of some planets at the time of their birth [16, 17]. The main problem is the weakness of the methodology with which they are constructed [50]. For instance, a recent study investigated the relationships between the intelligence and the date of birth and concluded negatively [23]. In fact, this is not at all the purpose of astrology to determine the level of intelligence and, consequently, astrology is not tested in this study. The authors also tested the personality reducing it to extraversion and neuroticism: again their results were negative but personality was reduced to very specific aspects which could possibly appear in a very detailed astrological chart requiring also the location of birth and a much deeper analysis! One may only conclude that such a study was poorly designed to assess astrology. Reading such paper is more or less equivalent to read an astrological report in a newspaper. Contrary to this, when astrological information is used as a whole, it is possible to discriminate psychological profiles based upon astrological differences [14]. Indeed, astrology is a complex combinations of many elements (combinations of the locations of celestial bodies and their possible meanings): as a result, interpreting all these features is an art, more or less as making a diagnosis for a physician [15]. Notice that there is a strong difference between these two arts since the latter is associated with a huge corpus of reproducible studies and the former not. Moreover, it should be noticed that self-attribution (activated when a clue for astrology is given) plays a clear influence on people's self-concept [51]. Today astrology does not have the status of a science just because there is not a corpus of reliable studies.

Astrology is based on three types of elements as follows.

1. The zodiacal sign associated with the location of the sun at the day of birth: it is longitude dependent. It is associated with the influence of the sun. With the moon, the sun is the most important object in the solar system, the zodiacal sign is the most important astrological element. As the solar contribution does not explain everything in the planetary motions, it does not provide the whole profile of humans.
2. The second level of features is provided by the locations of planets and of the moon at the time of birth. Each of these celestial bodies is considered as having an influence on a specific component of human activity as reported in Table A.1. Are particularly important the bodies in conjonction with the ascendant sign (AS) corresponding to the part of the sky rising on the eastern horizon at the time of birth (accurate knowledge of time and place of birth is therefore crucial). Bodies in conjonction with the mid heaven (MH, the zenith of the solar diurn motion) are

also considered as well as bodies in conjonction with the descendant (DS), and the background of heaven (BH, nadir), respectively. Planets and moon are ranked by relevance according to their location in the sky with respect of the time and place of birth as follows [14]

$$\mathrm{AS} > \mathrm{MH} > \mathrm{DS} > \mathrm{BH}\,.$$

3. The twelve houses are defined as the angular sectors corresponding to two hours, house H_I being associated with the first two hours after the time T_b of birth (here expressed in hour). The ith house thus corresponds to the time interval $[T_b + 2(i-1); T_b + 2i]$. Due to the inclination of the earth rotation axis, the angular sectors are not equal and are strongly dependent on the latitude. The twelve houses were introduced by Ptolemy, developed by the Hebrew astrologer Abraham Ibn Ezra (1089–1167) from Tudela (Spain) and finally popularized by Placidus de Titis (1603–1668) [32] who was a professor of mathematics, physics and astronomy at the university of Pavia (Italy). The importance of a given house is related to the number of celestial bodies from the solar system located in its associated angular sector at the time of birth.

Astrology is strongly influenced by the Pythagorean view of the world. Two relevant figures are the equilateral triangle and the square, designated as a trigone and a quadrature in astronomy. The regular polyhedrons are related to the four elements (earth, water, air and fire). Since $4 \times 3 = 12$, there are twelve zodiacal signs. The astrology is thus based on an hexadecimal system. These three numbers can be combined in various ways.

The twelve zodiacal signs are visited in one year (by the course of the sun). For obvious reasons, at least in the mediterranean basin, the cycle starts with Spring, that is, with April and Aries. The four seasons correspond roughly to a quarter of the lifetime, more from a psychological point of view than from a given duration. The meaning of the houses are reported in Table A.1.

Ptolemy introduced the four elements in astrology [34]. Today, they are interpreted and associated with the zodiacal signs as follows [4].

Earth:	realism, tenacity, practical mind	Taurus, Virgo, Capricorn
Water:	sensitivity, memory	Cancer, Scorpio, Pisces
Air:	flexibility, adaptability	Gemini, Libra, Aquarius
Fire:	initiative, will, growth	Aries, Leo, Sagittarius

The dominant element may be determined by summing the weights associated with the celestial bodies and two key points of the cardinal cross (ascendant and mid of heaven) located in each zodiacal signs. The weights of these components are reported in Table A.2 [4]: the heavier ones are the luminar bodies (the sun and the moon), then the closest planets (known from the Ancient astronomy) and then the more distant ones which were only discovered since the 19th century. Elements are associated with zodiacal signs by using trigone (Fig. A.1).

Table A.1 Possible interpretation of different elements used in astrology

Month	Sign		Statistical interpretation
April	Aries	♈	Impetus, energy, action, will
	Mars	♂	Energy, rivality, aggressiveness
	House I	H_I	Self, external appearance
May	Taurus	♉	Patience, materiality, fecundity
	Venus	♀	Love, beauty, arts
	House II	H_{II}	Value
June	Gemini	♊	Adaptability, mobility, communication
	Mercury	☿	Reactivity (communication)
	House III	H_{III}	Communication
July	Cancer	♋	Sentimentality, family, imagination
	Moon	☽	Sensitivity, feelings
	House IV	H_{IV}	Home and family
August	Leo	♌	Creation, power
	Sun	☉	Deep "Me", will, skills
	House V	H_V	Pleasure
September	Virgo	♍	Method, analysis, work, science
	Mercury	☿	Reactivity (analysis, science)
	House VI	H_{VI}	Health
October	Libra	♎	Justice, harmony, peace
	Venus	♀	Love, beauty, arts
	House VII	H_{VII}	Partnerships
November	Scorpio	♏	Secret, hidden power, aggressiveness
	Pluto	♇	Hidden things, marginality, subconscious
	House VIII	H_{VIII}	Self-transformation
December	Sagittarius	♐	Open mind, ideal, trip
	Jupiter	♃	Autority, unfolding, socio-professional realization
	House IX	H_{IX}	Philosophy
January	Capricorn	♑	Serious, work, ambition
	Saturn	♄	Reflexion, solitude
	House X	H_X	Social status
February	Aquarius	♒	Future, social relationships, independency
	Uranus	♅	Change, originality, independency
	House XI	H_{XI}	Friendships
March	Pisces	♓	Dedication, irrationality, vulnerability
	Neptune	♆	Spirituality, intuition, dreams
	House XII	H_{XII}	Mysticism and introspection

Table A.2 Weight of the components taken into account for determining the dominant elements by counting the components in each zodiacal sign, and then summing the weights with respect of the associated modality (see above)

Weight	Components
3	Sun, Moon, Ascendant
2	Mercury, Venus, Mars, Jupiter, Saturn
1	Uranus, Neptune, Pluto, Mid of Heaven

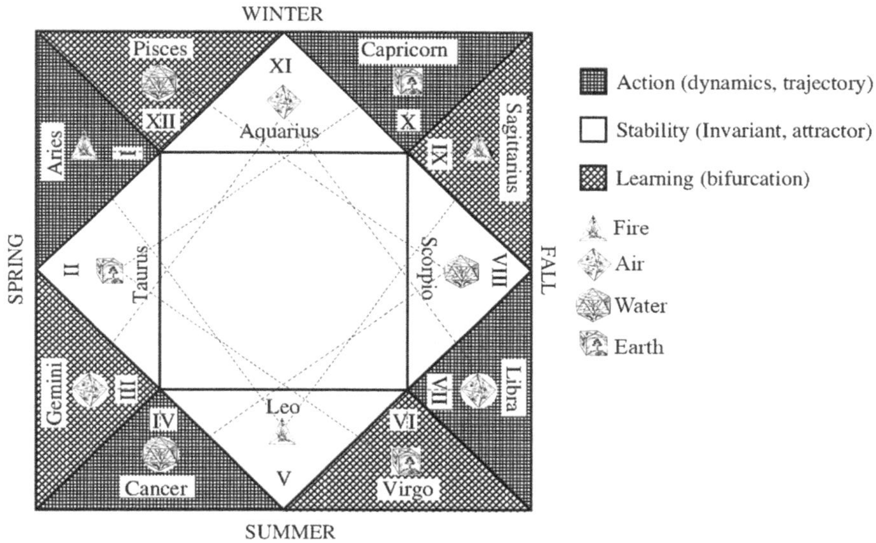

Fig. A.1 Astrological modalities (Action, Stability and Learning) and the elements (earth, water, air and fire) reported in our own geometrical (Pythagorean inspired) sketch of relationships with the planets and the houses

There are also astrological modalities which can be easily translated into concepts from the dynamical systems theory as follows.

Action	dynamics, trajectory	Aries, Cancer, Libra, Capricorn
Stabilization	Invariant sets, attractors	Taurus, Leo, Scorpio, Aquarius
Learning	bifurcation	Gemini, Virgo, Sagittarius, Pisces

Notice that Gaston Bachelard claims that when a new concept is learnt or understood, there is a *mutation* [5], that is, for a dynamicist, a *bifurcation*. The astrological modalities are related to zodiacal signs by using a square (Fig. A.1) depending on the location of celestial bodies. The first stage in life has to be dynamical, to draw a path: this is the first modality, *action*. Then an attractor can be evidenced and visited, a profile can be drawn: this is the second modality, *stability*. This is followed by some bifurcations (learning, maturation) to enrich the dynamics and the personality. This is also a stage preparing for a switch to a new phase in life: this is the third modality, *learning*. These three modalities are visited during each quarter of the lifetime.

The purpose of astrology is not necessarily to make predictions but rather to propose a global (psychological) analysis of the person. Trying to prove astrology by checking predictions seems to be vain, as already pointed by Ptolemy and Kepler.

According two proeminent contributors to the development of modern astronomy, Ptolemy and Kepler, astrology concerns the influence of celestial bodies on humans who live on earth. The purpose is to interpret these influences and not to explain the planetary motions which is the purpose of astronomy. If there is a crucial advantage to use the heliocentric system in astronomy, this is not the case in astrology, mostly because the focus point is in the earth. Ptolemy and Kepler recognized many weaknesses in the foundations of astrology although they did not denied its interest [6]. For instance, Kepler wrote [24]:

> Truly astrologers consider as causes of their predictions, to be sure, some that are physical, and some that are political, but when they yield the pen to enthusiasm, [they consider causes that are] in greater part inadequate, for the most part imaginary, foolish, and false, and finally altogether worthless. If they sometimes speak the truth when they are carried away by their enthusiasm, it ought to be marked up to good luck, and it must not be thought that it issues, more often and for the most part, from some higher and occult instinct.

These two astronomers considered astrology as profane, utilitarian and non rigorous while astronomy was sacred, abstract and rigorous. Kepler, as he did in his new astronomy [25], tried to evidence some physical causes to explain the influence of planets and of the moon, mostly using the light as a physical cause [24]. Kepler tried to impose the modern demand of causal determinism on archaic conceptions but with this requirement, he clearly constributed to discredit astrology [47]. Our purpose is not to convince the reader that astrology is well or badly founded but only to draw the astrological chart of Otto E. Rössler and to interpret it for providing a kind of psychological portrait with the help of Otto's recollections.

A.2 A Sketch of Otto Rössler's Personality

Otto E. Rössler is born on May 20, 1940 in Berlin. He is the son of Otto Rössler, born on February 1907 (by then Kismarton, Hungaria, now Eisenstadt, Austria), and Marianne Huth, born on February 6, 1909 (Bonn, Germany). Otto's grand-mother (who had the big table clock) was Louise Huth, born Schoepwinkel (Figs. A.2 and A.3).

Otto's father was a linguist working on the language of the Easter Islands' previous inhabitants. During the second world war, he received the scientific task to linguistically study an intact Jewish community: he refused, risking his life thereby. Otto remember to have attended, when he was a charming 3-years old boy, to a meeting between his father and his superiors (Fig. A.4): he still has the image of his father with a gun pointed at the temple: most likely, he helped to save him from life-threatening

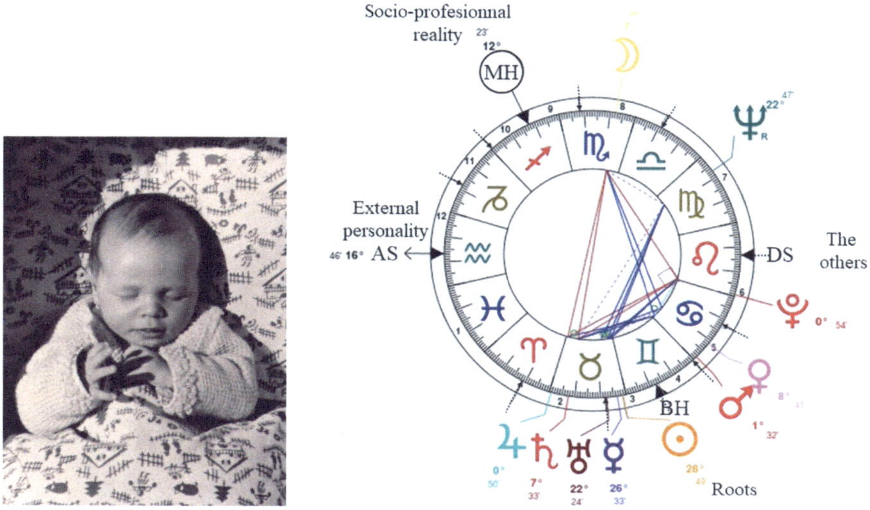

Fig. A.2 Astrological birth chart of Otto E. Rössler born on May 20, 1940 in Berlin, at 1:00 am. AS = Ascendant, DS = Descendant, MH = Mid of Heaven, BH = Background of Heaven. Zodiacal signs are ordered according to the earth revolution around the sun. Houses are ordered according to the rotation of the earth. In this chart, there is a quadrature between Jupiter, Pluto and the Moon (in red)

Fig. A.3 Otto with his father and (left) his sister, Ingrid, born on December 27, 1941 in Tübingen, and (right) with his mother, Marianne Huth

punishment for disobedience. From that period, Otto also remembers a bombing night in a cellar where he allegedly said "*But I do not want to die already, Mom.*" After the war, Otto's father introduced the concept of "Semito-Hamitic" group of languages, including, for example, Hausa in West Africa [37, 38, 39]. His students were often visiting him at home.

Fig. A.4 A house near the schloss of Tübingen where Otto's father bring him to a critical meeting in 1943

Otto got his first important meeting with religion when he was 13 years old[1]:

> I should say that at age 13, I was in a "Colonnie de vacances" near Mont Saint-Michel for four weeks at Granville, and became the privileged child who was allowed to ride on a moped in solitary excursions with the leading priest, Abbé Maurice Chaisnot (1921-2011). This was only few weeks before the shocking transition of puberty. The next year I felt much too mature to go there again. But I continued to like Aimé Duval's[2] ingenious religious songs (like "*Par la main* (Tout au long des longues longues plaines)" and "*Le seigneur reviendra*" and "*Le ciel est rouge il fera beau*").

In 1965, Otto recommended Duval's songs to Konrad Lorenz who had no understanding for that. As a reply, Konrad told Otto the story of his cousin in World War II in the trenches in Eastern France, who on a Sunday had to patrol the two-meters-high corridors, and suddenly while opening up his butter bread on turning around a corner stood 20 cm before a French soldier. Baffled, he re-directed the motion of his arm to offer the bread to the other man. They both could no longer be used in their armies getting imprisoned for "fraternisation," to become lifelong friends.

[1]Otto to Christophe, Message on August 23, 2019.

[2]Aimé Lucien Duval (1918–1984) was a French jesuit priest who was also a singer-composer and guitar player. He got a rather wide hit in the 1950s and 1960s.

Fig. A.5 Otto with his father and the Teddy Bear he wanted to baptize. Summer 1942

Fig. A.6 Maurice Chaisnot, priest in Granville (France) during the 1950s

Otto was baptized several times[3] (Figs. A.5 and A.6):

The first time, a protestant clergyman came to our flat to baptize me and my sister Inge (born one and half years after me). I was very much looking forward to the occasion because I had carefully prepared my teddy-bear to be baptized along with the "water of life" so he would be made alive, of course. But then I spotted that the clergyman went to the kitchen sink to fetch the water from there. I was terribly disappointed from that moment on for realizing that there was no chance the teddy would be awakened. I was about 6 at the time (or 5 1/2 ?). Later, my mother switched to catholicism, so I was sent to a catholic school. There, all new pupils got baptized together in a ceremony "to make sure." At age 52, I learned from my father on his deathbed that his mother was Jewish. I tried to switch but was not admitted.

---

[3]Message from Otto to Christophe, August 8, 2019.

Fig. A.7 Otto's radio amateur licence obtained in 1957

When Otto was 13, he became fascinated in telephones and built a connection between the floor on which his parents lived with his room under the roof so that he could listen while the other side could not (with three wires instead of two—this could have been more elegant, to Otto's opinion). Then this gradually led to repairing and then building radios, and radio emitters. He got a radio amateur's license DL9KF when he was 17 (Fig. A.7). He still owns it. His radio amateur call was "*CQ80—CQ80 DL9KF*." During his years at highschool at Tübingen, Otto, got acquainted with electronics. He never forgot his first interests in electronics and read the book written by Andronov's school [3] and a little book entitled "*Measuring-signal generators, Frequency measuring devices and multivibrators* from the radio-amateur library [49]. Otto then switched to getting interested in ethics, via a school assignment on poet Schiller's dramas.

A friend of the Rössler family was Pater Guido, a Franciscan monk who made Otto a member of the Third Order[4] at age 17 as a Secular. He then decided to become a begging monk, but his father forbade it because he was not yet legally mature below 21 years of age. He remembers that he "*was half-relieved because he did not know whether he would be strong enough for a life on the streets as a begging monk.*" He also had a bad conscience. Not sure that today Otto has any regret, just because religion can cause such unnecessary internal quarrels. Nevertheless, today Otto remains a proud member of the Third order of Saint-Francis. For him, religions never contradict each other. Nor does secularism.

[4]The Third Order of Saint Francis is made of Regulars (religious congregations of men and women) and the Seculars (the members do not wear a religious habit, take vows, or live in community). The First Order corresponds to friars and the Second Order to nuns.

In 1958, pushed by his father, and as a substitute for becoming a monk, Otto studied in medicine, out of the same idealistic motivation. Then in the second year, he got attracted by scientific medical books (histology, biochemistry) and began a dissertation working with mice in a long-term immunization study under the supervision of Erich Letterer (1895–1982) [40]. But unexpectedly, mices became immune-tolerant to the bovine serum albumin used in the immunization. Letterer was very disappointed. This effect, theoretically predicted by Macfarlane Burnet (1899–1985) [9], was observed by the group of Peter Medawar (1915–1987) in the 1950s [7]. Both received the Nobel prize in 1960. Moreover, Letterer had little patience with Otto's experimental skill. As a result, the thesis has only 24 pages.

During the last years of his medical studies, Otto read the book *The history of nature* published in 1949 by Carl Friedrich von Weizsäcker (1912–2007) [53]. He then attempted to get scientific contacts (and perhaps recognition) by writing letters to Weizsäcker. Otto started to bombard him with letters as a response to his book (Fig. A.8). After few exchanges, Otto had been invited to meet Carl in his home. Carl recommended Otto to complete his medical studies and to work as a scientist afterwards. After Otto's medical studies, he even secured for him a first scientific position (a stipend)—at the Max Planck Institute for Behavioural Physiology in Biocybernetics with Horst Mittelstaedt (1923–2016) at Seewiesen.[5] During this year, Otto, who was already exchanging letters with Konrad Lorenz, developed a big friendship in very long discussions at his home. Otto's ideas being taken seriously by Lorenz with the biogenesis theory and the ensuing deductive biological approach was an unexpected stimulation in this time window. According to Otto's recollections, they discovered deductive brain theory together, as later formulated [42, 44, 46].

Otto is Taurus ascendent Aquarius. The first specificity of the astrological birth chart of Otto is that there is no planet in the ascendant and the sun is in the background of the heaven: this could be related to a tendency to introversion and to favour his private life. He may also have a touch of egocentricity. This chart presents a "cup" shape, all the planets being located in the hemisphere centered at the descendant. Commonly, peoples with such a profile are very atypical, powerful and conscious of their skills. In Otto's case, the hemisphere associated with the social relationships is empty: Otto has a certain restraint while communicating with others, but when he exchanges, he is always driven by benevolence. Discussing with Otto immediately immerges in the world of ideas, of concepts: quite detached from real cases. He is very much attracted by theories and, perhaps, quickly afraid by the reality of the world. With no planet in the first house, his self-confidence and his ability to promote himself are under developed. For instance, in spite of a quite significant popularity in the early 1980s, the present book was rejected by Springer by then: Otto never submitted it elsewhere and the manuscript was left in a drawer until 2010 when Reimara Rössler showed it to Christophe.

[5] Seewiesen (Upper Bavaria), is located between the lakes Starnberger See and Ammersee and was the site of the Max Planck Institute for Behavioural Physiology from 1954 to 1999. It was the workplace of Konrad Lorenz for many decades. On April 1, 1954, a resolution of the Senate of the Max Planck Society appointed Erich von Holst (1908–1962) and Lorenz as Directors at the Institute.

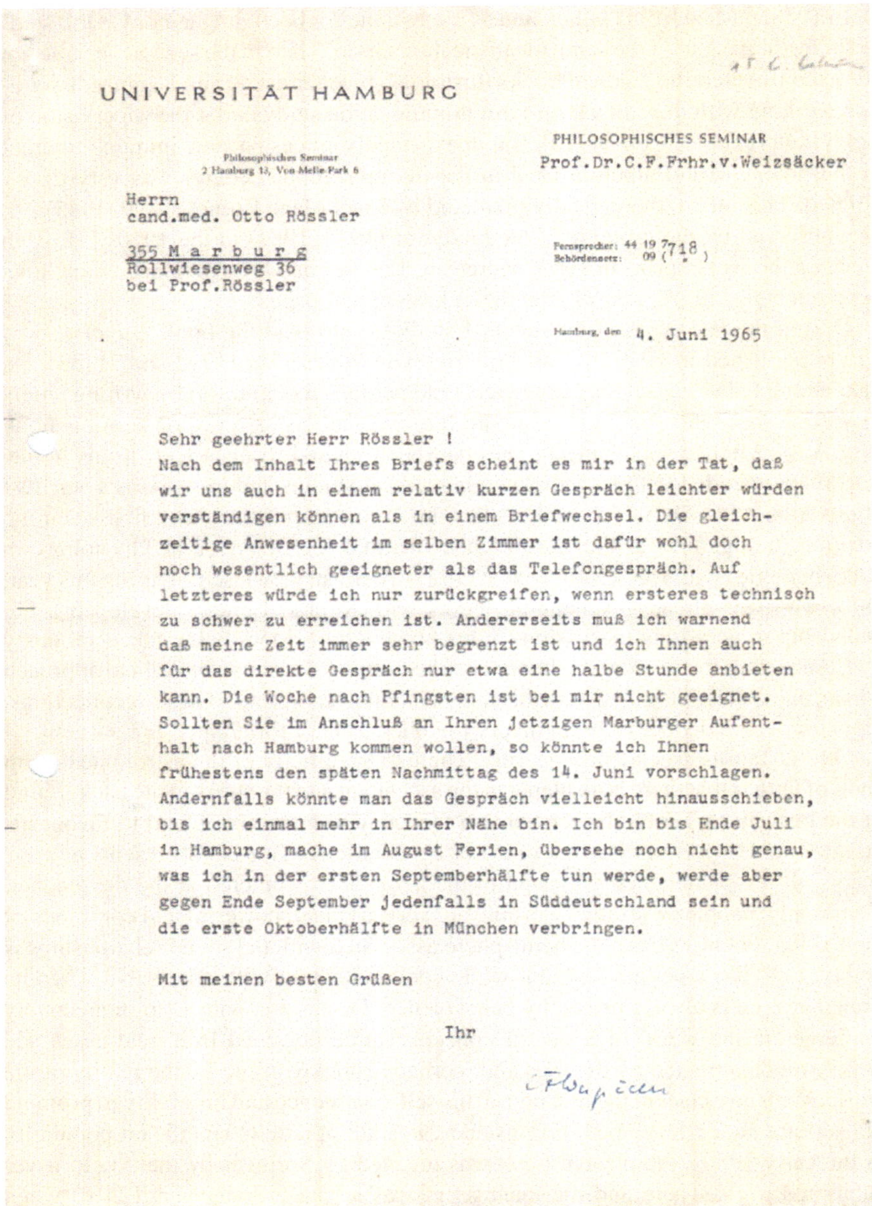

UNIVERSITÄT HAMBURG

PHILOSOPHISCHES SEMINAR
Prof.Dr.C.F.Frhr.v.Weizsäcker

Philosophisches Seminar
2 Hamburg 13, Von-Melle-Park 6

Herrn
cand.med. Otto Rössler

355 Marburg
Rollwiesenweg 36
bei Prof.Rössler

Fernsprecher: 44 19 7718
Behördennetz: 09 ( )

Hamburg, den 4. Juni 1965

Sehr geehrter Herr Rössler !
Nach dem Inhalt Ihres Briefs scheint es mir in der Tat, daß wir uns auch in einem relativ kurzen Gespräch leichter würden verständigen können als in einem Briefwechsel. Die gleichzeitige Anwesenheit im selben Zimmer ist dafür wohl doch noch wesentlich geeigneter als das Telefongespräch. Auf letzteres würde ich nur zurückgreifen, wenn ersteres technisch zu schwer zu erreichen ist. Andererseits muß ich warnend daß meine Zeit immer sehr begrenzt ist und ich Ihnen auch für das direkte Gespräch nur etwa eine halbe Stunde anbieten kann. Die Woche nach Pfingsten ist bei mir nicht geeignet. Sollten Sie im Anschluß an Ihren jetzigen Marburger Aufenthalt nach Hamburg kommen wollen, so könnte ich Ihnen frühestens den späten Nachmittag des 14. Juni vorschlagen. Andernfalls könnte man das Gespräch vielleicht hinausschieben, bis ich einmal mehr in Ihrer Nähe bin. Ich bin bis Ende Juli in Hamburg, mache im August Ferien, übersehe noch nicht genau, was ich in der ersten Septemberhälfte tun werde, werde aber gegen Ende September jedenfalls in Süddeutschland sein und die erste Oktoberhälfte in München verbringen.

Mit meinen besten Grüßen

Ihr

Fig. A.8 Letter from Carl F. Weizsäcker to Otto E. Rössler, dated on April 6, 1965

Otto has a natural tendency to develop new ideas, to work at the interface between different disciplines. With a background in electronics, completed studies in medicine, he spent his post-doc to continue to enrich his pool of knowledge. He was not afraid to trigger exchanges with very well-known peoples: Carl von Weiszäcker, Konrad Lorenz, Ludwig von Bertalanffy, Robert Rosen, Art Winfree, Eberhard Hopf...He wrote to many others as Stephen Smale, Robert May, Philip Holmes, John Guckenheimer...He had no fear for writing them, most likely considering these exchanges as "private". With the recognition from some of these "authorities", Otto found the needed legitimacy for pushing his own ideas. When he was challenged in 1975 by Art Winfree to produce a chemical chaotic reaction scheme, he received then very positive comments from him about his first results. He was already combining the topological analysis in terms of branched manifold and first-return maps. He submitted his work but...Otto did that to the *Zeitschrift für Naturforschung* and not in a more appropriate journal for such a kind of work: he did that just because he was used to publish in this journal, full of recognition for the editor-in-chief, Alfred Klemm.[6] He had the feeling to be welcome.

In that spirit, if his two most well-known papers (quoted 3816 and 1364 times, respectively) were published in a journal with a wider audience, *Physics Letters A*, this is due to the stimulation of Hermann Haken who himself published his paper reporting the analogy existing between a laser system and the Lorenz system in that journal [22]. It is quite noticeable that his most quoted papers are certainly the less rich, and the most conventional among the long series of contributions to chaos theory he had written during eight years. There are also the two papers which are written in the standard format, without any "off-the-wall" terminology or pictures.

For Otto, there is a tendency to consider that he is badly understood, kept at distance by the others and has an affection deficiency. According to the astrological chart (Fig. A.2), this is quite prominent since the domain visited by the celestial bodies is bounded by three of them which are in quadrature, namely Jupiter, Pluto and the Moon. The influence of these three bodies is thus preponderant. Jupiter is associated with personal unfolding. This could correspond to a tendency presented by Otto to be provocative, out of the stream. As an example, in his first paper about chaos [43], Otto inserted a sketch of Dali's soft watch. There is no scientific reason for that and Otto justified this insertion for preventing the possibility to be considered too seriously. He added that he was uncomfortable with his lack of mathematical background as most of students in medicine. However, Otto was radio-amateur in his teenage years and constructed himself his radio emitter-receiver: he had therefore a quite solid background in electronics (with a minimum of mathematics, see [29] for details). In spite of this, Otto needed to protect himself, in the case he would be wrong! With the present book which was written in the early 1980s, we saw that his understanding of most of the abstract concepts was rather accurate. He was even able to combine them for constructing a full paradigm with many examples and many connections with the real world. If the door was largely opened by Poincaré whose legacy which was

[6]With Hans Friedrich-Freksa, Alfred Klemm founded the *Zeitschrift für Naturforshung* in 1946, to provide a German journal publishing their colleagues' scientific results.

efficiently synthetized by Lorenz [31], the American meteorologist did not promote this background as Otto did. Otto proposed various examples of different dynamics, most often interpreted in terms of branched manifold ("paper sheet model") and first-return map, sometimes combined with the Li and Yorke theorem [30]. For Lorenz, "chaos" was a tool to explain the unavoidable limitation in weather predictions. For Otto, "chaos" is a field in itself, with its own concepts, results and procedures. Otto spent his "chaos years" to transform the "chaos revolution" into normal science, if one wants to speak in the way of Thomas Kuhn [28]. In the view of his contributions, Otto's lack of self-confidence seems to be rather non justified and could be perhaps explained by his own history.

As indicated by the quadrature with Pluto, Otto may spend a lot of energy for defending his ideas against all odds: this is not in contradiction with his lack of self-confidence. These two aspects tend to be counter-productive, the latter annihilating the former. In astrology, Pluto is associated with transformation, deep metamorphosis. This is one of the amazing character revealed by Otto's path. He started by studying electronics, completed his studies in medicine but was never an active physician. He then spent a post-doc to learn about socio-biology with Konrad Lorenz. He traveled in US for a second post-doc focused on theoretical biology with Robert Rosen (Buffalo, today the State University of New York) where he learned how to write differential equations for reproducing observed dynamics [36]. At his return to Germany, once he attended to an analog computer course provided by the *Electronic Associates Inc.*, he started his academic career in teaching computer programming at the department of theoretical chemistry at the university of Tübingen, thanks to his background in electronics.

The house whose angular sector contains the largest number of celestial bodies is house H_{III} (Table A.3) which is associated with intellectual activities: With such a house, Otto has skills in communication, relationships, and learning. Otto is a theoretician (as he revealed himself with his poor skills in conducting his experiments during his Ph.D. thesis). His contributions to chaos were very often the results of interactions with others. For instance, his very first contribution to chaos theory was stimulated by Art Winfree [29]. During the four months after his meeting with Winfree in Vienna (September 1975), Otto acquired most of the topological concepts needed for describing chaotic attractors with the inheritance from Poincaré. First he took from his background in electronics the concept of slow-fast manifold developed in Andronov's book [3]. He got the use of first-return map from the 1963 paper by Edward Lorenz [31]. He developed his paper-sheet model from the works on branched manifold not only by Lorenz (who used isopleths to construct such a branched manifold) but also by Stephen Smale [48], and later by John Franks [13] and Robert F. Williams [54]. The third ingredient Otto used to prove that his system produces chaos was the Li–Yorke theorem "period-3 implies chaos" [30]. These four concepts were mastered in the last four months of 1975 as revealed by his first paper on chaos [43]. All of these concepts were new for him and at that time, none combined them. There is one noticeable exception with the work by John Guckenheimer, George Oster and A. Ipaktchi [21] which was existing as a preprint by the end of

Table A.3 The different houses ranked according to the physical influence of the celestial bodies at Otto's birth time

House	Statistical interpretation	Celestial bodies
H_{III}	Communication	Uranus, mercury, sun
H_{VIII}	Self-transformation	Moon
H_{II}	Value	Jupiter, saturn
H_V	Pleasure	Mars, venus
H_{VII}	Partnerships	Neptune
H_{VI}	Health	Pluto
H_I	Self, external appearance	
H_{IV}	Home and family	
H_{IX}	Philosophy	
H_X	Social status	
H_{XI}	Friendships	
H_{XII}	Mysticism and introspection	

1975. Otto has thus an obvious skills to assimilate new concepts and to produce something new with them.

With Uranus in house III, Otto would have a certain sense of rebellion that he expresses by his provocative touch. In his papers on chaos, mostly published between 1976 and 1983, Otto tried to develop a hierarchy of chaos, a kind of classification of chaotic attractors according to their topological characteristics [45]. For that purpose, he built various sets of differential equations producing different types of chaotic attractors, always investigated with topological tools as first-return map to a Poincaré section or branched manifold (that he called "paper-sheet model").[7] Rather than developing a mathematically-based series of names for designating this different types of chaos, Otto used names imported from the daily life, as walking-stick map, folded-towel map, noodles map, pancake map, sandwich chaos, etc. This was the second stage of his off-the-wall humour after the insertion of Dali's soft-watch in his first paper on chaos [43]. Such a drift in the terminology was not noticed before 1976. Combined with his powerful creativity, Otto constructed many different types of chaos, most often with a very suggestive and detailed analysis of the underlying structure.

More recently, Otto stimulated an opposition to some experiments conducted at CERN to replicate conditions that existed just a fraction of a second after the

[7]There is even a book on that topic which was edited by Gerold Baier and Michael Klein, both from the Universität of Tübingen (Germany) for Otto's 50th birthday with a very interesting contribution by these two scientists [26]. Such a complete classification of chaotic attractors based on topological invariants remains to construct today. An updated version of it can be found in Appendix B.

Table A.4 Location of celestial bodies in the astrological signs. The ascendant and the mid of heaven are also reported. The element and the contributing weight are indicated

Sign	Contributing components	Element	Weight
Taurus	Jupiter, Saturn, Uranus, Mercury, Sun	Earth	10
Cancer	Mars, Venus	Water	4
Scorpio	Moon	Water	3
Aquarius	Ascendant	Air	3
Leo	Pluto	Fire	1
Sagittarius	Mid of heaven	Fire	1

Big Bang because it may create a mini-black hole that could tear the earth apart.[8] Academic publications concluded to a very low risk for such a catastrophic feature [27]. Where is the fear that inhibited him to be considered too seriously in his first paper on chaos? Or perhaps, he had no fear because he knew that many peoples will not consider him too seriously…

In his astrological chart (Fig. A.2), the zodiacal signs are visited as reported in Table A.4. The dominant elements are Earth (45.5%) and Water (31.8%). Air and Fire are rather lacking. This would mean that Otto has some tenacity in accomplishing his goal. He is very sensitive to the others; for Otto, "dialog is the fountain spring of new ideas."[9] Otto's house is full of books, everywhere, but "books are important only when someone has opened up to you for a specific reason."[10] According to Otto, previous knowledge is not important but experience is, particularly "personal experiences made oneself." Here Otto follows in fact Poincaré who promotes, in the scientific methodology, to select the facts which something hidden behind and to recognize which law is hidden, to perceive the possibility of a generalization, to attach importance to the elegance of the method, which is connected to the economy of thought, but to not count too much on the memory, on the initial knowledge [33].

Although memory is not primordial in his research, Otto has a good one. While preparing this book, he only rarely answered "*I forgot*" or "*I don't know*," remembering in which book or paper, the answer could be find.

Otto may present a lack of flexibility and of will. This could be paradoxal; this is in fact not. Until the flow does not affect his own objective, his mind, Otto is very flexible and is always happy to follow the flow. His lack of adaptability emerges when something would affect his own path, only related to quite limited topics. Opened to others, Otto always tries to be benevolent, to please others.

[8]This opposition was covered by many newspaper, for instance, *The Telegraph*, Legal bid to stop CERN atom smasher from 'destroying the world' (August 30, 2008), *The Guardian*, Will the world end on Wednesday? (September 8, 2008), etc.

[9]Otto to Christophe, message dated on September 24, 2019.

[10]Otto, September 24, 2019, *Ibid.*

In Otto's astrological chart (Fig. A.2), there are five celestial bodies associated with the astrological modality of stability: Jupiter, Saturn, Mars, Venus and the Moon. Four of them are associated with the modality of *learning*: Uranus, Mercury, the Sun and Pluto. With a dominant stability, Otto is goal-oriented and is willing to do whatever it takes to accomplish what he has set out to do. With a learning modality also well visited, particularly by the Sun, Otto is rather often able to follow the flow. He sees all sides of a situation. By nature, he is communicative: his wife, Reimara, is used to say that he would have been a bad physician, just because "*he was talking too much to his patients*." His energy must be misconstrued as being provocative. In fact, Otto is flexible until his goal is in question: he receives positively any suggestion but when a proposition does not match with what he is or what he wants, it is nearly impossible to change his mind. For Otto, the way things are done is not too important, only the global picture has value.

The dominant tendency to establish, to stabilize some concepts, is rather obvious with his contribution to chaos, which appeared like a mushroom! He spent his teenage years as a radio-amateur, acquainting some background in electronic circuits: in particular, he learnt to master the ingredients required to produce oscillatory behaviors. Inspired by Andronov's school, he focused his attention on Abraham and Bloch's multivibrator [2], known to produce non-trivial oscillations: at that time, the resulting behavior was supposed to be periodic with many harmonics. Otto was then interested in producing a chemical multivibrator [41]. He quickly started to learn from quite mathematical papers, mostly following the way opened by Rosen during his post-doc at Buffalo. When Art Winfree challenged him for designing a chaotic chemical reaction, he spent four months to find a convincing example: his first paper contains a correct—and new at that time—picture of what a topological characterization of chaos should be. Clearly, once the first task completed, his goal was clearly to design various types of chaos and to establish a hierarchy among them [45]. To sum up, Otto spent four months for learning and seven years for enriching his set of examples, for exhibiting the important features to retain for distinguishing them. He thus contributed to develop the topological characterization of chaotic attractors, an approach which was then developed in the 1990s by Robert Gilmore (Drexel University) and co-workers [18, 19]. Otto's contribution is quite well illustrated by the present book. In Otto's mind, the program he developed in 1976 was achieved in 1983 with his hierarchy of chaos: it was time to switch to another program. It was time for a new bifurcation!

References

1. C.G. Abbot, Solar variation, a weather element. Proc. Natl. Acad. Sci. **56**(6), 1627–1634 (1966)
2. H. Abraham, E. Bloch, Mesure en valeur absolue des périodes des oscillations électriques de haute fréquence. Annales de Physique **9**(1), 237–302 (1919)
3. A.A. Andronov, A.A. Vitt, S.E. Khaikin, *Theory of Oscillators* (1937); Translated in English by F. Immirzi, Dover (1966)
4. C. Aubier, *Dictionnaire pratique de l'astrologie*, Editions Solar (1989)
5. G. Bachelard, *La formation de l'esprit scientifique*, Vrin (1938)

6. J.B. Backenridge, A. Rossi, Johannes Kepler's on the more certain fundamentals of astrology. Proc. Amer. Philos. Soc. **123**(2) (1979)
7. R.E. Billingham, L. Brent, B. Medawar, 'Actively acquired tolerance' of foreign cells. Nature **172**, 603–606 (1953)
8. D.B. Boivin, Influence of sleep-wake and circadian rhythm disturbances in psychiatric disorders. J. Psychiatry Neurosci. **25**(5), 446–458 (2000)
9. F.M. Burnet, F. Fenner, *The Production of Antibodies*, 2nd edn. (Macmillan, New York, 1949)
10. W.M. Elsasser, Induction effects in terrestrial magnetism Part I: theory. Phys. Rev. **69**, 106–116, (1946)
11. W.M. Elsasser, Induction effects in terrestrial magnetism Part II: the secular variation. Phys. Rev. **70**(3–4), 202–212 (1946)
12. W.M. Elsasser, Induction effects in terrestrial magnetism Part II: the secular variation. Phys. Rev. **72**(9), 821–833 (1947)
13. J. Franks, The structure of smale diffeomorphisms. *Lecture Notes in Mathematics*, vol. 668 (1978), pp. 117–126
14. S. Fuzeau-Braesch, An empirical study of an astrological hypothesis in a twin population. Pers. Individ. Differ. **13**(10), 1135–1144 (1992).
15. S. Fuzeau-Braesch, *Pour l'astrologie — Réflexions d'une scientifique* (Albin Michel, Paris, 1996)
16. M. Gauquelin, *Les hommes et les astres* (Denoël, Paris, 1960)
17. M. Gauquelin, *L'hérédité planétaire: Notre naissance et l'horloge cosmique* (Planète, Paris, 1969)
18. R. Gilmore, Topological analysis of chaotic dynamical systems. Rev. Mod. Phys. **70**(4), 1455–1529 (1998)
19. R. Gilmore, M. Lefranc, *The Topology of Chaos* (Wiley, New York, 2002)
20. L.J. Gray, J. Beer, M. Geller, J.D. Haigh, M. Lockwood, K. Matthes, U. Cubasch, D. Fleitmann, G. Harrison, L. Hood, J. Luterbacher, G.A. Meehl, D. Shindell, B. van Geel, W. White, Solar influences on climate. Rev. Geophys. **48**, RG4001 (2010)
21. J. Guckenheimer, G.F. Oster, A. Ipaktchi, Periodic solutions of a logistic difference equation. J. Math. Biol. **4**, 101–147 (1977)
22. H. Haken, Analogy between higher instabilities in fluids and lasers. Phys. Lett. A **53**(1), 77–78 (1975)
23. P. Hartmann, M. Reuter, H. Nyborg, The relationship between date of birth and individual differences in personality and general intelligence: a large-scale study. Pers. Individ. Differ. **40**(7), 1349–1362 (2006)
24. J. Kepler, On the fundamentals of astrology: a new brief dissertation regarding cosmic theory with a physical prognosis of the coming year 1602 A.D. written for philosophers (Schuman Press, Prague, 1602). English translation by A. Rossi, J. Bruce Backenbridge (1979)
25. J. Kepler, *Astronomia Nova* (1609). Translated by W.H. Donahue, Cambridge University Press (1992)
26. M. Klein, G. Baier, Hierarchies of dynamical systems, in *A Chaotic Hierarchy* (World Scientific Publishing, Singapore, 1991), pp. 1–24
27. B. Koch, M. Bleicher, H. Stöcker, Exclusion of black hole disaster scenarios at the LHC. Phys. Lett. B **672**(1), 71–76 (2009)
28. T. S. Kuhn, *The Structure of Scientific Revolutions* (University of Chicago Press, Chicago, 1962)
29. C. Letellier, V. Messager, Influences on Otto E. Rössler's earliest paper on chaos. Int. J. Bifurc. Chaos **20**(11), 3585–3616 (2010)
30. T.Y. Li, J.A. Yorke, A period-three implies chaos. Amer. Math. Mon. **82**, 985–992 (1975)
31. E.N. Lorenz, Deterministic nonperiodic flow. J. Atmos. Sci. **20**, 130–141 (1963)
32. P. de Titis, *Astronomy and Elementary Philosophy* (1789)

33. H. Poincaré, *Science et méthode* (1908), Translated in English by B. Russell, *Science and Method* (Nelson & Sons, London, 1914)
34. C. Ptolemy, *Tetrabiblos*, English version by J.H. Ashmand, *Ptolemy's Tetrabiblos or quadripartite being four books of the influence of stars newly translated from the Greek paraphrase of Proclus*, Davis and Dickson, London (1822).
35. I.P. Rolf, *Rolfing and Physical Reality* (Inner Traditions/Bear, Rochester, 1990)
36. R. Rosen, *Dynamical System Theory in Biology* (Wiley, New York, 1970)
37. O. Rössler, Libyca. Wiener Zeitschrift für die Kunde der Morgenländer **49**, 282–311 (1942)
38. O. Rössler, Der semitische Charakter der libyschen Sprache. Zeitschrift für Assyriologie **16**, 121–150 (1952)
39. O. Rössler, Libysch-Hamitisch-Semitisch. Oriens **17**, 199–216 (1964)
40. O.E. Rössler, *Dauerimmunisierung von Albinomäusen mit Rinder-γ-Globulin* [Long-term Immunization of Albino Mice with Bovine-γ-Globulin] (University of Tübingen, Tübingen, 1966)
41. O.E. Rössler, Basic circuits of fluid automata and relaxation systems. Zeitschrift für Naturforschung B **27**, 333–343 (1972)
42. O.E. Rössler, Adequate locomotion strategies for an abstract organism in an abstract environment — A relational approach to brain function, in *Lecture Notes in Biomathematics*, vol. 4 (1974), pp. 342–369
43. O.E. Rössler, Chaotic behavior in simple reaction system. *Zeitschrift für Naturforschung A* **31**, 259–264 (1976)
44. O.E. Rössler, An artificial cognitive-plus-motivational system, in *Progress in Theoretical Biology*, vol. 6 (1981), pp. 147–160
45. O.E. Rössler, The chaotic hierarchy. *Zeitschrift für Naturforschung A* **38**, 788–801 (1983)
46. O.E. Rossler, L.-R. Vial, F. Kuske, A. Nitschke, T. Ikegami, A. Ujica, Brain equation and personogenesis. Clin. Pediatr. **2**, 1011 (2019)
47. G. Simon, L'astrologie de Kepler: le sens d'une réforme, L'Astronomie **86**, 325–336 (1972)
48. S. Smale, Differentiable dynamical systems. I. Diffeormorphisms. Bull. Amer. Math. Soc. **73**, 747–817 (1967)
49. H. Sutaner, *Meßsender Frequenzmesser und Multivibratoren*. Radio-Praktiker Bucherei, vol. 128/130 (Franzis Verlag, Munich, 1966)
50. P.R. Thagard, Why astrology is a pseudoscience. Proc. Bienn. Meet. Philos. Sci. Assoc. **1**, 223–234 (1978)
51. J.J.F. Van Rooij, Introversion-extraversion: astrology versus psychology. Pers. Individ. Differ. **16**(6), 985–988 (1994)
52. J.A. Van Allen, Energetic particles in the earth's external magnetic field. Hist. Geophys. **7**, 235–251 (1997)
53. C.F. von Weizsäcker, *Die Geschichte der Natur*, Göttingen (1948) — English translation, *The History of Nature* (University of Chicago Press, Chicago, 1949)
54. R. Williams, The Lorenz Attractor. *Lecture Notes in Mathematics*, vol. 615 (1977), pp. 94–112
55. G.A. Zherebtsov, V.A. Kovalenko, S.I. Molodykh, K.E. Kirichenko, Solar variability manifestations in weather and climate characteristics. J. Atmos. Sol.-Terr. Phys. **182**, 217–222 (2019)

Appendix B
An Updated Hierarchy of Chaos

by Christophe Letellier and Otto E. Rössler

B.1 Introduction

There are various type of dynamical behaviors which can be observed. Their nature strongly depends on the dimension m of the state space $\mathbb{R}^m$. Otto E. Rössler ended his program of researches on chaos with a first attempt to construct a hierarchy of chaos, providing examples (many of them are discussed in the present book) [43]. For his 50th birthday, Mikael Klein and Gerold Baier proposed a second version, with a possible classification based on the spectrum of Lyapunov exponents [10]. Thirty years later, we here develop an updated version. Although still based on the spectrum of Lyapunov exponents, the main difference with the Klein–Baier hierarchy consists in distinguishing the different types of chaos by using the first-return map or the Poincaré section, in the spirit of Otto's works. The present classification is a direct continuation of these two initial contritutions.

We will limit ourselves to the cases of continuous dynamical systems. This is mainly justified because a first-return map to a Poincaré section of a continuous system provides discrete system. As already mentioned, the first ingredient for constructing a hierarchy of chaos is the number of positive and null Lyapunov exponents λ_i. Let be p and q these two numbers, respectively. Oscillations require at least one null exponent which is in the direction of the flow ($q \geq 1$) and a state space whose dimension $m \geq 2$. To avoid too specific cases, we will only consider those where there is at least one negative Lyapunov exponent. When the considered system is dissipative, the nature of the attractor $\mathscr{A}$ does not change when the number q of negative Lyapunov exponents is increased. Consequently, it is sufficient to describe those which are characterized by a single negative Lyapunov exponent.

In dynamical systems, there is an important property related to the volume of the state space visited by the flow. If the trace of the jacobian matrix of the system is negative, then the volume visited by the trajectory shrinks to zero: this is the Liouville theorem. The system is said to be *dissipative*. When the trace is null, the volume visited is an invariant and there is neither attractor nor transient regime: the

O. E. Rössler and C. Letellier, *Chaos*, Understanding Complex Systems,
https://doi.org/10.1007/978-3-030-44305-4

system is commonly said to be *conservative*. In this latter case, the system is also most often invariant under time reversal ($t \mapsto -t$) [36]. Although there are some rare counter-examples, we will keep the common definition for conservative system, that is, a null trace and invariant under time reversal.

In order to characterize the stability of periodic orbits, Henri Poincaré introduced the concept of cross-section transverse to the flow, commonly designated as Poincaré section [34]. It was shown by Edward Lorenz that a first-return map to a Poincaré section can be used for characterizing chaotic attractors [26]. This was then extensively used by Rössler [37, 38, 40]. First-return maps are rather useful for low-dimensional chaotic attractor $\mathscr{A}$. When the dimension is greater than 3 or when the system is conservative, a Poincaré section should be preferred, particularly when the regime is toroidal-chaos: the structure of the attractor $\mathscr{A}$ is better characterized by a two-dimensional projection of the Poincaré section. By definition, the Poincaré section of a m-dimensional dynamical system has a dimension $m - 1$. Consequently, when $m \geq 4$, the Poincaré section is at least three-dimensional and it is already quite difficult to describe its structure. We here recover the important gap which exists between three- and four-dimensional space to develop topological characterization [12]. In a talk given at the University of Utah from June 12 to June 23, 1978,[11] Otto suggested a "hierarchy theorem" as follows.

> Every dimension carries a less trivial type of attractor.
>
> - Dimension 1: Point attractor
> - Dimension 2: Periodic attractor
> - Dimension 3: Chaotic attractor
> - Dimension 4: Hyperchaotic attractor
> - Dimension 5: Hyper^2 chaotic attractor
> - Dimension $n \to \infty$: Limit attractor ($n - 2$ [positive Lyapunov exponents])
>
> The maximally extended attracting invariant sets are ω-stable, but their internal flow in almost all cases are not.

He then distinguished minimal chaos with a single positive Lyapunov exponent from maximal (full) chaos with $m - 2$ positive Lyapunov exponents [43]. Of course, for a three-dimensional system, minimal chaos is also maximal. Thus, when $m > 3$, maximal chaos is the hyperchaotic C^{m-2} case and it is also possible to have toroidal chaos C^{m-2}-T^2 but such a configuration seems to be rather unstable. According to the hierarchy theorem, to each additional dimension, it should be possible to find at least one new type of chaos, not seen in smaller dimensional state space. We will here limit ourselves to explicit the different possibilities in a space whose dimension is up to 4.

---

[11] 10th Summer Seminar on Applied Mathematics, organized by Frank C. Hoppensteadt. Some proceedings were published with contributions by L. N. Howard, C. Steele, A. S. Winfree, D. Ludwig, F. C. Hoppensteadt, O. E. Rössler, J. K. Hale and S. Guckenheimer. Otto's paper [41] only combines some hyperchaotic system which are in this book (see Sect. 6.6.).

When the dimension of the dynamical system is 1, the solution either settles down onto a stable singular point or is ejected to infinity. We will limit ourselves to consider only bounded attractor. When the dimension is equal to two, the solution is either a point or a periodic orbit. Here we have to distinguish dissipative from conservative systems. In the former, the solution converges—after a transient regime—to an asymptotic set, called the attractor $\mathscr{A}$, and latter system produces marginally stable solution. For instance, in a three-dimensional state space, dissipative as well as conservative chaos are characterized by a single positive and a single null Lyapunov exponents ($p = q = 1$). Strongly dissipative systems are easy to describe by using a first-return map [14]. Conservative systems are far more complex to describe in spite of the KAM theory showing that the state space of a system with n degrees of freedom is foliated by invariant tori T^n [1, 11, 31]. In fact, a chaotic sea is rather difficult to describe or to characterize. Of course, there are intermediary situations with weakly dissipative systems. We will here focus our attention on "school cases", which are good representatives for clear given types.

B.2 Two-Dimensional Oscillators

Let us start with the simple harmonic oscillator

$$\ddot{x} + x = 0\,. \tag{B.1}$$

This second-order differential equation is rewritten into the set of two ordinary differential equations

$$\left\{ \begin{array}{l} \dot{x} = y \\ \dot{y} = -x \end{array} \right. \tag{B.2}$$

where the state space is explicitly spanned by the two variables x and y. This system is conservative and produces periodic orbits whose amplitudes are dependent on the initial conditions (Fig. B.1a).

When a damping switch function is added, the conservative oscillator (B.2) is transformed into a dissipative one, that is,

$$\left\{ \begin{array}{l} \dot{x} = y \\ \dot{y} = -x + \rho \left(1 - x^2\right) y\,. \end{array} \right. \tag{B.3}$$

This is the well-known van der Pol oscillator [51] which produces a limit cycle (Fig. B.1b) which does not depend on the initial conditions.

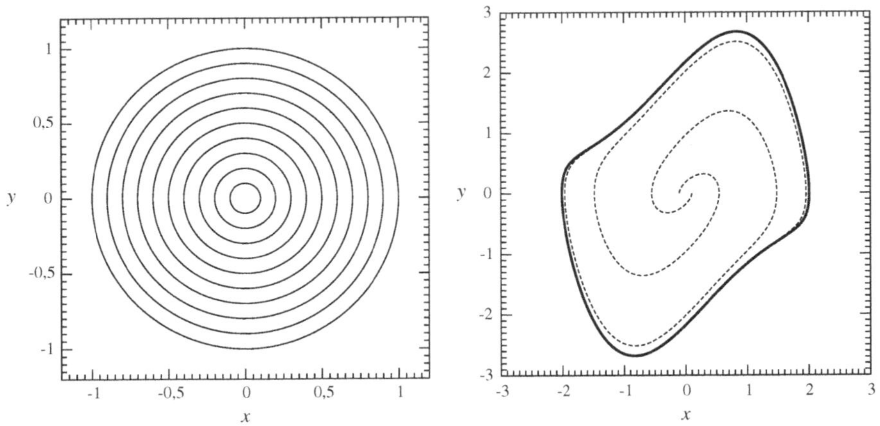

(a) Conservative case ($\rho = 0$) (b) Dissipative case ($\rho = 1$)

Fig. B.1 State portrait of the conservative oscillator (B.2) for ten initial conditions (**a**). Its dissipative analog (B.3) produces a limit cycle (**b**): in the latter case, two transient regimes are plotted in dashed lines. The state portrait can be also produced with system (B.3) with $\rho = 0$

B.3 Quasi-periodic Regimes and Toroidal Chaos

It is possible to couple each of these oscillators with a switch function, leading to

$$\begin{cases} \dot{x} = y \\ \dot{y} = -x + \rho \left(1 - x^2\right) y - z \\ \dot{z} = a \left(1 - x^2\right) \end{cases} \tag{B.4}$$

This system is conservative when $\rho = 0$ and dissipative when $\rho > 0$. In the former case, this system produces a trajectory visiting the surface of a torus T^2, characterized by two incommensurable frequencies. Such a torus is evidenced with a Poincaré section (for instance, defined as $y_\mathrm{p} = 0$) which has the shape of a closed loop (Fig. B.2a). When a damping nonlinearity is added ($\rho > 0$), it is possible to get toroidal chaos which, unfortunately, exists only for a finite duration. Such a regime was termed metastable by James Yorke [52]. We propose to designate toroidal chaos by the symbol C-T^2, meaning that the main structure is a torus T^2 and "C" stands for its chaotic nature. We add a tilde to specify that the behavior is metastable, thus leading to $\tilde{\mathrm{C}}$-T^2. The Poincaré section of the metastable toroidal chaos is quite similar to chaotic sea as observed in conservative systems as exhibited by Michel Hénon and Carl Heiles [9]. Few island of regular motion can be detected in the upper part of the Poincaré section. The system is in fact only slightly dissipative (we were not able to find toroidal chaos with larger ρ-value) and these islands are reminiscent from the quasi-periodic regimes observed in the conservative case. This $\tilde{\mathrm{C}}$-T^2 regime is thus of the "sea" type (see Sect. B.6).

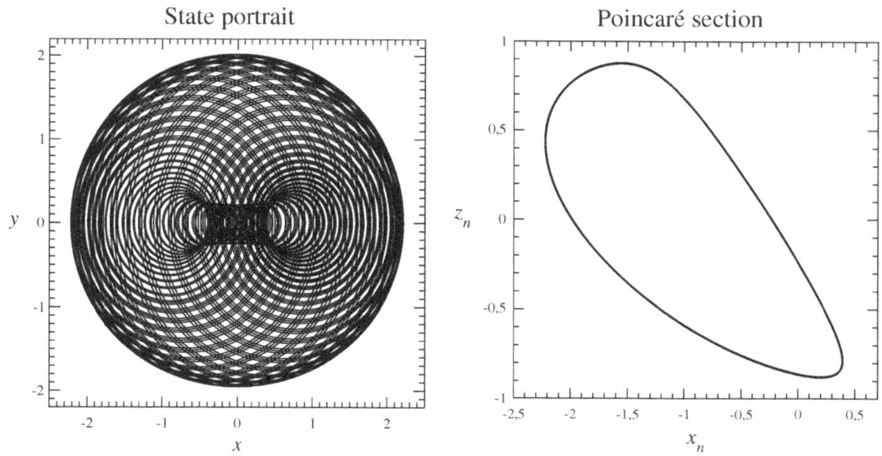

(a) Conservative case ($a = 0.02$ and $\rho = 0$): Quasi-periodic T^2

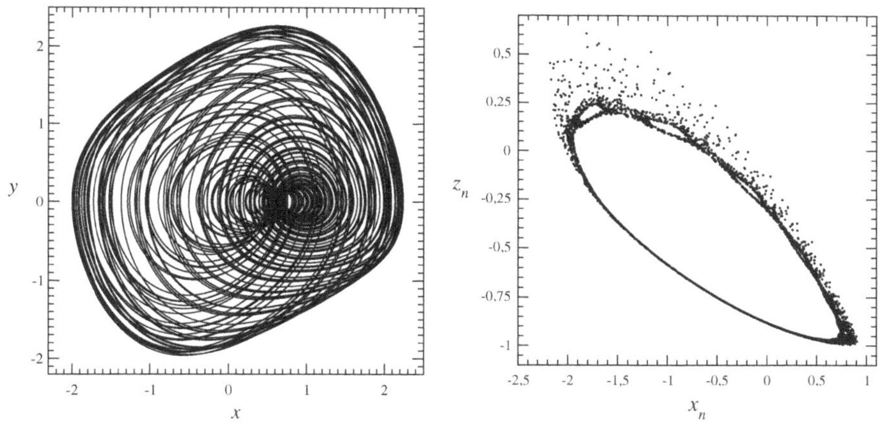

(b) Dissipative case ($a = 0.07$ and $\rho = 0.3$): metastable toroidal chaos $\tilde{C}$-T^2

Fig. B.2 State portrait of the three-dimensional oscillator (B.4) resulting from the combination of the 2D linear oscillator (B.2) with a switch function. Initial conditions: $x_0 = y_0 = z_0 = 0.1$ for the conservative case and $x_0 = 1.5$, $y_0 = 1.1$, and $z_0 = 0.1$ for the dissipative case

Finding a stable toroidal chaos C-T^2 in a three-dimensional state space produced by a continuous system is rather difficult and most often, when there is a chaotic regime, it is metastable: see for instance, the case of Eq. (6.17) and (6.18) which both produces metastable toroidal chaos $\tilde{C}$-T^2 of the sea type (Figs. 6.10b and 6.18b). A few examples of toroidal chaos C-T^2 are known. For instance, there is one produced by a modified Lorenz system [25, 22] and one produced by a system proposed by Bo Deng [6]. The latter is the rather complicated system

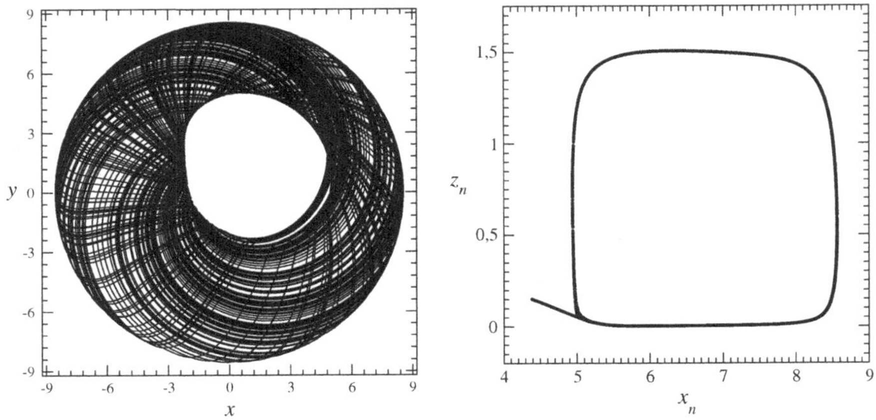

(a) State portrait (b) Poincaré section

Fig. B.3 Toroidal chaos C-T^2 produced by the three-dimensional dissipative system (B.5). Its Poincaré section presents a single folding. Parameter values: $a = 3$, $b = 0.8$, $c = 1$, $d = 0.1$, $m = 0.05$, $\eta = 3.312$, $R = 10$, $\alpha = 2.8$, $\beta = 5$, $\varepsilon = 0.1$, $\lambda = -2$ and $\mu = 0.415$. Initial conditions: $x_0 = y_0 = z_0 = 0.1$

$$\left\{\begin{array}{l} \dot{x} = z\left(\lambda x - \mu y\right) + (2-z)\left[\alpha x\left(1 - \dfrac{x^2+y^2}{R^2}\right) - \beta y\right] \\ \dot{y} = z\left(\mu x + \lambda y\right) + (2-z)\left[\alpha y\left(1 - \dfrac{x^2+y^2}{R^2}\right) + \beta x\right] \\ \dot{z} = \dfrac{z\left[(2-z)\left(a\left(z-2\right)^2 + b\right) - dx\right]\left[z + m\left(x^2+y^2\right) - \eta\right] - \varepsilon c\left(z-1\right)}{\varepsilon} \end{array}\right. \tag{B.5}$$

The chaotic nature of the toroidal behavior (Fig. B.3a) produced by this system is clearly evidenced with a Poincaré section which reveals the closed loop shape characteristic of the toroidal behavior and a folding (left bottom part of Fig. B.3b) ensuring the mixing properties. Such a folding was associated with chaotic properties by James Curry and James Yorke [5]. Since there is a single folding, this toroidal chaos C-T^2 is unimodal.

An extension to the fourth dimension is possible by transforming the switch function into a nonlinear oscillator. The conservative form of system (B.4) is thus

$$\left\{\begin{array}{l} \dot{x} = y \\ \dot{y} = -x - z - w \\ \dot{z} = a\left(1 - x^2\right) - cw \\ \dot{w} = cz\,. \end{array}\right. \tag{B.6}$$

As expected, there is now three frequencies in the Fourier spectrum (Fig. B.4b) computed from variable x for the toroidal behavior thus produced. They are $f_1 = 0.0008$ Hz, $f_2 = 0.020333$ Hz, and $f_3 = 0.159304$ Hz, respectively. They are incom-

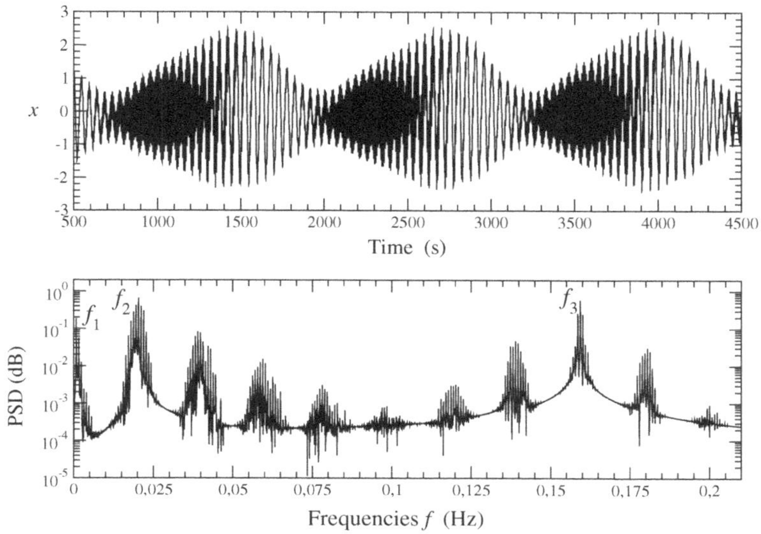

Fig. B.4 Time series of variable x and its related Fourier spectrum for the toroidal chaos C-T^3 produced by four-dimensional conservative system (B.6). Parameter values: $a = 0.03$ and $c = 0.1313$. Initial conditions: $x_0 = 0.6173$, $y_0 = -0.1368$, $z_0 = -0.1231$, and $w_0 = -0.0078$

mensurable. The spectrum also reveals a band of emerging frequencies: this is a signature of chaos.

The Poincaré section (defined as $y_{\rm p} = 0$) shows the structure of a torus T^2 (Fig. B.5). The toroidal chaos is thus structured around a torus T^3. It is then possible to compute a second-order Poincaré section [33] defined as

$$\mathscr{P}_2 \equiv \left\{ (x_n, w_n) \in \mathbb{R}^2 \mid y_n = 0, -\varepsilon < z_n < +\varepsilon, \dot{z}_n > 0 \right\} \tag{B.7}$$

where ε is here equal to 0.04: this is a band selected in the plane projection of the Poincaré section shown in Fig. B.5. This allows to "unfold" or to "reconstruct" a high-dimensional Poincaré section. Such a technique is very demanding in the number of intersections in the Poincaré section. In the present case, to get 32,000 points in the second-order Poincaré section $\mathscr{P}_2$ requires a numerical integration of the trajectory over a duration of $4.2 \cdot 10^6$ s which corresponds to $6.4 \cdot 10^5$ intersections with the Poincaré section.

The second-order Poincaré section reveals a closed-loop shape which is characteristic of a toroidal structure (Fig. B.6). This loop has a thickness which may depend on the ε-value used for its construction. Nevertheless, in the left top part of it, there is at least five islands of regular motion which confirms its reality: this toroidal chaos C-T^3 is therefore of the "sea" type.

Unfortunately, it is not always possible to exhibit the annular structure of a conservative toroidal chaos C-T^2. For instance, the four-dimensional system [27]

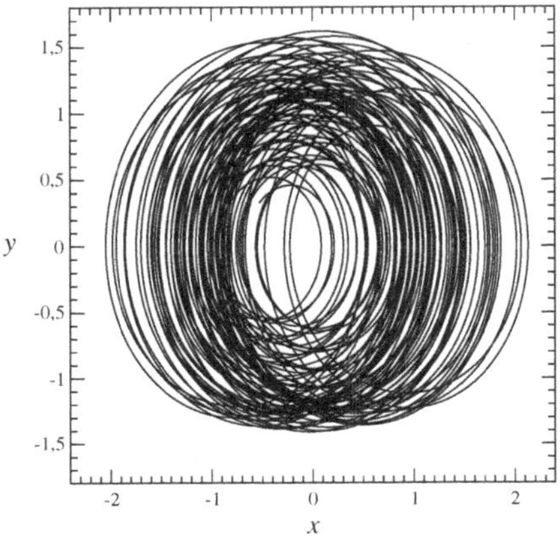

(a) State portrait

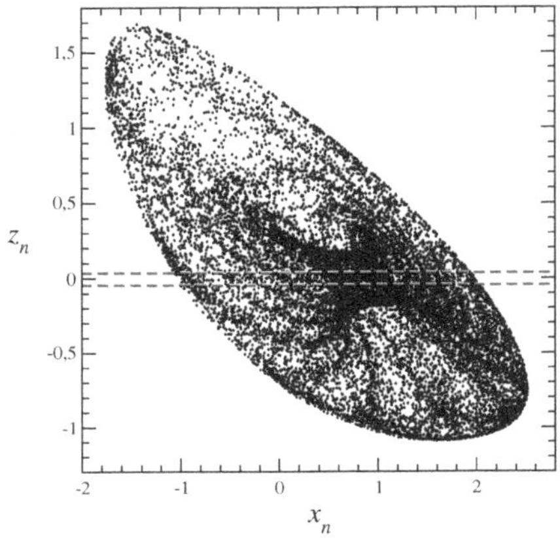

(b) Poincaré section

Fig. B.5 Toroidal chaos C-T^3 produced by the four-dimensional conservative oscillator (B.6). The location of the second-order Poincaré section $\mathscr{P}_2$ is indicated with dashed lines. Same parameter as in Fig. B.4

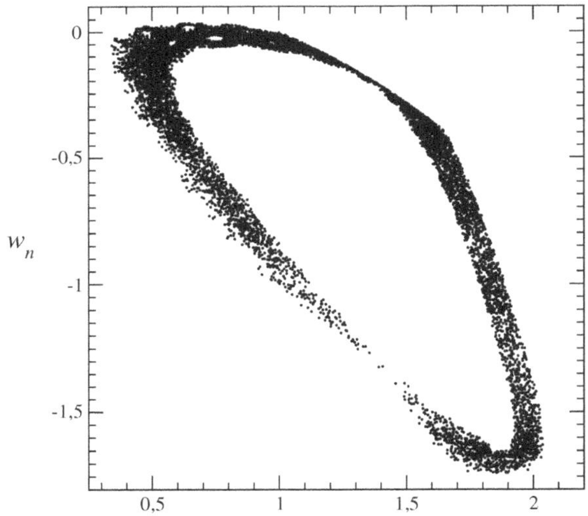

Fig. B.6 Second-order Poincaré section $\mathscr{P}_2$ of the toroidal chaos C-T^3 produced by the four-dimensional conservative oscillator (B.6). Same parameter values as in Fig. B.4

$$\begin{cases} \dot{u}_1 = av_1 \\ \dot{v}_1 = b\left(u_1^2 - u_2^2\right) - c\left(u_1^3 - u_1u_2^2\right) \\ \dot{u}_2 = av_2 \\ \dot{v}_2 = -2bu_1u_2 - c\left(u_2^3 + u_2u_1^2\right) \end{cases} \tag{B.8}$$

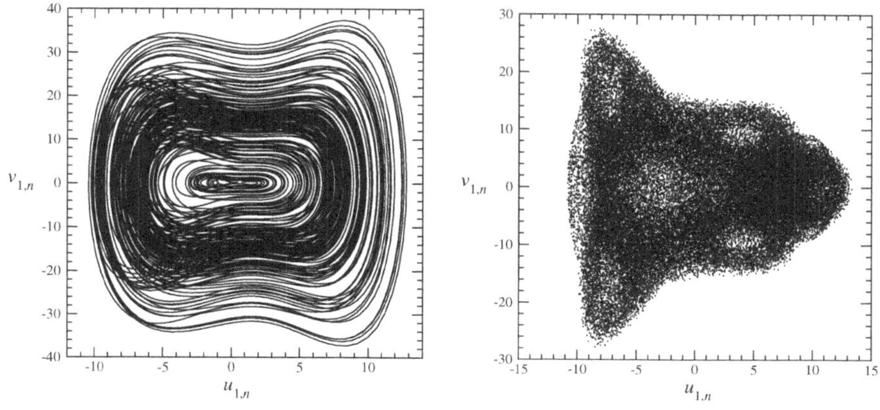

(a) State portrait (b) Poincaré section

Fig. B.7 Conservative toroidal chaos C-T^2 produced by the conservative four-dimensional system (B.8). Parameter values: $a = 10, b = 3$ and $c = 2$. Initial conditions: $u_1(0) = u_2(0) = 0.5$, $v_1(0) = 1$, and $v_2 = 0$

produces a conservative toroidal chaos C-T^2 characterized by the Lyapunov exponents

$$\begin{cases} \lambda_1 = +0.57 \\ \lambda_2 = 0 \\ \lambda_3 = 0 \\ \lambda_4 = -0.57 \end{cases} \tag{B.9}$$

which confirm the toroidal nature of this chaos (Fig. B.7). A Poincaré section (defined as $u_2 = 0$) reveals a chaotic sea, more or less suggesting a "filled" torus. Computing a second-order Poincaré section (defined as $-0.4 < v_1 < +0.4$) does not allow to exhibit an annular section. This indicates that the conservative toroidal chaos is structured in a different way. It is also possible that the chaotic sea is more spread out than in the previous case shown in Fig. B.5. It is still a tough challenge to characterize such chaos.

B.4 Dissipative Chaos

It is possible to construct a simple dissipative version of the oscillator (B.4) by adding the term $-cz$ to the third equation; we thus have

$$\begin{cases} \dot{x} = y \\ \dot{y} = -x - z \\ \dot{z} = a\left(1 - x^2\right) - cz\,. \end{cases} \tag{B.10}$$

This dissipative system produces a Rössler attractor, that is, a chaos C (Fig. B.8a). Its first-return map to a Poincaré section (defined by $y_p = 0$) is made of two foliated monotone branches (Fig. B.8b). We would like to have a more dissipative system to remove the foliation. We were not able to obtain such an idealized map by varying the parameter values of system (B.10). We therefore rewrote the Rössler system [39] in a form which as close as possible to the previous one, that is, as

$$\begin{cases} \dot{x} = y + ax \\ \dot{y} = -x - z \\ \dot{z} = b + yz - cz \,. \end{cases} \tag{B.11}$$

For appropriate parameter values, a chaos C with a smooth unimodal map is obtained (Fig. B.8b). The critical point between the increasing and the decreasing branch is differentiable: it is smooth. The Rössler attractor is therefore a smooth unimodal chaos C.

Strictly speaking, two attractors are topologically equivalent when (i) their two branched manifolds (templates) are the same and (ii) their populations of unstable periodic orbits are equal. In practice, it is rather rare that the two populations of periodic orbits are exactly the same. This is not so relevant since, most often, they can be easily adjusted by tuning one of the parameter values. This is why it may be considered that two attractors are topologically equivalent when they are characterized by the same template. Sometimes, the number of branches differs; in that case, when the smaller template is included in the larger one, we say that the two attractors are topologically *compatible* [28].

When the first-return map has a significant thickness or when they are foliated, there are more than one periodic orbit associated with a given symbolic sequence: it is thus possible to have two different linking numbers between orbits characterized by the same symbolic sequences [28]. We say that the two attractors are ε-topologically equivalent, meaning that when the thickness is removed, a branched manifold can be constructed. Here, the two attractors shown in Fig. B.8 are ε-topologically equivalent. It is therefore not necessary to distinguish them in a classification of chaotic attractors.

As we saw in this book, there are many types of chaotic attractors which are not toroidal. Otto named them screw, funnel, sandwich, Lorenzian…He always provided the first-return map to a Poincaré section to support their characterization. With two monotone branches, there is a single critical point which can be differentiable as we saw with smooth unimodal chaos C. It is possible to have a non-differentiable critical point characterized by a cusp point: this is the Lorenz map, introduced by Edward Lorenz to characterize his attractor [26]. With Peter Ortoleva (Indiana University), Otto proposed the rational system [40]

$$\begin{cases} \dot{x} = ax + by - cxy - \dfrac{(dz + e)x}{x + K_1} \\ \dot{y} = f + gz - hy - \dfrac{jxy}{y + K_2} \\ \dot{z} = k + lxz - mz \end{cases} \tag{B.12}$$

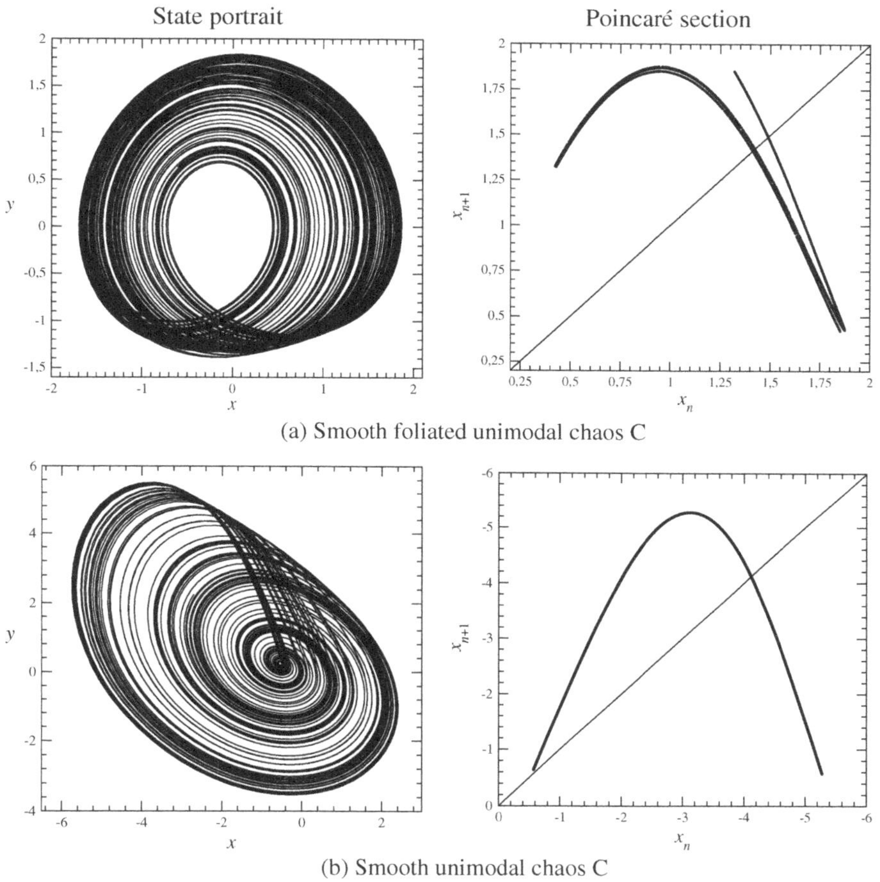

Fig. B.8 Smooth foliated unimodal chaos C (**a**) produced by the three-dimensional dissipative oscillator (B.10) and smooth unimodal chaos C (**b**) produced by the three-dimensional dissipative oscillator (B.11). Parameter values: **a** $a = 0.27675$ and $c = 0.2$; **b** $a = 0.432125$, $b = 2$, and $c = 4$

with such a map but without the rotation symmetry $\mathcal{R}_z$ around the z-axis that the Lorenz system has. With appropriate parameter values, this system produces a unimodal chaos C with a tearing mechanism leading to a "Lorenz" map (Fig. B.9a): the Poincaré section is defined as $x_p = 116$ ($\dot{x}_p > 0$). We will designate such a map as a torn unimodal map. This behavior is thus a torn unimodal chaos C. From the branched manifold (template) point of view, this attractor is not topologically different from the smooth unimodal chaos C. What differs is the sequence with which periodic orbits are created or removed when a parameter value is varied [8]. The tearing mechanism also offers more possibilities for constructing a branched manifold with two stripes than a folding associated with a smooth unimodal chaos C [4].

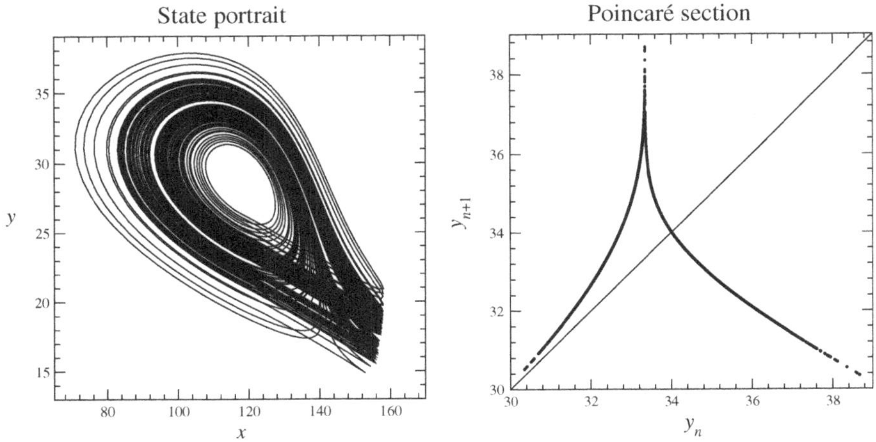

(a) System (B.12) proposed by Rössler and Ortoleva

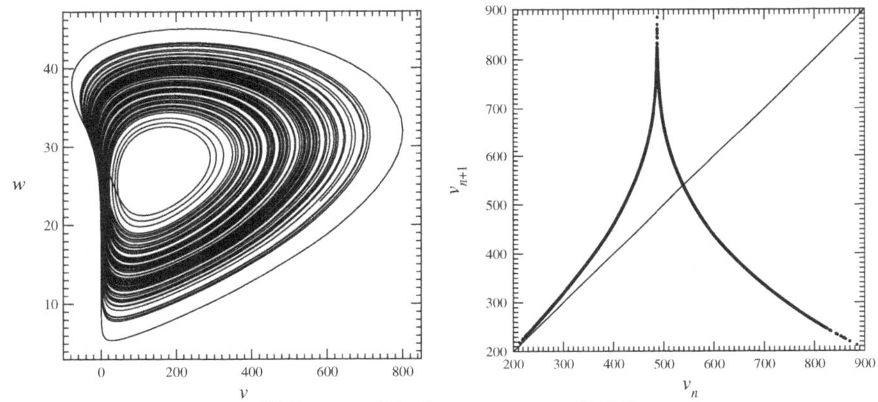

(b) Image of the Lorenz system (B.15)

Fig. B.9 Torn unimodal chaos C (**a**) produced by the three-dimensional system (B.12) and by the image of the Lorenz system (**b**). Parameter values for system (B.12): $a = 33$, $b = 150$, $c = 1$, $d = 3.5$, $e = 4815$, $f = 410$, $g = 0.59$, $h = 4$, $j = 2.5$, $k = 2.5$, $l = 5.29$, $m = 750$, $K_1 = 0.01$, and $K_2 = 0.01$. Initial conditions: $x_0 = 142.611$, $y_0 = 15$, and $z_0 = 0.05$. Parameter values for the image of the Lorenz system: $R = 28$, $\sigma = 10$, and $b = \frac{8}{3}$

The torn unimodal is not natural in the Lorenz attractor. It was shown that symmetric attractors with n scrolls (or leaves) should be characterized by using a n-component Poincaré section [13, 15, 16]. This was generalized with the concept of bounding tori by Tsvetelin Tsankov and Robert Gilmore [47, 48]. An attractor can be enclosed in a bounded surface which takes the form of a torus with a genus g (the number of holes the torus presents). All the attractors previously discussed in this appendix are bounded by genus-1 tori. The Lorenz system is bounded by a genus-3 torus: one hole per singular point. An attractor bounded by a genus-g torus must be characterized by a $(g-1)$-component Poincaré section [48]. The Lorenz system must be therefore characterized with a two-component Poincaré section [13, 15, 16]. Consequently, when correctly computed, the first-return to a Poincaré section of the Lorenz attractor has four branches and, consequently, three critical points [4]: this is a torn three-modal chaos C. The Lorenz system has a rotation symmetry: as a result, there are two copies of a so-called fundamental domain [13, 16]. It is possible to modded out the symmetry to obtain a representation of the symmetric attractor—the image attractor—without any residual symmetry [7, 15, 30]. The Lorenz map is the natural map for the image of the Lorenz attractor.

Let us construct the image of the Lorenz system. Starting from the Lorenz equation [26]

$$\left\{ \begin{array}{l} \dot{x} = -\sigma x + \sigma y \\ \dot{y} = Rx - y - xz \\ \dot{z} = -bz + xy \end{array} \right. \tag{B.13}$$

and using the coordinate transformation

$$\Psi = \left| \begin{array}{l} u = \Re\,(x+i\,y)^2 = x^2 - y^2 \\ v = \Im\,(x+i\,y)^2 = 2xy \\ w = z\,, \end{array} \right. \tag{B.14}$$

one get the proto-Lorenz system [30]

$$\left\{ \begin{array}{l} \dot{u} = (-\sigma - 1)u + (\sigma - R)v + vw + (1-\sigma)\rho \\ \dot{v} = (R-\sigma)u - (\sigma+1)v - uw + (R+\sigma)\rho - \rho w \\ \dot{w} = -bw + \dfrac{1}{2}v \end{array} \right. \tag{B.15}$$

where $\rho = \sqrt{u^2+v^2}$. This system produces a genus-1 attractor characterized by a torn unimodal map (Fig. B.9b). Since by simply using coordinate transformation, it is possible to inject or to modded out symmetry properties [7, 17], a classification of chaotic attractors should focus on attractors without symmetry properties [23]. Specifying the nature of the symmetry in addition to the type of chaos of its image would avoid to distinguish types of chaos which are in fact equivalent modulo a symmetry. For instance, the Lorenz attractor is a torn unimodal chaos C with a rotation symmetry $\mathcal{R}_z$.

There is another fundamental possibility to construct a torn unimodal map: rather than having one increasing and one decreasing branches, the two branches may have slopes of the same sign by twisting by π one of the branches. This was called "sandwich" chaos by Otto but let us designate it as twisted torn unimodal chaos. A possible way to construct such an attractor is to break the symmetry in a Lorenz-like system, to get [40]

$$\left\{ \begin{array}{l} \dot{x} = -dx + ay + e \\ \dot{y} = -cx + y - yz \\ \dot{z} = -bz + y^2\,. \end{array} \right. \tag{B.16}$$

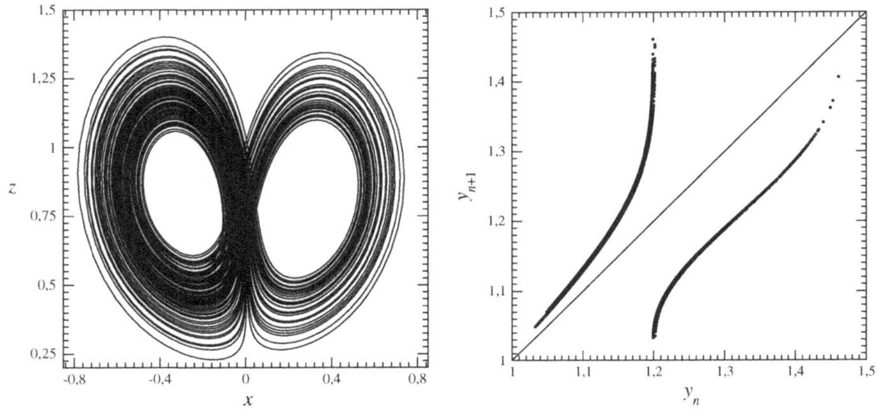

(a) State portrait (b) Poincaré section

Fig. B.10 Twisted torn chaos C produced by the perturbed Lorenz-like system (B.16). Parameter values: $a = 0.08$, $b = 0.1$, $c = 1$, $d = 0.42$, and $e = 0.0015$. Initial conditions: $x_0 = 0.125$, $y_0 = 0.2$, and $z_0 = 0.4$

It can be shown that the resulting attractor (Fig. B.10a) is bounded by a genus-1 torus [19] and a twisted torn unimodal chaos C is obtained as exhibited by the first-return map to a Poincaré section (defined as $x_\text{p} = -0.3$, $\dot{x}_\text{p} > 0$) shown in Fig. B.10b. With this last example, we exhausted the possibilities to construct unimodal map.

More developed chaos can be constructed by adding new branches with folding or tearing mechanisms. It is also possible to add full torsion in the attractor: a number of half-twists is added to all the branches, without modifying the general shape of the first-return map [2, 3]. This produces new types of branched manifold but this would not change drastically the relative organization of unstable periodic orbits.

B.5 New Features Allowed by a Fourth Dimension

With system (B.6), we got a toroidal chaos which can be seen as a torus T^3 with a small chaotic sea. Such a behavior is the natural chaos for a conservative system. Other forms of toroidal chaos are possible, particularly because a fourth dimension opens possibilities non offered in a three-dimensional state space.

The first way to construct a four-dimensional system producing toroidal chaos is to start with one of the variant of the van der Pol oscillator, namely

$$\begin{cases} \dot{x} = y \\ \dot{y} = \rho \left(1 - \gamma x^2\right) y - x^3 + u \\ \dot{u} = v \\ \dot{v} = -\omega^2 u \end{cases} \tag{B.17}$$

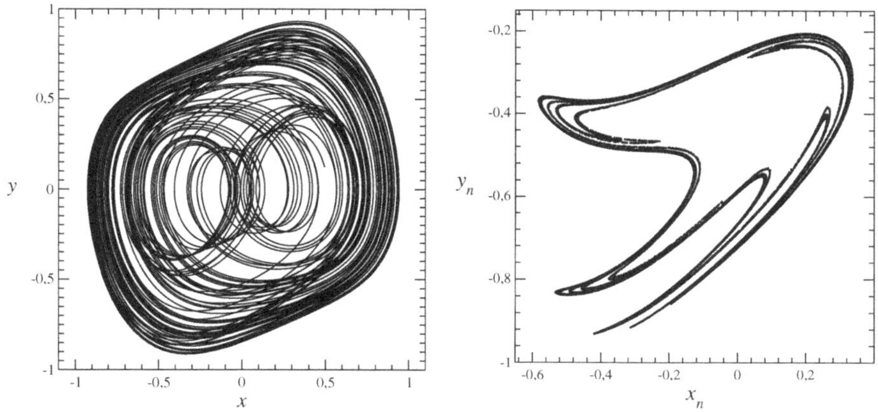

(a) State portrait (b) Poincaré section

Fig. B.11 Smooth (most likely) three-modal toroidal chaos C-T^2 produced by the driven van der Pol system (B.17). Parameter values: $\rho = 0.2$, $\gamma = 8$, and $\omega = 1.02$. Initial conditions: $x_0 = 0.2$, $y_0 = 0.2$, $u_0 = 0.35$, and $v_0 = 0$

as investigated by Yoshisuke Ueda in his 1965 Ph.D. thesis [49] but only published in 1981 [50]. Compared to system (B.6), the linear term $-x$ is removed from the second equation. This system is clearly made of two oscillators, the first one being driven by the second one which is an harmonic oscillator. The driven oscillator is dissipative while the second is conservative: the four-dimensional (autonomous) system is semi-dissipative (or semi-conservative) [29]. The amplitude of the driving signal is governed by the initial conditions u_0 and v_0 as seen in Fig. B.1a: it is thus not dependent on a parameter!

For appropriate parameter values, there is toroidal chaos (Fig. B.11). At least three foldings (Fig. B.11b, left bottom), with perhaps two others located in the upper part of the section, inside the torus, are observed in the Poincaré section (defined as $u_n = 0$). There is no tearing in this attractor: it is thus a smooth (most likely) five-modal toroidal chaos C-T^2. This toroidal chaos emerges according to the Curry–Yorke scenario [5] and a period-doubling cascade [21].

Two dissipative oscillators, bidirectionally coupled, can be obtained from system (B.6) as follows [10]

$$\left\{ \begin{array}{l} \dot{x} = y \\ \dot{y} = -x - az - bw \\ \dot{z} = d\left(1 - x^2\right) - cw \\ \dot{w} = cz - ew\,. \end{array} \right. \tag{B.18}$$

For appropriate parameter values, and when the d-value is increased, there is a period-doubling cascade on the torus T^2: this means that the Poincaré section

$$\mathscr{P} \equiv \left\{ (y_n, z_n, w_n) \in \mathbb{R}^3 \mid w_n = 2x_n - 1, \dot{x}_n > 0 \right\} \tag{B.19}$$

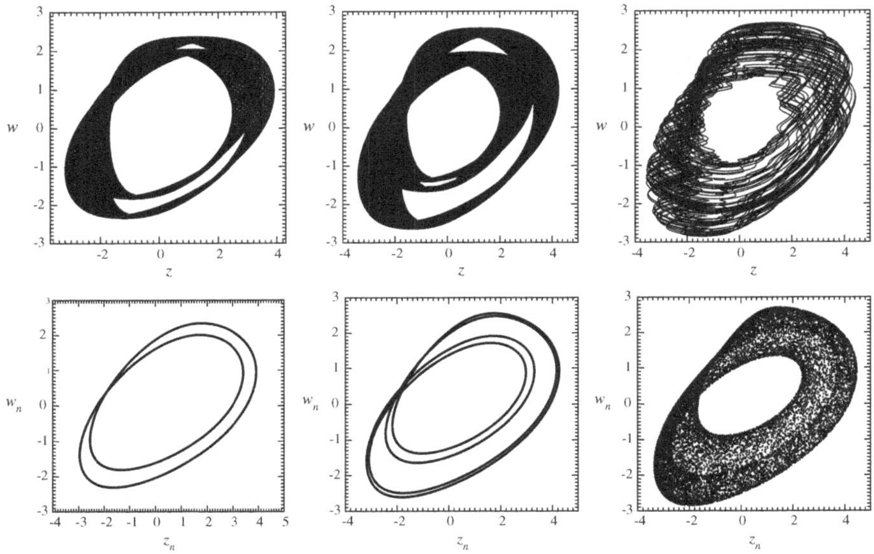

(a) $d = 0.37$: $P_2 \otimes T^2$ (b) $d = 0.383$: $P_4 \otimes T^2$ (c) $d = 0.3922$: $C \otimes T^2$

Fig. B.12 Conjugated chaos $C \otimes T^2$ produced by the dissipative four-dimensional system (B.18). Plane projections of the Poincaré section $\mathscr{P}$ are also shown. Other parameter values: $a = 0.15$, $b = 0.25$, $c = 0.1$, and $e = 0.05$. Initial conditions: $x_0 = -0.043$, $y_0 = -0.157$, $z_0 = -3.113$, and $w_0 = -1.826$

of the torus reveals a period-1 orbit ($d = 0.36$), a period-2 orbit ($d = 0.37$, Fig. B.12a), a period-4 orbit ($d = 0.383$, Fig. B.12b), and a chaotic attractor $d = 0.3922$, Fig. B.12c) which has the structure of Rössler attractor. Such a cascade was found by Klein and Baier [10]. It was also observed in two different laser systems [18, 20].

A period-doubling cascade on the torus can only be observed in a state space whose dimension is at least equal to four. A torus T^2 is conjugated with a smooth unimodal chaos: the chaotic regime shown in Fig. 6.18c is therefore termed conjugated smooth toroidal chaos $C \otimes T^2$. We were unable to extract a smooth unimodal first-return map to a second-order Poincaré section. It seems that the relationships between the three-dimensional smooth unimodal chaos C and this four-dimensional one is quite complex and remains to be further investigated.

All the dynamics discussed above were characterized by a single positive Lyapunov exponent. Otto introduced in 1979 the four-dimensional system [42]

$$\left\{ \begin{array}{l} \dot{x} = y + ax + w \\ \dot{y} = -x - z \\ \dot{z} = b + zy \\ \dot{w} = -cz + dw \end{array} \right. \tag{B.20}$$

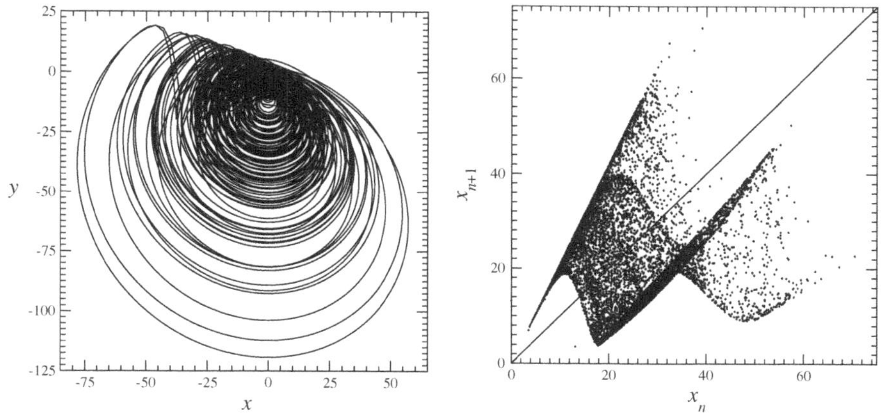

(a) State portrait (b) Poincaré section

Fig. B.13 Smooth hyperchaos chaos C^2-T^2 produced by the four-dimensional system (B.20). Parameter values: $a = 0.25$, $b = 3.0$, $c = 0.5$, and $d = 0.05$. Initial conditions: $x_0 = -10$, $y_0 = -6$, $z_0 = 0$, and $w_0 = 10.1$

producing a chaotic regime associated with two positive Lyapunov exponents. He called *hyperchaotic* this regime. Hyperchaos remains poorly characterized, mostly because the approach based on linking numbers and branched manifolds does no longer work. Moreover, there is no universality exhibited in hyperchaotic systems as encountered with the smooth unimodal chaos C. The example here proposed is interesting due to its simplicity and because, by contruction, it is a direct extension of the smooth unimodal chaos C. As suggested by Klein and Baier, we proposed to designate by C^k a chaotic attractor characterized by k positive Lyapunov exponents (Fig. B.13).

The first-return map to a Poincaré section (defined as $\dot{x} = 0$ and $\ddot{x} > 0$) seems to have two increasing branches (perhaps a third one on the right bottom side) blurred by a thickness "linking" them. It is rather hard to provide a more accurate description.

Notice that we were not able to find a quasi-periodic behavior T^3 in the previous models, thus confirming the result by Ruelle and Takens that a dynamical behavior structured around a torus T^k is most often chaotic when $k \geq 3$ [44].

B.6 Conservative chaos

Conservative chaos can be found in the three-dimensional system proposed by Shuichi Nosé [32] and William Hoover [35]. It was later rediscovered by Julian Sprott [45]. The system reads as

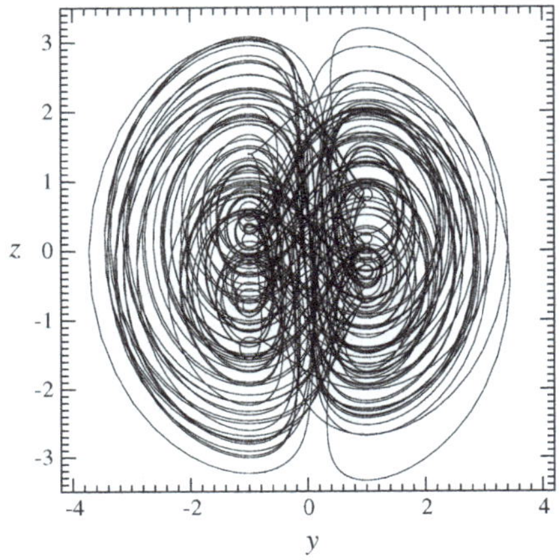

(a) State portrait

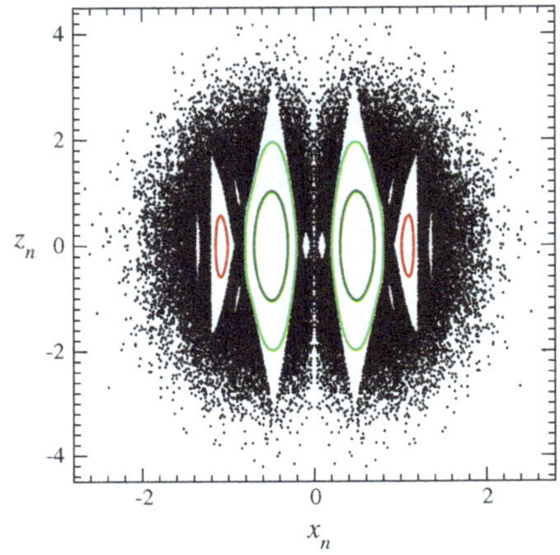

(b) Poincaré section

Fig. B.14 Sea chaos C produced by the conservative three-dimensional system (B.21). The Poincaré section is defined as $y_p = 0$. Parameter values: $a = 0.3$. Initial conditions for the chaotic sea: $x_0 = 1.5$, and $y_0 = z_0 = 0$

$$\begin{cases} \dot{x} = ay \\ \dot{y} = -x + yz \\ \dot{z} = 1 - y^2 \end{cases} \tag{B.21}$$

which is a harmonic oscillator coupled in a nonlinear way with a switch function. For appropriate parameter values, this system produces a chaotic regime (Fig. B.14a) which is characterized by a chaotic "sea" organized around some tori T^2 (a few of them are also shown in Fig. B.14b). Identifying the mechanisms responsible for the chaotic regime is not yet possible: it is therefore difficult to determine whether this chaos is unimodal or not.

In order to produce a conservative hyperchaos C^2-T^2, the most obvious configuration would require two positive Lyapunov exponents balanced by two negative ones, plus one null along the trajectory: the system would be thus five-dimensional. With this perspective, the system

$$\begin{cases} \dot{x} = ay + w^2 \\ \dot{y} = -ax \\ \dot{z} = xv \\ \dot{v} = bw - xz \\ \dot{w} = -bv - xw \end{cases} \tag{B.22}$$

was proposed [53]. Nevertheless this system has a jacobian matrix $\mathcal{J}$ whose trace is $\text{Tr}(\mathcal{J}) = -x$. This system can be globally conservative if

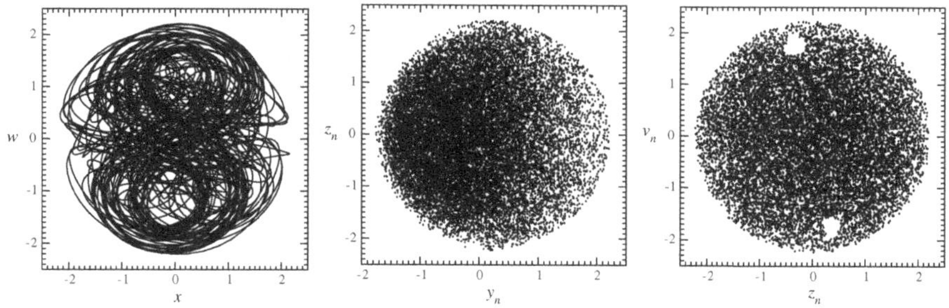

(a) State portrait (b) 1st-order Poincaré section (c) 2nd-order Poincaré section

Fig. B.15 Hyperchaotic sea chaos C^2 produced by the weakly-dissipative five-dimensional system (B.22). Parameter values: $a = 1$ and $b = 1$. Initial conditions: $x_0 = y_0 = z_0 = v_0 = w_0 = 1$

$$\overline{\text{Tr}(\mathcal{J})} = \frac{1}{T} \int_{t=0}^{T} x \, \text{d}t = 0 \,. \tag{B.23}$$

Unfortunately, this is not the case here since, for the retained parameter values, the averaged trace is slightly negative. Strictly speaking, this system in not conservative and a transient regime may be observed before reaching the attractor. In the other hand, the Lyapunov exponents [53]

$$\left\{ \begin{array}{l} \lambda_1 = +0.836 \\ \lambda_2 = +0.045 \\ \lambda_3 = 0 \\ \lambda_4 = -0.045 \\ \lambda_5 = -0.836 \end{array} \right. \tag{B.24}$$

with the initial requirement. Most likely, this system is weakly dissipative and the toroidal chaos is indeed hyperchaotic.

The chaotic behavior takes the form of a complex torus. A Poincaré section (defined as $x_p = 0$) reveals a "filled" torus (Fig. B.15b). The "sea" type can be glimpsed with a second-order Poincaré section (defined as $y_p = 0$): two small regular islands can be observed (Fig. B.15c). Once again, new techniques would be required to provide more insights about the topological structure of this attractor.

B.7 A Possible Classification of Attractors

With the set of examples of different types of chaos, it is possible to determine the sensitive properties for a possible classification. As specified in the beginning of this appendix, typically there is a single negative Lyapunov exponent in the retained examples to avoid to duplicate similar (if not topologically equivalent) attractors.

Table B.1 Nature of the possible attractor $\mathcal{A}$ produced by m-dimensional dissipative dynamical system. The spectrum of Lyapunov exponents λ_i is reported as well as the range of the fractal dimension d. Only cases where there is a single negative Lyapunov exponent are reported

m	Nature of $\mathcal{A}$	Sgn(λ_i)	$d_{\mathcal{A}}$	Map	Type
1	Singular point	$-$	0	∅	
2	Limit cycle	0 $-$	1	•	
3	Torus T^2	0 0 $-$	2	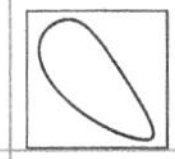	
	Toroidal chaos C-T^2	+ 0 $-$	$2 < d \leq 3$		Sea
				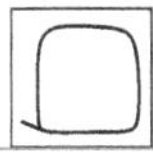	Unimodal toroidal
	Chaos C^1	+ 0 $-$	$1 < d \leq 3$		Smooth unimodal
				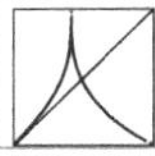	Torn unimodal
				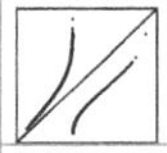	Twisted torn unimodal
	Conservative chaos C^1	+ 0 $-$	$d \approx 3$	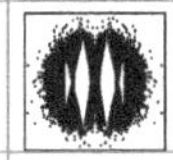	Sea
4	Torus T^3	0 0 0 $-$	3		
	Toroidal chaos C-T^2	+ 0 0 $-$	$2 < d \leq 3$		Smooth five-modal
	Toroidal chaos C⊗T^2	+ 0 0 $-$	$2 < d \leq 3$		Conjugated
	Toroidal chaos C-T^3	+ 0 0 $-$	$2 < d \leq 3$		Sea
	Chaos C^2	+ + 0 $-$	$2 < d \leq 4$		Smooth
	Conservative toroidal chaos C-T^2	+ 0 0 $-$	$d \approx 3$		Sea
> 4	Torus T^{m-1}	0^{m-1} $-$	$m - 1$		
	Toroidal chaos C^p-T^q	$+^p$ 0^q $-$	$2 < d \leq m - 1$		
	Chaos C^{m-2}	$+^{m-2}$ 0 $-$	$2 < d \leq m$		

Most of the examples here discussed thus correspond to maximal (full) chaos ($m - 2$ positive Lyapunov exponents for a m-dimensional systems). Thus, a single positive Lyapunov exponent is associated with minimal chaos (minimal chaos is also maximal when $m = 3$) [43]. Toroidal chaos requires two null Lyapunov exponents. Dynamical regimes are mainly characterized by their numbers p and q of positive and null Lyapunov exponents, respectively. The next ingredient is topological and provided by either a Poincaré section or a first-return map to a Poincaré section.

Mixing properties are associated with either folding or tearing mechanisms [4]. When there is a single mechanism, the system is said to be unimodal. We showed that there is at least four types of unimodal chaos: smooth, torn, twisted torn, and toroidal. While the three formers are well characterized with a branched manifold (template), characterizing accurately the latter remains an open question. All these chaotic regimes were produced by dissipative systems. When the system is conservative, typically the Poincaré section does not present any obvious structure and we describe such a chaos by a "sea". To the best of our knowledge, it is not known how many modes (folding and/or tearing) could be associated with a chaotic sea. Of course, new situations arise with the fourth dimension. For now, we do not pretend that the examples here discussed cover all possible types of chaos. The most surprising chaos is certainly the conjugated toroidal chaos $C \otimes T^2$. An updated hierarchy of chaos can be thus proposed as reported in Table B.1.

B.8 Conclusion

The proposed classification allows to describe the main different types of chaos which may be observed. It also allow to suggest a certain hierarchy between the different types of chaos by using the number of positive and null Lyapynov exponents. Such a hierarchy remains to be refined by using some markers as entropies and complexity measures [24].

References

1. V.I. Arnold, Instability of dynamical systems with many degrees of freedom. Doklady Akademii Nauk SSSR **156**, 9–12, 1964.
2. G. Boulant, M. Lefranc, S. Bielawski, D. Derozier, Horseshoe templates with global torsion in a driven laser. Phys. Rev. E **55**(5), 5082–5091 (1997)
3. G. Boulant, M. Lefranc, S. Bielawski, D. Derozier, A non-Horseshoe template in a chaotic laser model. Int. J. Bifur. Chaos **8**(5), 965–976 (1998)
4. G. Byrne, R. Gilmore, C. Letellier, Distinguishing between folding and tearing mechanisms in strange attractors. Phys. Rev. E **70**, 056214 (2004)
5. J. Curry, J.A. Yorke, A transition from Hopf bifurcation to chaos: computer experiments with maps on $\mathbb{R}^2$. *Lecture Notes in Mathematics*, vol. 668 (1978), pp. 48–66
6. B. Deng, Constructing homoclinic orbits and chaotic attractors. Int. J. Bifur. Chaos **4** (1994), pp. 823–841

7. R. Gilmore, C. Letellier, *The Symmetry of Chaos* (Oxford University Press, Oxford, 2007)
8. R. Gilmore, Explosions in Lorenz maps. Chaos Solitons Fractals **76**, 130–140 (2015)
9. M. Hénon, C. Heiles, The applicability of the third integral of motion: some numerical experiments. Astron. J. **69**(1), 73–79 (1964)
10. M. Klein, G. Baier, Hierarchies of dynamical systems, in *A Chaotic Hierarchy* (World Scientific Publishing, Singapore, 1991), pp. 1–24
11. A.N. Kolmogorov, On the conservation of conditionally periodic motions under small perturbation of the Hamiltonian. Doklady Akademii Nauk SSSR **98**, 527–530 (1954)
12. M. Lefranc, Reflections from the fourth dimension, in *Topology and Dynamics of Chaos*, ed. by C. Letellier (World Scientific Publishing, Singapore, 2013), pp. 205–226
13. C. Letellier, P. Dutertre, G. Gouesbet, Characterization of the Lorenz system taking into account the equivariance of the vector field. Phys. Rev. E **49**(4), 3492–3495 (1994)
14. C. Letellier, P. Dutertre, B. Maheu, Unstable periodic orbits and templates of the Rössler system: toward a systematic topological characterization. Chaos **5**(1), 271–282 (1995)
15. C. Letellier, G. Gouesbet, Topological characterization of a system with high-order symmetries: the proto-Lorenz system. Phys. Rev. E **52**(5), 4754–4761 (1995)
16. C. Letellier, P. Dutertre, J. Reizner, G. Gouesbet, Evolution of multimodal map induced by an equivariant vector field. J. Phys. A **29**, 5359–5373 (1996)
17. C. Letellier, R. Gilmore, Covering dynamical systems: two-fold covers. Phys. Rev. E **63**, 16206 (2001)
18. C. Letellier, Modding out a continuous rotation symmetry for disentangling a laser dynamics. Int. J. Bifur. Chaos **13**(6), 1573–1577 (2003)
19. C. Letellier, E. Roulin, O.E. Rössler, Inequivalent topologies of chaos in simple equations. Chaos Solitons Fractals **28**, 337–360 (2006)
20. C. Letellier, M. Bennoud, G. Martel, Intermittency and period-doubling cascade on tori in a bimode laser model. Chaos Solitons Fractals **33**, 782–794 (2007)
21. C. Letellier, V. Messager, R. Gilmore, From quasi-periodicity to toroïdal chaos: analogy between the Curry-Yorke map and the van der Pol system. Phys. Rev. E **77**(4), 046203 (2008)
22. C. Letellier, R. Gilmore, Poincaré sections for a new three-dimensional toroidal attractor. J. Phys. A **42**, 015101 (2009)
23. C. Letellier, L.A. Aguirre, Required criteria for recognizing new types of chaos: application to the "cord" attractor. Phys. Rev. E **85**, 036204 (2012)
24. C. Letellier, I. Leyva, I. Sendiña-Nadal, Dynamical complexity measure to distinguish organized from disorganized dynamics, Physical Review E, **101**, 022204, (2020)
25. D. Li, A three-scroll chaotic attractor. Phys. Lett. A **372**(4), 387–393 (2007)
26. E.N. Lorenz, Deterministic nonperiodic flow. J. Atmos. Sci. **20**, 130–141 (1963)
27. G.M. Mahmoud, M.E. Ahmed, Analysis of chaotic and hyperchaotic conservative complex nonlinear systems. Miskolc Math. Notes **18**(1), 315–326 (2017)
28. S. Mangiarotti, I. Sendiña-Nadal, C. Letellier, Using global modeling to unveil hidden couplings in small network motifs. Chaos **28**, 123110 (2018)
29. O. Ménard, C. Letellier, J. Maquet, L. Le Sceller, G. Gouesbet, Analysis of a non synchronized sinusoidally driven dynamical system. Int. J. Bifur. Chaos **10**(7), 1759–1772 (2000)
30. R. Miranda, E. Stone, The proto-Lorenz system. Phys. Lett. A **178**, 105–113 (1993)
31. J.K. Moser, Convergent series expansions for quasi-periodic motions. Math. Ann. **169**(1), 136–176 (1967)
32. S. Nosé, A unified formulation of the constant temperature molecular dynamics methods. J. Chem. Phys. **81**(1), 511–519 (1984)
33. T.S. Parker, L.O. Chua, *Practical Numerical Algorithms for Chaotic Systems* (Springer, Berlin, 1989)
34. H. Poincaré, Sur le problème des trois corps et les équations de la dynamique. Acta Math. **13**, 1–270 (1890)

35. H.A. Posch, W.G. Hoover, F.J. Vesely, Canonical dynamics of the Nosé oscillator: stability, order, and chaos. Phys. Rev. A **33**(6), 4253–4265 (1986)
36. J.A.G. Roberts, G.R.W. Quispel, Chaos and time-reversal symmetry. Order and chaos in reversible dynamical systems. Phys. Rep. **216**(2–3), 63–177 (1992)
37. O.E. Rössler, Chaotic behavior in simple reaction systems. Zeitschrift für Naturforschung A **31**, 239–264 (1976)
38. O.E. Rössler, Different types of chaos in two simple differential equations. Zeitschrift für Naturforschung A **31**, 1664–1670 (1976)
39. O.E. Rössler, An equation for continuous chaos. Phys. Lett. A **57**(5), 397–398 (1976)
40. O.E. Rössler, P.J. Ortoleva, Strange attractors in 3-variable reaction systems. *Lecture Notes in Biomathematics*, vol. 21 (1978), pp. 67–73
41. O.E. Rössler, Chaotic oscillations: an example of hyperchaos. *Lecture Notes in Applied Mathematics*, vol. 17 (1980), pp. 141–156
42. O.E. Rössler, An equation for hyperchaos. Phys. Lett. A **71**, 155–157 (1979)
43. O.E. Rössler, The chaotic hierarchy. Zeitschrift für Naturforschung A **38**, 788–801 (1983)
44. D. Ruelle, F. Takens, On the nature of turbulence. Commun. Math. Phys. **20**, 167–192 (1971)
45. J.C. Sprott, Some simple chaotic flows. Phys. Rev. E **50**(2), 647–650 (1994)
46. R. Thomas, V. Basios, M. Eiswirth, T. Kruel, O.E. Rössler, Hyperchaos of arbitrary order generated by a single feedback circuit, and the emergence of chaotic walks. Chaos **14**(3), 669–674 (2004)
47. T.D. Tsankov, R. Gilmore, Strange attractors are classified by bounding tori. Phys. Rev. Lett. **91**(13), 134104 (2003)
48. T.D. Tsankov, R. Gilmore, Topological aspects of the structure of chaotic attractors in $\mathbb{R}^3$. Phys. Rev. E **69**, 056206 (2004)
49. Y. Ueda, Some problems in the theory of nonlinear oscillations. Ph.D. thesis (1965). Reprinted in *The Road to Chaos*, Aerial Press (1992)
50. Y. Ueda, N. Akamatsu, Chaotically transitional phenomena in the forced negative-resistance oscillator. IEEE Trans. Circ. Syst. **28**(3), 217 (1981)
51. B. van der Pol, On "relaxation-oscillations". Philos. Mag. VII **2**(11), 978–992 (1926)
52. J.A. Yorke, E.D. Yorke, Metastable chaos: the transition to sustained chaotic behavior in the Lorenz model. J. Stat. Phys. **21**(3), 263–277 (1979)
53. A. Wu, S. Cang, R. Zhang, Z. Wang, Z. Chen, Hyperchaos in a conservative system with nonhyperbolic fixed points. Complexity **2018**, 9430637 (2018)

Appendix C
Chaotic Realizations in the Real World

by Christophe Letellier

C.1 Belousov–Zhabotinski Reaction

Chaotic oscillations were observed in the Belousov–Zhabotinski reaction in the middle of the 70s by John Hudson's group in Charlottesville [25]. This group had observed aperiodic oscillations for few years but they had no explanations for them and left these results in a drawer.[12] This is only when Otto Rössler published his first paper about chaotic oscillations [63] that Hudson got a theoretical paradigm to explain these time series (potential from a platinium wire electrode) that were obviously aperiodic (see Fig. C.1) as follows [67].

> A crucial property is the geometric configuration of the manifold. Here, taking the geometric representation used in a paper by Rössler [169], we suppose that the manifold is an S-shaped folded sheet. *[...]* The key necessary features are a pseudo-steady-state manifold with a folded or overlapping portion.

The mechanism governing this dynamics was explained by Hudson's group [67] with pictures in the spirit of Fig. 1.1 and of the drawings (Figs. C.2) by Otto in his first paper [63].

The Bonhöffer–van der Pol circuit was important for Otto since it provides a simple model for relaxation oscillations. The fact Bonhoeffer proposed his model for explaining the nerve conduction [9] touched Otto's sensitivity to physiology. Otto was looking for a chemical Bonhöffer–van der Pol circuit [62] and saw the Belousov–Zhabotinski reaction as a natural realization of it. Having understood that the mechanism underlying chaotic oscillations requires state space representation to be explained, Otto and Klaus Wegmann plotted the electrochemical potential *versus* the potential of a bromide ion recorded in a well stirred Belousov–Zhabotinski reaction under isothermal open flow conditions [79]. They thus obtained a two-dimensional projection of the state portrait (Fig. C.3). These plots were the very first chaotic state portraits obtained from experimental data. They thus distinguished two

[12] J. Hudson to C. Letellier, private communication.

O. E. Rössler and C. Letellier, *Chaos*, Understanding Complex Systems,
https://doi.org/10.1007/978-3-030-44305-4

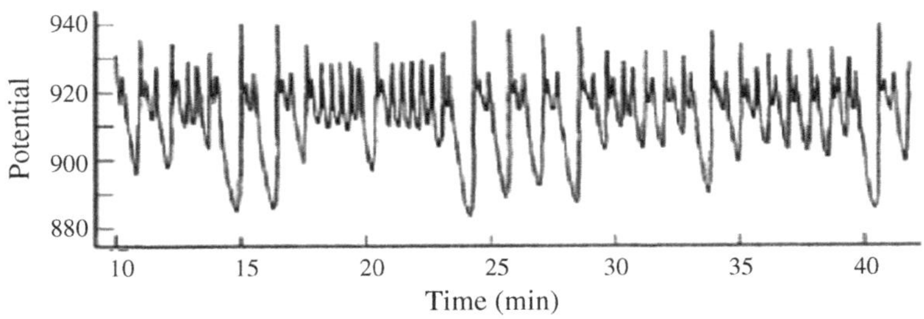

Fig. C.1 Time evolution of the potential across a platinum electrode inserted into the Belousov–Zhabotinsky reaction vessel. These were interpreted as chaotic oscillations by John Hudson and colleagues in 1977

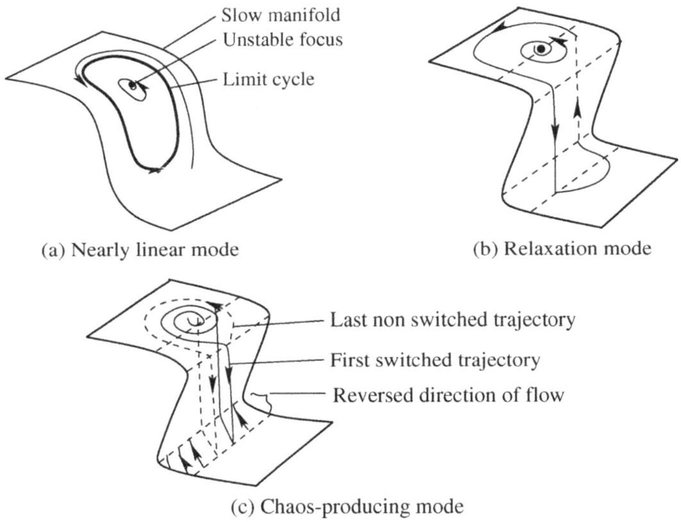

Fig. C.2 Sketches as drawn by Rössler for picturing the main trajectorial flow of a universal circuit in his first chaos paper. This type of drawings was used in 1977 by Hudson's group for explaining the aperiodic oscillations observed in the Belousov–Zhabotinski reaction

types of chaos, endogenous chaos without external noise from exogenous chaos with external noise.

These early observations were followed by those of Jean-Claude Roux, Annie Rossi, Sabine Bachelart and Christian Vidal in 1980 [66, 76]. Rather than measuring two different chemical variables, the Frenches "reconstructed" a three-dimensional state portrait from measurements of a single variable, the concentration in Ce^{4+} ions. Roux's team used derivative coordinates following a procedure adopted by Packard, Crutchfield, Farmer and Shaw, and that was submitted for publication at that time [59]. The principle was to "unfold" the time series by using delay coordinates, that is, by plotting the measured variable $x(t)$ versus the first derivative $\dot{x}$ and

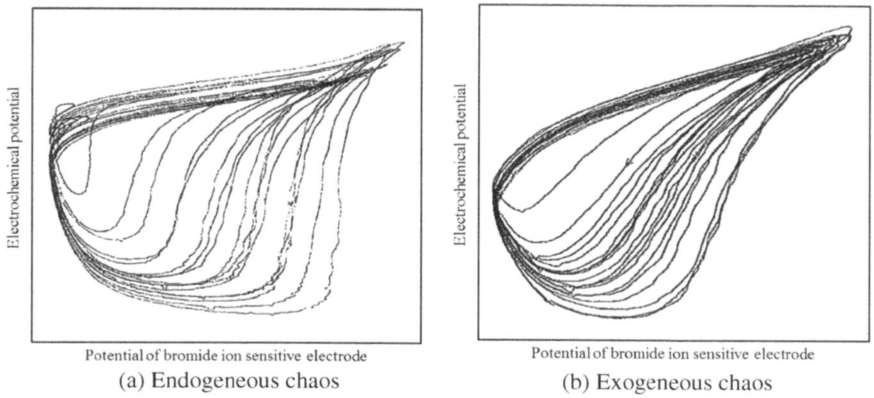

Fig. C.3 Two different types of chaos observed by Wegmann and Rössler in a Belousov–Zhabotinsky reaction in 1978

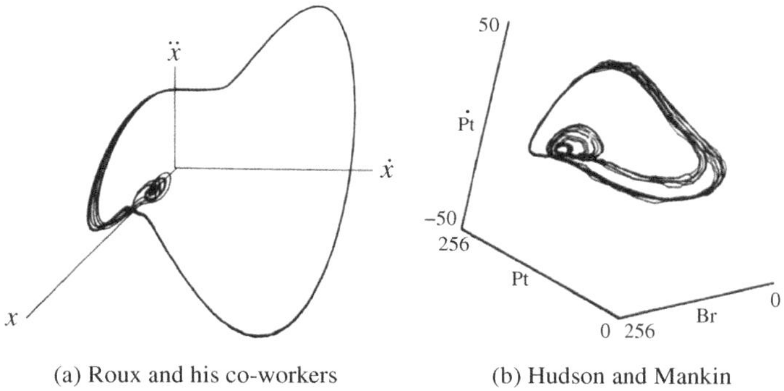

Fig. C.4 Reconstructions of state portraits using derivative coordinates from measurements in Belousov–Zhabotinski reaction. From the concentration [Ce^{4+}] by Roux et al. and from the platinium electrode and the bromide ions (as used by Rössler and Wegmann) by Hudson's group

the second derivative $\ddot{x}$. This approach which can be applied to any system was first suggested by David Ruelle in various seminars in American universities and then published by Packard and co-workers. The validity of this technique to any system was demonstrated by Floris Takens [71]. Roux et al. thus constructed a state portrait as shown in Fig. C.4a. The same year, also based on the work by Packard and his colleagues, John Hudson and Joe Mankin provided a similar chaotic attractor (Fig. C.4b). In the spirit of Otto's work, they also constructed a first-return map to a Poincaré section for characterizing the type of chaos and showed that it was a unimodal smooth map (walking-stick map) as found in the spiral attractor [32].

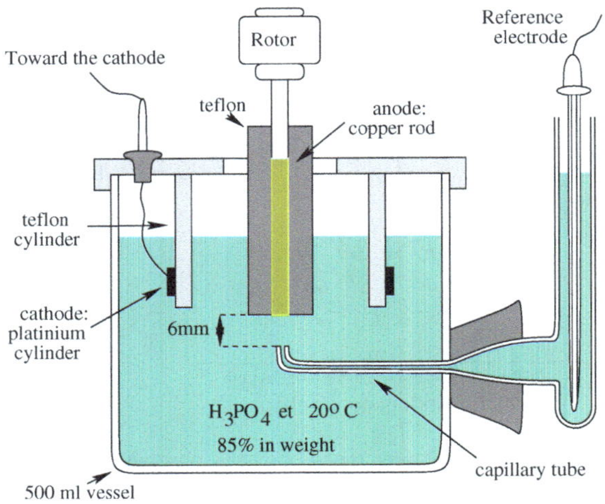

Fig. C.5 Schematic of the copper electrolysis experiment carried out by John Hudson and Zihao Fei

C.2 Electrodissolution

Another type of chemical experiments produce chaos: electrodissolution. Indeed, chaos was observed in copper electrodissolution, again by Hudson's group [5, 6]. The chemical dynamics was investigated from current measurements during potentiostatic electrodissolution of a rotating Cu electrode in phosphoric acid (H_3PO_4). The anode, a copper cylinder 8.26 mm in diameter inserted in a teflon cylinder 2 cm in diameter is rotated at a moderate speed ($\approx$700 rotations/minute) in an 85% (by mass) solution of phosphoric acid maintained at 20° by a heat bath (Fig. C.5): the rotation speed must be low enough to avoid creating turbulence but high enough to avoid reactants migrating too far. A cylindrical platinum net band (much larger than the disc) was put around the disk as a counter electrode to obtain uniform potential and current distributions. All these precautions are needed in order to get a sufficiently homogeneous accurate measurements of the time evolution of the current I.

The cell was a 500 ml flask with a side neck in which a capillary probe was fixed. The saturated calomel reference electrode was separated from the solution by a capillary. The distance between the disc surface and the tip of the capillary was about 6 mm. A potentiostat (Princeton Applied Research model 273) was used to regulate the potential of the working disc electrode with respect to the reference electrode and to monitor the current. The potential was equal to 689 mV. The data were recorded at a sampling frequency f_s of 1500 Hz.

The current I presents relatively large oscillations that could reach amplitudes as large as 15 mA while the average current was 40 mA [41]. These oscillations have an irregular amplitude while their period is relatively constant. Starting from the time evolution of the current I, a state portrait is reconstructed by using the derivative coordinates (Fig. C.6a). A first-return map to a Poincaré section is smooth and unimodal (Fig. C.6b). Its smooth nature is sufficient to guarantee that the electrodissolution

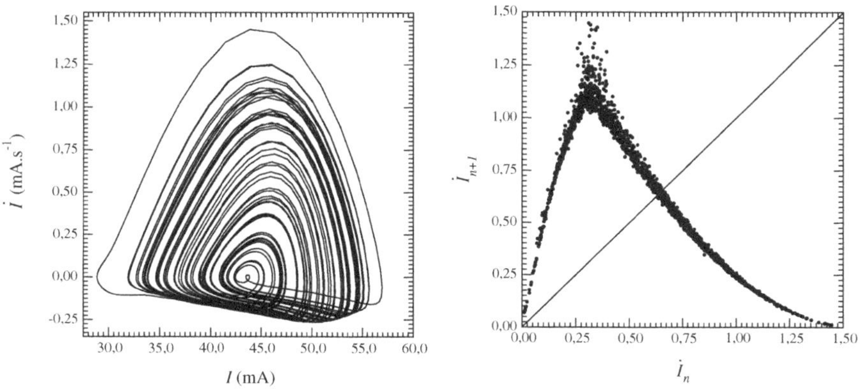

(a) Spiral chaotic attractor

(b) Smooth unimodal first-return map

Fig. C.6 Spiral chaos observed in the copper electrodissolution experiment. A smooth unimodal first-return map to a Poincaré section indicates that a period-doubling cascade can be observed

dynamics has the same type of chaos as the logistic map or the spiral attractor: as a result, a period-doubling cascade can be observed in this experiments [60].

These observations provide experimental confirmation for the existence of chaotic behavior that is compatible with very simple theoretical models, that is, first-return map belonging to the class of the logistic map. Using the horseshoe map (Fig. 7.10), it is possible to predict the relative organization of periodic orbits in the branched manifold (paper sheet model as shown in Fig. 2.5 [41]. This chaotic dynamics is a perfect example of experimental dynamics that presents a topological organization equivalent to the paper sheet model of the spiral attractor. The theory of chaos can be actually used as a paradigm for the study of phenomena that are governed by a deterministic dynamics.

C.3 Thermionic Diode

Electronics played a relevant role in the path followed by Rössler for bridging the gap between periodic and chaotic oscillations [51]. If there are few evidences for early observations of non-trivial behaviors in electronic circuits [53], chaos was observed in many of them in 1980s [48]. Among the experimental realizations of a van der Pol oscillator, there are circuits with a thermionic diode. Andronov's school investigated some of them [4]. A typical example is the circuit carried out by Thomas Klinger and his colleagues [57]. The diode was constructed with a tungsten filament (cathode) placed in a 15 cm diameter cylinder in which a plate of stainless steel (anode) was inserted at a distance of 10 cm from the cathode (Fig. C.7). The diode is driven by a static potential on the top of which is superposed a sinusoidal component.

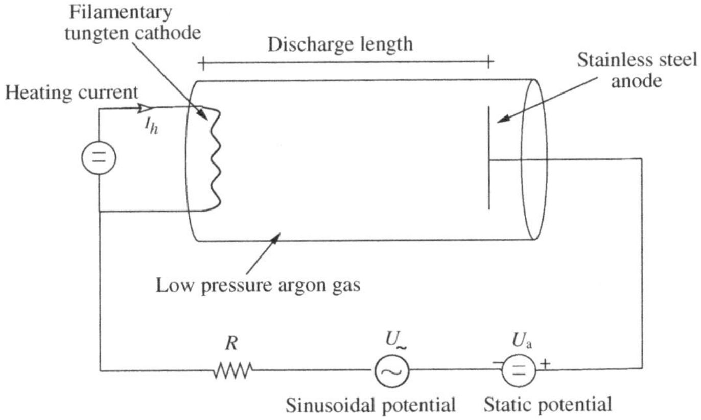

Fig. C.7 Block diagram of the circuit investigated by Klinger et al. The external resistance R is equal to 100 Ω and the heating current $I_h = 4.2$ A. The cylinder was filled with argon to a pressure of 0.4 Pa

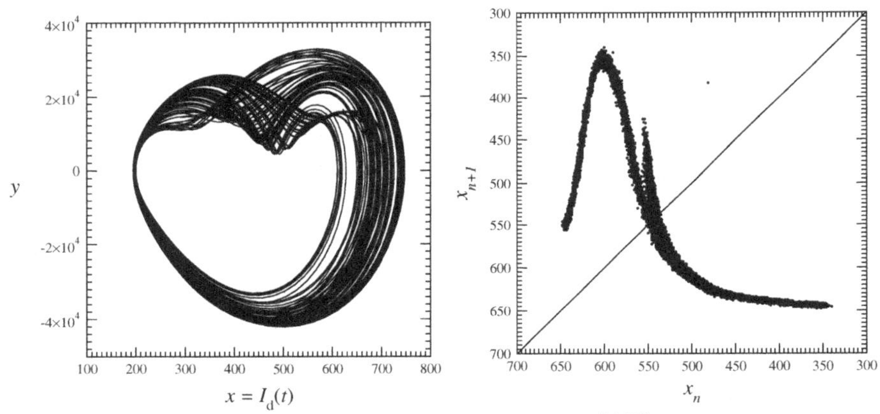

(a) State portrait (b) First-return map

Fig. C.8 Chaotic behavior produced by the circuit with a thermionic diode. The state portrait is reconstructed using the derivative coordinates. The first-return map to a Poincaré section is smooth and unimodal (with a slightly foliated decreasing branch)

From the discharge current I_d, a state portrait is reconstruced by using derivative coordinates (Fig. C.8a): it is characterized by a smooth unimodal first-return map to a Poincaré section (Fig. C.8b) [43]. Most likely this chaotic attractor is of a funnel type. A very similar chaotic regime can be produced by a driven van der Pol oscillator [53].

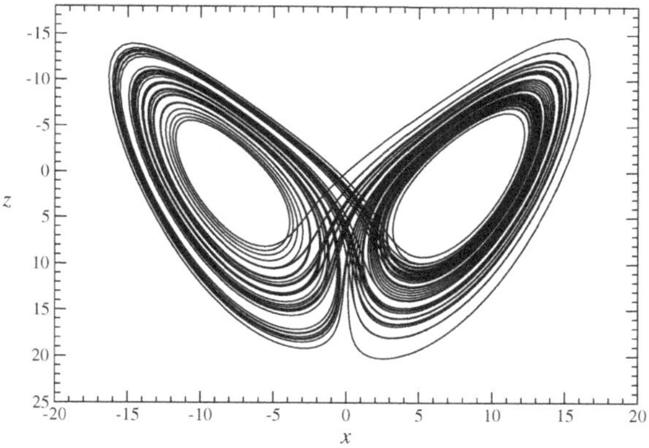

Fig. C.9 Lorenz attractor produced by the Haken model (C.1). Parameter values: $R = 25$, $\sigma = 10$ and $b = 8/3$

C.4 Laser

There is a field from physics which provided many examples of chaotic oscillations: laser physics. If oscillations were seen in the early 1960s, it seems that one of the first contributions to chaotic oscillations was a theoretical work by Arkadii Grasiuk and Anatoly Oraevsky [23, 24]. At that time, Ernest Buley and Frederick Cummings investigated a model of an idealized laser considered as a population of two-level atoms interacting strongly with a single-radiation field mode and mentioned that plotted state portraits of limit cycles or transient regimes; they also mentioned a "series of almost random spikes" [8]. A decade later, Hermann Haken evidenced an analogy between the Lorenz system and a laser system [27]. Beginning with the Maxwell–Bloch equations, and under certain conditions, Haken was able to derive the equations

$$\left\{ \begin{array}{l} \dot{x} = \sigma(y - x) \\ \dot{y} = -y + xz \\ \dot{z} = bR - bz - xy\,. \end{array} \right. \tag{C.1}$$

producing a Lorenz attractor (Fig. C.9a). Due to this analogy, the Lorenz system was more important for laser physics than the Rössler attractor.

The experimental development of a laser with Lorenz dynamics was finally realized 1986 by Carl O. Weiss and J. Brock [78]. Their laser used a gas, ammonia (NH_3). The variable measured was the light intensity, which is related to the electric field strength. It is possible to compare the experimental dynamics[13] with the model one (Fig. C.10) when $I = x^2 + y^2$ is used. The most striking feature is the first-return map which is characterized by the cusp between the two monotone branches: this is very specific to the Lorenz attractor. The foliated structure of this map results from a slight symmetry breaking in the experimental system. With all of these experimental

[13]The data are from Dingyuan Tang's Ph.D. thesis [14] and were kindly provided by Carl Weiss.

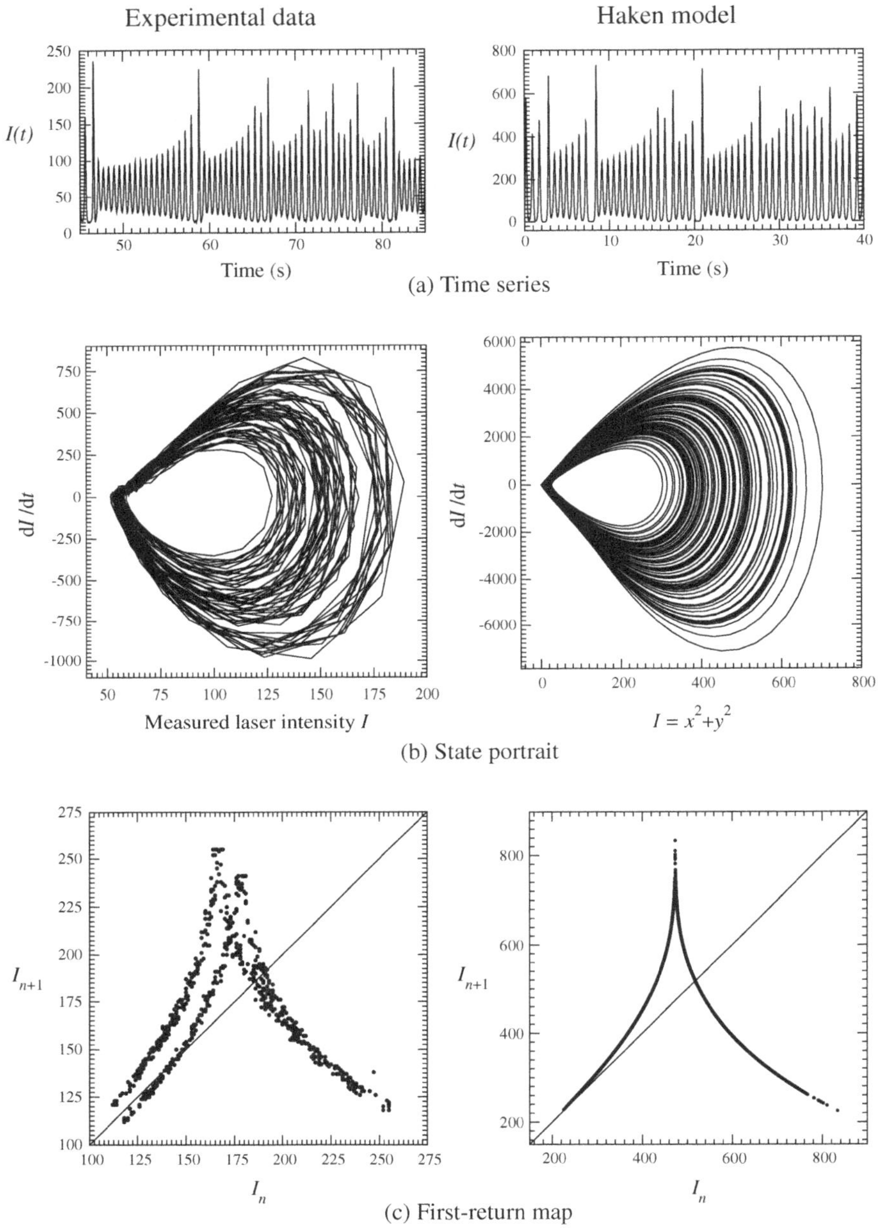

Fig. C.10 Comparisons between the NH_3 laser conducted by Tang and Weiss, and the Haken model (C.1). The laser has a pump intensity of 6 W/cm^2, a gas pressure of 10 Pa and a pump frequency of 20 MHz. Same parameter values as in Fig. C.9

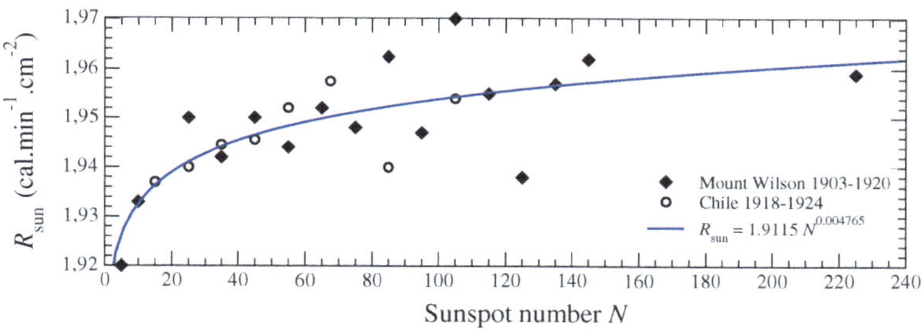

Fig. C.11 Solar constant values versus sunspot numbers. These data were investigated by Abbot from more than 2,000 observations collected at the Mount Wilson observatory (Pasadena, CA) and the Montezuma solar observatory (Chile). A power law was fitted to Abbot's data. $1.96\,\text{cal.min}^{-1}.\text{cm}^{-2} \approx 1.366 \cdot 10^3\ \text{W.m}^{-2}$

confirmations, the Lorenz system became a reference model for demonstrating the possibility of chaotic behavior in real physical systems.

Many other types of chaotic attractors were then observed in laser systems. For instance, homoclinic chaos was observed in a single-line, single-mode, homogeneously broadened infrared CO_2 laser containing an intracavity low-pressure absorber gas [31, 39]. Chaotic attractors with different topologies were observed by Guillaume Boulant et al. in this type of laser [10, 11].

C.5 Solar Activity

Solar activity is important to investigate since solar wind—stream of charged particles released from the upper atmosphere of the sun—interacts with the terrestrial atmosphere and may have effects in its local space environment, by producing ionizing radiation which may damage some communication satellite. These solar winds are strongly dependent on the solar activity and it thus useful to be able to predict the maximum amplitude of solar cycles [38, 30].

As in many stars, there is a magnetic field produced by the streams of charged particles flowing within the sun. It presents some fluctuations directly resulting from the oscillations in the solar activity. These oscillations can be investigated from the sunspot numbers observed in the solar disk. These dark spots correspond to the convection which is blocked by strong magnetic lines. The more numerous sunspots are, the more active the sun is. This is only in the early 20th century that the sunspot numbers were correlated to the solar constant R_{sun} by Charles G. Abbot (Fig. C.11) [1].

These solar cycles are known since the 19th century but more or less reliable observations can be already found in the 17th century (Fig. C.12).[14]

[14]Data may be downloaded from http://www.sidc.be/silso/datafiles#total.

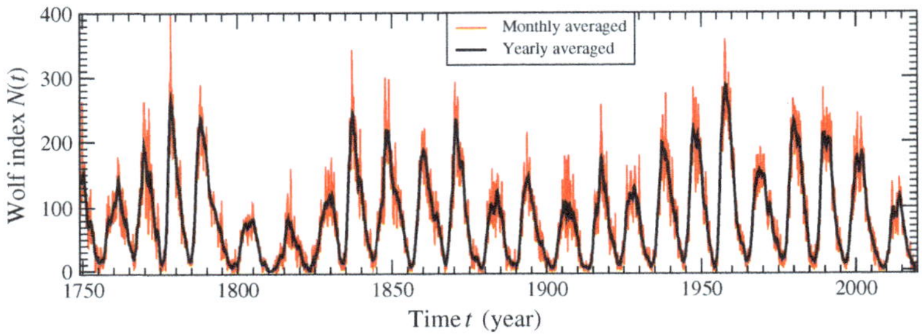

Fig. C.12 Sunspot number N from the data collected since 1749. The cycles before 1850 were reconstructed by Rudolf Wolf from published observations [82]. The monthly and yearly averaged sunspot numbers are plotted

The modulation of the solar cycle is aperiodic and could have a deterministic origin. Unfortunately, the available data are not sufficiently accurate to determine with common signal analysis whether they are governed by a deterministic chaotic process or a random one. An alternative is to find a set of differential or discrete equations that reproduce accurately the dynamics underlying the data: this is called *global modelling* [3].

The time series of the sunspot numbers (Fig. C.12) presents an almost constant period, but large amplitude variations. These variations in the sunspot numbers reflect the fact that the solar magnetic field significantly affects this behavior. This is known since Heinrich Schwabe that the period of the solar activity is about eleven year [68]. George Hale later evidenced that the magnetic field reverses at each cycle, so that the periodicity is in fact about 22 year [29].

After 300 years of observations, 23 cycles of data are available (Fig. C.12): this is not so much, particularly for classical (statistical) data analysis. Moreover, data before 1850 have been reconstructed by Wolf and a bias may be suspected since a quite different dynamics underlying these data has been evidenced [13]. Consequently, only a ten of reliable solar cycles remain for a global modelling.

Since we are primarily interested in long term variations in the solar dynamics, that is, fluctuations from one cycle to another, monthly averaged sunspot numbers allowed to remove high-frequency fluctuations. A state portrait (Fig. C.13a) reconstructed using delay coordinates from the Wolf index $N(t)$ where a time delay τ is equal to 15 months. The reconstructed trajectory (Fig. C.13a) still shows fluctuations at frequencies much higher than the eleven year cycle: the structure of this portrait is therefore blurred. The data are therefore smoothed with a filtering process which does not significantly affect the long term dynamics (Fig. C.13b).

Based on Hale's observation that the polarity of the solar magnetic field reverses at each solar cycle, Ronald Bracewell introduced an inversion of the sunspot numbers at each minimum for taking into account the reversals of the solar magnetic field [12]. In place of a period of 11 years as shown in Fig. C.12, the period of the sunspot cycles

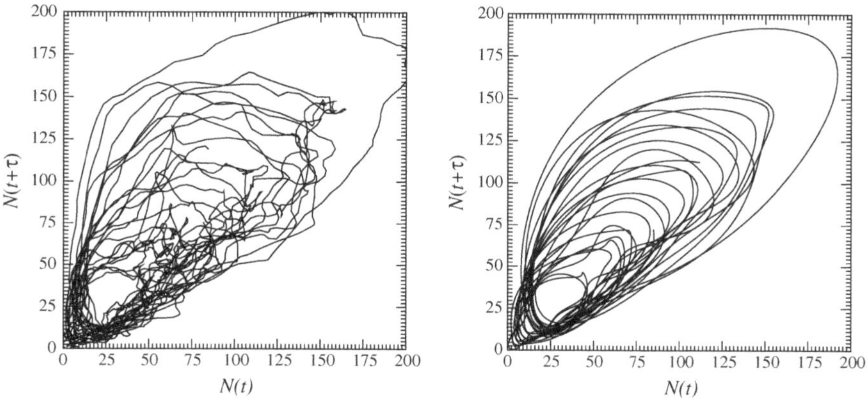

(a) Monthly averaged (b) Smoothed data

Fig. C.13 Attractors reconstructed using delay coordinates $N(t)$ and The $N(t+\tau)$ where the time delay τ is 15 months. The 23 solar cycles observed since 1749 are presented here

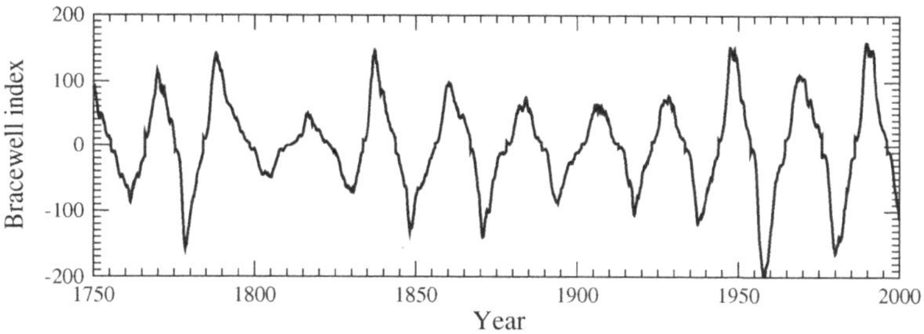

Fig. C.14 Evolution of the number of sunspots (Wolf index), taking account of the magnetic field inversions after each cycle, as suggested by Bracewell

would thus be about 22 years. He alternately plotted the sunspot cycles, positive and negative, as represented in Fig. C.14.

Unfortunately, discontinuities induced at each reversal prevent to get an accurate global model. As a result, it was useful to introduce symmetry properties to properly model the magnetic field inversion. Let us start with the Lorenz system which has a rotation symmetry by π radians about the z-axis: the attractor is invariant under the change of coordinates $(x, y, z) \mapsto (-x, -y, z)$.

By using the change of coordinates

$$\Psi = \left| \begin{array}{l} u = \text{Re}\,(x+iy)^2 = x^2 - y^2 \\ v = \text{Im}\,(x+iy)^2 = 2xy \\ w = z \end{array} \right. \tag{C.2}$$

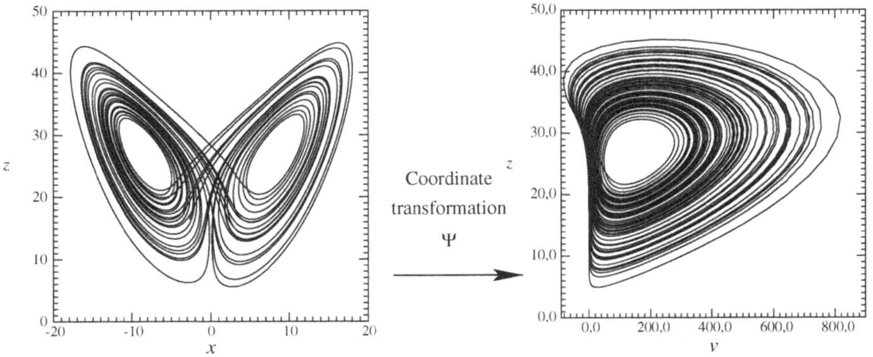

(a) Modding out the symmetry in the Lorenz attractor

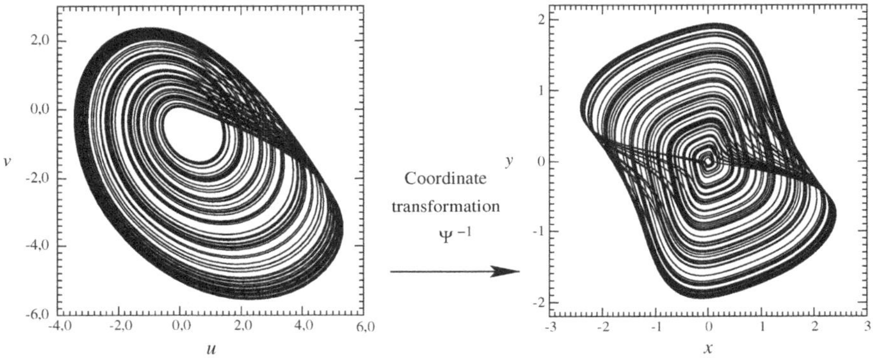

(b) Inserting a symmetry in the Rössler attractor

Fig. C.15 Modding out (**a**) and injecting a rotation symmetry by using an appropriate coordinate transformation (Ψ and Ψ^{-1}, respectively)

where $i = \sqrt{-1}$ (variable z is unchanged since it corresponds to the rotation axis), it is possible to construct a representation of this attractor without symmetry [44, 20]. The chaotic attractor without symmetry (Fig. C.15a, right) is nearly dynamically equivalent to the Lorenz attractor with symmetry (Fig. C.15a, left).

This is reversible, and symmetries can be injected by applying the inverse transformation Ψ^{-1}. A Rössler attractor (initially without symmetry) with a rotation symmetry around the z-axis can be produced as shown in Fig. C.15b. The folding characteristic of the Rössler attractor is now found twice in the "symmetrized" Rössler attractor. Placing the location of the rotation axis in the non-visited zone in the middle of the attractor reconstructed from the sunspot number (Fig. C.16a), the symmetric attractor (Fig. C.16b) presents an inversion of the magnetic field at each cycle, corresponding to what is observed in the solar dynamics [47]. An interesting outcome of the insertion of a rotation symmetry is that it allows the trajectory to remain sufficiently far from the origin (where the variables assume values close to

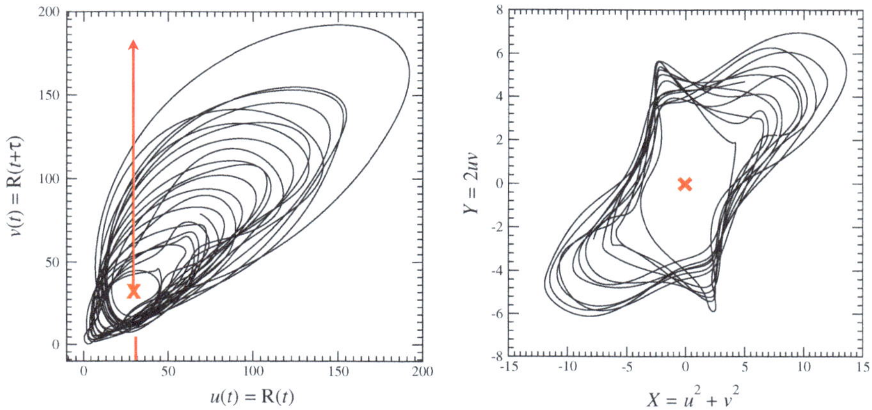

(a) Attractor without symmetry (b) Symmetric attractor

Fig. C.16 When the rotation axis (indicated by a cross) is located inside the non-visited zone in the middle of the attractor without symmetry, the sign of the variables describing the symmetric attractor is inverted after each eleven year

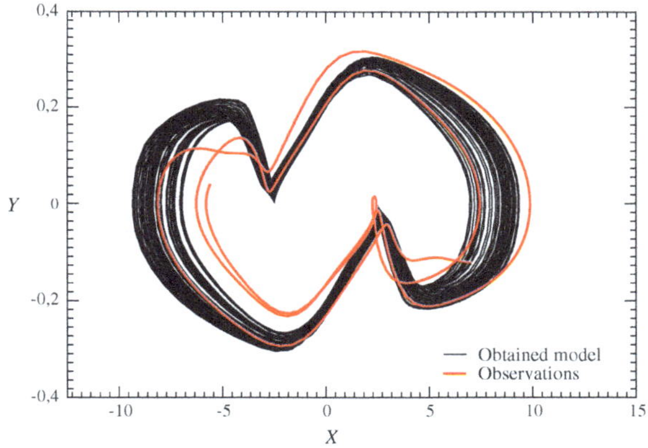

Fig. C.17 Chaotic attractor obtained from a three-variable global model, starting from "symmetrized" sunspot data

zero and where, as a result, they are very sensitive to noise and measurement uncertainties). Beginning with the first five cycles observed after 1850, a set of ordinary differential equations reproducing the dynamics underlying the long-term evolution of sunspots was obtained [2]. The chaotic attractor produced by this global model is compared in Fig. C.17 with the fifty years of data used for this global modelling procedure. The attractor has a structure that closely follows the observational data. Regrettably, the available observational data set is too short to allow a complete topological characterization of the attractor.

The global model can be numerically integrated for times much longer than the 23 cycles between 1750 and 2000. The "numerical data" can be used for a finer characterization of the dynamics. The attractor without symmetry produced by the

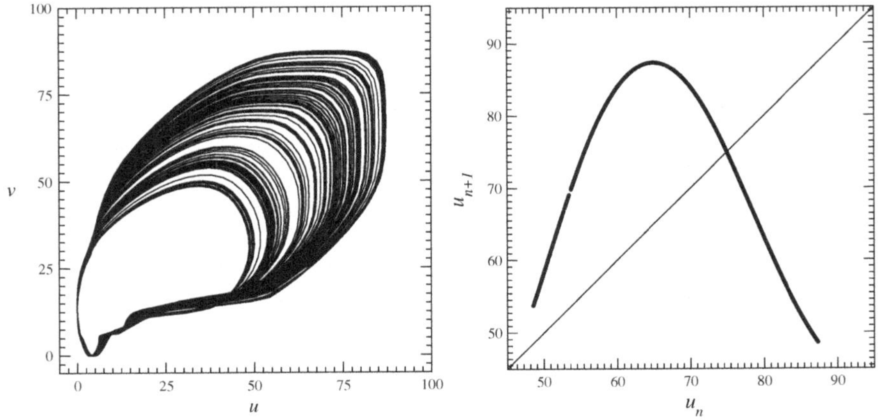

(a) Chaotic attractor (b) First-return map

Fig. C.18 Chaotic attractor (**a**) produced by the global model when the symmetry is modded out. A smooth unimodal first-return map (**b**) confirms the relation with the Rössler attractor

global model (Fig. C.18a) has the same branched manifold as the Rössler attractor. In particular, the first-return map to a Poincaré section is a smooth unimodal map (Fig. C.18b). As a result, the idea of using the Rössler system as a paradigm for studying solar activity [35, 33] is therefore justified.

The global model does not completely reproduce the observed dynamics. In particular, large oscillations are not reproduced by this model. This frequently happens for models obtained using a global modelling procedure since fluctuations resulting from high-dimensional dynamics, appearing as random, cannot be reproduced by low dimensional deterministic models [49]. With such a model, the long term dynamics underlying solar activity is shown to possess a deterministic low-dimensional component. Difference equations, obtained by Luis A. Aguirre from the same data set, were used for prediction [2]. These equations are easier to use for forecasting solar cycles because they take information (in the data) up to a given time and they map such information (predict) in the future. Of course, the newly predicted value can be injected into the model to predict even further, and so on. In particular, the difference equations take information spread over the last 44 months and predicts 4 months into the future. A key point in our experience is that both modeling and prediction of the sunspot time series has been greatly facilitated if performed in the transformed (symmetrical) space rather than directly in the original space. After performing analysis and prediction in the transformed space, the results should be mapped back into the original space. Prediction results (performed in 2005) are shown in Fig. C.19.

On June 5, 2012, the current prediction by the NASA for the next sunspot cycle maximum gave a smoothed sunspot number maximum of about 60 in the Spring of

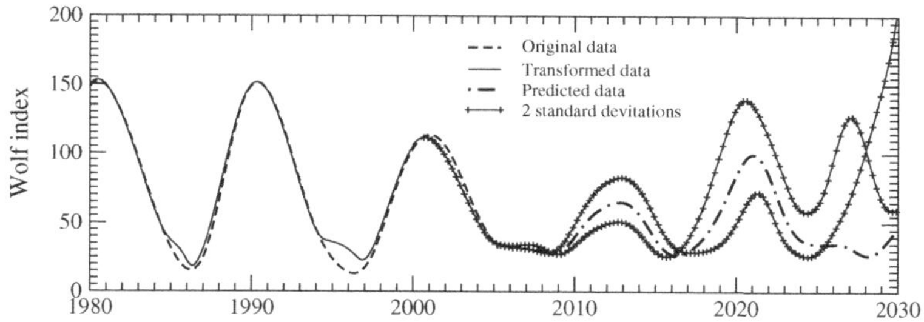

Fig. C.19 This plot shows the original sunspot numbers with dashed line up to 2005. From 2005, the predictions transformed back to the original space are shown

2013.[15] With our global model, we proposed two different predictions (performed in slightly different ways) they were a peak at about 65 in September or October 2012, or a peak at about 87 in between June and December of 2013. The maxima was finally observed in April 2014 with a peak at 81.8. We therefore quite correctly predicted the amplitude (87 *versus* 82) but we slightly anticipated the occurrence of that maximum (June to December *versus* April 2014, that is, a relative error between 2.7 and 6.7%). The fact that the solar cycle 24 was unusually long and weak [7] could explain why our prediction anticipated its maximum. The data we had in 2005 when we made this prediction did not allow us to identify the unusually long minimum with which this cycle started. In spite of this, we predicted rather accurately its amplitude, thus confirming the reliability of our deterministic model and, consequently, evidencing that there is indeed a deterministic component underlying the solar cycle.

C.6 Food Chains

Many studies have been carried out for more than a century on the evolution of population chains. Nevertheless, ecologists have great problems finding observational data of sufficient quality for convincingly showing chaotic evolution of real species. As a result, the possibility of chaotic evolution in food chains is even now still in debate although there are plenty of models producing chaos. This is due to the fact that, in order to prove the existence of chaos in ecosystems, it is necessary to plot the trajectory in some state space, and this implies long data sets. To be specific, if we assume that it takes at least fifty oscillations sampled at the rate of about fifty points per oscillation, this means that if an animal that breeds once per year is being studied, it is necessary to record the population once every week for at least fifty years! As a result, only experiments with insects (which reproduce very rapidly) can realistically be carried out to identify chaotic tendencies.

[15]http://solarscience.msfc.nasa.gov/predict.shtml.

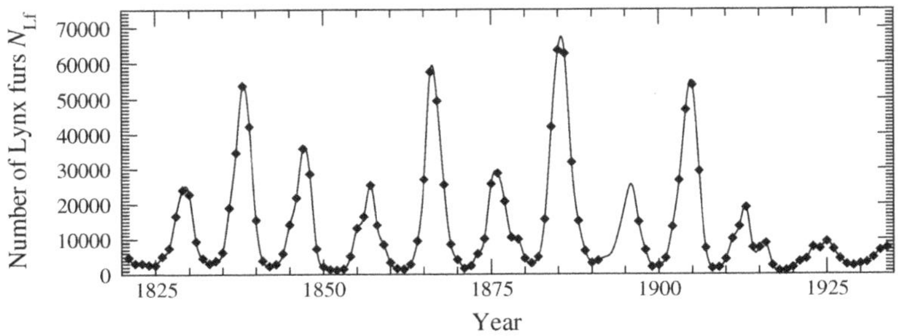

Fig. C.20 History of the number of lynx pelts sold in the different regions of Canada between 1821 and 1934

The earliest evidence for fluctuations comes from records of the Hudson Bay Company regarding the populations of lynxes and hares. This company bought and sold lynx furs. Their records (Fig. C.20)—studied by Charles Sutherland Elton and Mary Nicholson [15]—show that the number of animals fluctuates with a period of about nine years, so that about a dozen cycles were observed in the data during the century. The amplitude of these cycles varies significantly from year to year, suggesting chaos.

As for the copper electrodissolution (see Sect. C.2), the goal is to find a set of ordinary differential equations that reproduces the dynamics underlying the Elton-Nicholson data. The derivative coordinates were used to reconstruct the state space trajectory (Fig. C.21a). The attractor thus obtained[16] is quite similar to a Rössler attractor reconstructed from the derivatives of the z-variable; unfortunately, this variable is known to offer a poor observability of the dynamics [42] and a global model is quite difficult to obtain [36].

Elton and Nicholson found that the lynx population depends not only on the hare population but also on other preys as snowshoe rabbits, muskrats and salmons. Further, the lynx is not the top predator since it is hunted by trappers [56]. Quite surprisingly, these additional species are not taken into account in most of the models proposed to explain these data. Consequently, in order to get a model reproducing the variations in the lynx population in a realistic way, one might need to use at least a four-dimensional model. In fact, a rather satisfactory model in the form

$$\begin{cases} \dot{X} = Y \\ \dot{Y} = Z \\ \dot{Z} = W \\ \dot{W} = F(X, Y, Z, W) , \end{cases}$$

where $X = N_{\mathrm{Lf}}$ was found [55]. Here Y, Z, and W are the first three derivatives of the number of Lynx furs N_{Lf} and F is a polynomial function which is rather complex. The difficulty with this ecosystem is the limited amound of available data (there are

[16] Data available at http://www.atomosyd.net/spip.php?article39.

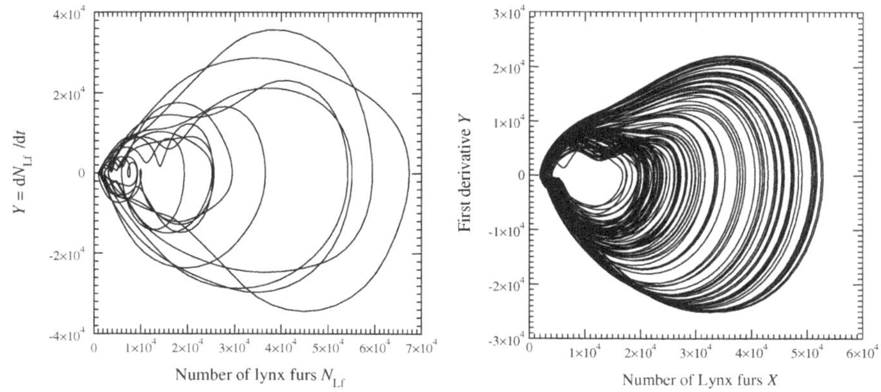

(a) Original data

(b) Synthetic data

Fig. C.21 Different state portraits for the lynx population **a** spanned by the records of the Hudson Bay Company and **b** produced by the four-dimensional global model obtained from that records

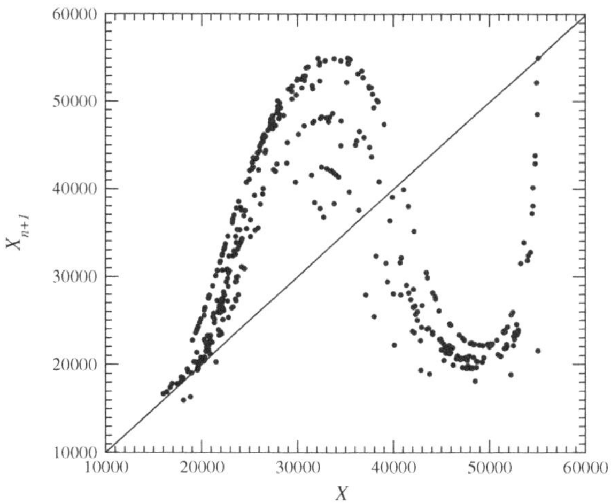

Fig. C.22 First-return map to a Poincaré section of the metastable chaos produced by the four-dimensional model obtained from the data collected by Elton and Nicholson in the records of the Hudson Bay Company

only 10 cycles). Unfortunately, this obtained model is not completely stable. It was integrated over 420 cycles (this corresponds to about 4,000 years) before an ejection of the trajectory to infinity. This lack of stability could be a consequence of two factors: (i) the measured time series (Fig. C.20) looks like variable z of the Rössler system which is known to provide a poor observability of the dynamics and (ii) there is only one point per year, that is, roughly, ten points per oscillations. A data set is considered to be good when it has at least about fifty points per oscillation [49]; such a requirement ensures a reliable computation of the successive time derivatives of the recorded time series.

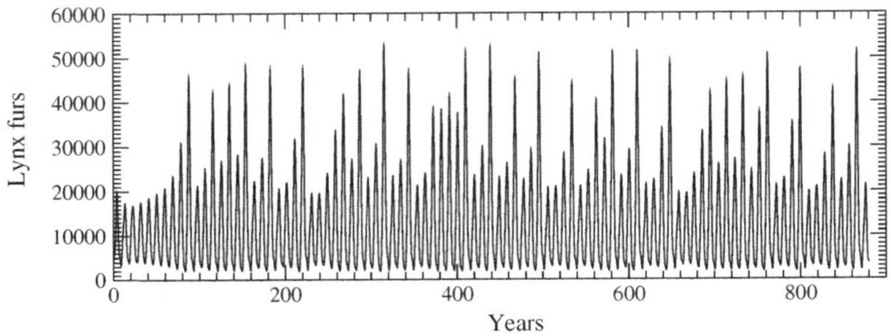

Fig. C.23 Evolution of the lynx population obtained by numerically integrating the global model. The 9 year period and large fluctuations in amplitude are observed as in the original data

The model produces a metastable chaos (Fig. C.21b) characterized by a first-return map to a Poincaré section with three monotonic branches whose structure is foliated (Fig. C.22). The first increasing branch is nearly tangent to the diagonal: this is a signature of "lethargy"—long minima around zero—when there are very few lynxes. This model can be integrated over 4,000 years, clearly for a larger duration than the one over which one could expect a stationary dynamics: the hunters were most likely quite rare before 1565, that is, only 500 years ago. In spite of this, the simulated data clearly present a time period of about 9 years and large fluctuations in amplitude as observed in the measured data. Such a model proves that there is a low-dimensional dynamics explaining a significant part of the data. It is a reliable proof that there is an underlying determinism in the evolution of this particular ecosystem (Fig. C.23).

C.7 Heart Variability

Since Balthazar van der Pol and Jan van den Mark showed that realistic electrocardiogram could be produced by three coupled nonlinear oscillators [75], many attempts to construct a chaotic model producing realistic electrocardiograms were proposed. Otto E. Rössler, with Reimara Rössler and Herbert D. Landahl suggested the driven Bonhöffer-van der Pol model[17]

$$\left\{ \begin{array}{l} \dot{x} = x - \dfrac{x^3}{3} + (y - d) - u \\ \dot{y} = \frac{1}{c} \left[a - x - b(y - d) \right] \\ \dot{u} = v \\ \dot{v} = -\omega^2 u \end{array} \right. \tag{C.3}$$

[17] The model here reported was slightly modified from the original version according to a suggestion provided by Otto in Tübingen, October 29, 2010 which was in agreement with the draft of the abstract shown in Fig. C.24.

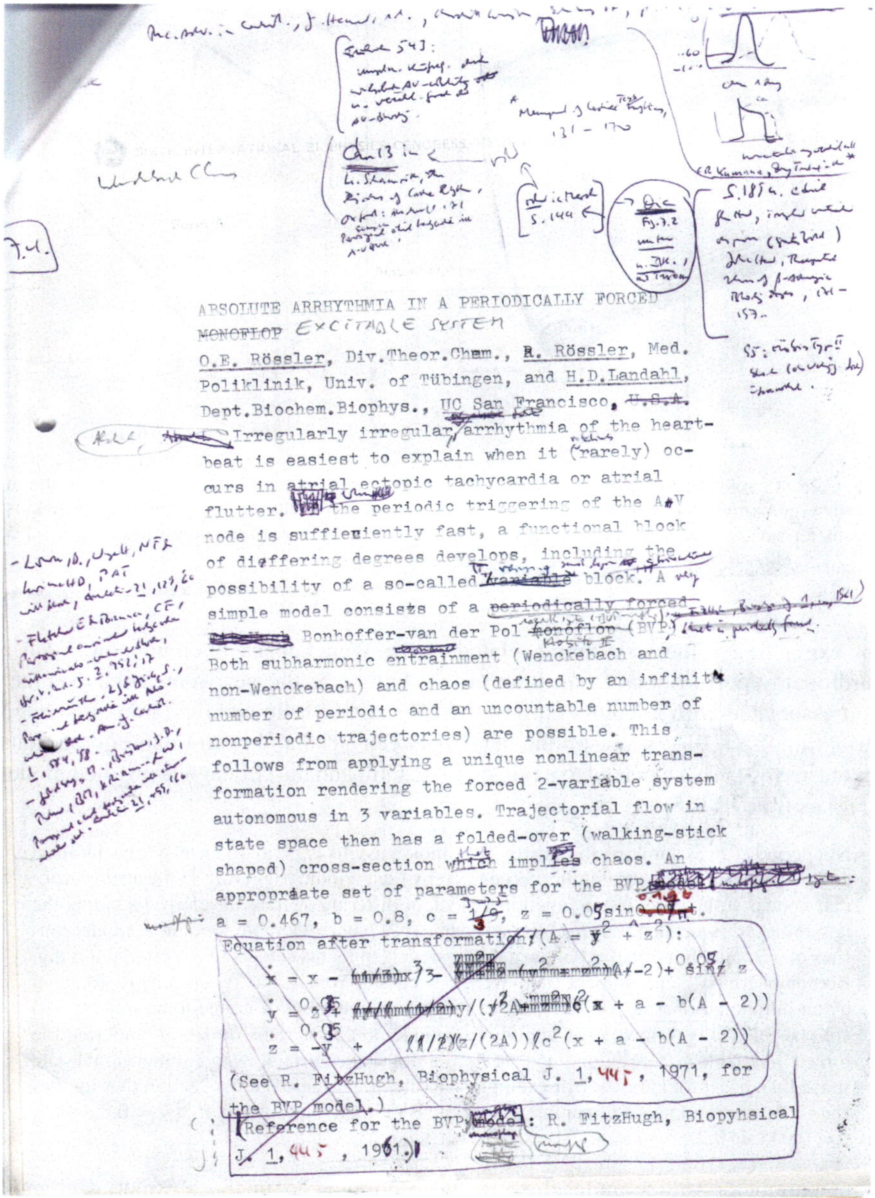

ABSOLUTE ARRHYTHMIA IN A PERIODICALLY FORCED MONOFLOP EXCITABLE SYSTEM

O.E. Rössler, Div.Theor.Chem., R. Rössler, Med. Poliklinik, Univ. of Tübingen, and H.D.Landahl, Dept.Biochem.Biophys., UC San Francisco, U.S.A.

Irregularly irregular arrhythmia of the heart-beat is easiest to explain when it (rarely) occurs in atrial ectopic tachycardia or atrial flutter. The periodic triggering of the A-V node is sufficiently fast, a functional block of differing degrees develops, including the possibility of a so-called variable block. A very simple model consists of a periodically forced Bonhoffer-van der Pol monoflop (BVP). Both subharmonic entrainment (Wenckebach and non-Wenckebach) and chaos (defined by an infinite number of periodic and an uncountable number of nonperiodic trajectories) are possible. This follows from applying a unique nonlinear transformation rendering the forced 2-variable system autonomous in 3 variables. Trajectorial flow in state space then has a folded-over (walking-stick shaped) cross-section which implies chaos. An appropriate set of parameters for the BVP: a = 0.467, b = 0.8, c = 1/3, z = 0.05 sin 0.4 t.

Equation after transformation (A = y² + z²):

(See R. FitzHugh, Biophysical J. 1, 445, 1971, for the BVP model.)

Reference for the BVP: R. FitzHugh, Biophysical J. 1, 445, 1961.

Fig. C.24 Draft of the abstract presented by Otto E. Rössler at the Sixth International Biophysics Congress held in Kyoto (Japan), September 3–9, 1978

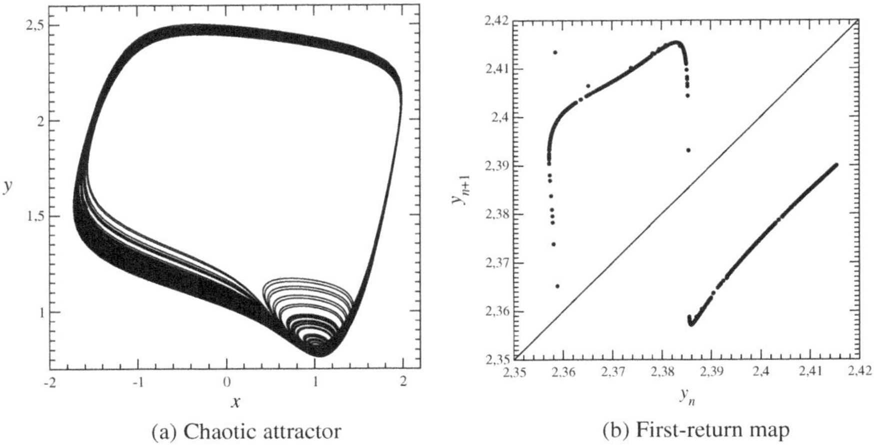

Fig. C.25 Chaos solution to the Bonhoeffer-van der Pol system (C.3) to model cardiac arrhythmia. The first return map to a Poincaré section reveals two increasing branches: this is a twisted torn chaos. Parameter values: $a = 0.4691$, $b = 0.8$, $c = 3$, $d = -1.5$, $\omega = 0.2$. Initial conditions: $x_0 = -1.5$, $y_0 = 1.5$, $u_0 = 0.045$ and $v_0 = 0$

for explaining some arrhythmia related to the Wenckeback phenomenon. Such a cardiopathy corresponds to an intermittent failure of the atrioventricular conduction associated with a lengthening of the PR interval followed by a non-conducted atrial impulse [19]. As a result the delay between atrial and ventricular contractions increases [80]. This can lead to a missed beat. Otto and his colleagues justified model (C.3) as follows [65].

> Nonperiodic arrhythmia of the heartbeat is most easy to explain when it occurs in atrial ectopic tachycardia or atrial flutter. When hereby the periodic triggering of the atrioventricular node is sufficiently fast, a functional block of differing degrees develops, including the possibility of a so-called 'varying' second degree A-V block. A simple continuous model consists of a Bonhoeffer-van der Pol excitable system [16] that is subjected to periodic forcing. Both subharmonic entrainment (non-Wenckebach and Wenckebach) and 'chaos' (defined by an infinite number of periodic plus an uncountable number of nonperiodic trajectories) are possible. This follows from applying a unique nonlinear transformation rendering the forced 2-variable system autonomous in 3 variables. Trajectorial flow in 3-dimensional state space then has a folded-over (for example walking-stick shaped) cross section that implies chaos. An appropriate set of parameters for the BVP monoflop: $a = 0.467$, $b = 0.8$, $c = 3$, $z = 0.045 \sin 0.2t$.

With parameter values quite close to those announced in the abstract, a chaotic behavior—close to be a twisted torn unimodal type—was in fact observed (Fig. C.25).

Many works tried to identify chaos in the heart rate. Most of them used geometric arguments as correlation dimension, Lyapunov exponents [69, 26]. If these markers are consistent with a chaotic behavior, they cannot be conclusive [34, 26]. Indeed, it is rather tough to get a definite answer for identifying a chaotic behavior. It is for instance difficult to distinguish some colored noise from purely deterministic chaotic dynamics. Chaos means "fluctuations produced by deterministic rules that nevertheless lead to irregular, long-term unpredictable, dynamics" [21]. Therefore, prior to asserting that a behavior is chaotic, there should be a clear evidence that deterministic equations govern its dynamics. Identifying a determinism from heart rates is still an open question [22, 81, 17].

If global modelling is one of the best techniques to evidence a determinism underlying measured data [3], however, a lack of global model from electrocardiogram or tachogram does not prove that there is not a deterministic component governing the heart rate. Global modelling technique just offers a conclusive answer when a satisfactory model is obtained, but such a positive answer strongly depends on the observable and not only on the underlying dynamics [42, 45]. Consequently, there is no proof for a determinism underlying heart rates but this does not mean that there is no structure in the cardiodynamics which can be used for discriminating different cardiopathies.

For not breaking the causality chain, arrhythmias were not removed from the measurements. To avoid long term drift, the cardiodynamics can be investigated from [18]

$$\Delta \mathrm{RR}_n = \mathrm{RR}_{n+1} - \mathrm{RR}_n \, . \tag{C.4}$$

Using data collected from the PHYSIONET website, RR-intervals—the duration between two successive R-waves in an electrocardiogram—are investigated for healthy subjects and patients with various pathologies. Three typical types of cardiodynamics can be evidenced with the help of first-return maps built from the $\Delta \mathrm{RR}_n$ [18]. The healthy subjects are characterized by a small cloud of points centered at the origin of the map (Fig. C.26a): only limited fluctuations are observed and there is no obvious structure within this cloud of points. Patients with congestive heart failure are characterized by a first-return map with four well marked segments (Fig. C.26b) which are associated with ectopic beats. An ectopic beat is commonly characterized by a delay τ by which the beat is anticipated with respect to the sinusal rhythm. The sequence of $\Delta \mathrm{RR}_n$ is thus $-\tau$, $+2\tau$ and $-\tau$. This feature explains that the first-return map presents four well-defined segments as follows.

- Segment 1 is defined as $\Delta \mathrm{RR}_n = 0 \quad (\Delta \mathrm{RR}_n < 0)$
- Segment 2 is defined as $\Delta \mathrm{RR}_{n+1} = -2\Delta \mathrm{RR}_n \quad (\Delta \mathrm{RR}_{n+1} > 0)$
- Segment 3 is defined as $\Delta \mathrm{RR}_{n+1} = -0.5\Delta \mathrm{RR}_n \quad (\Delta \mathrm{RR}_{n+1} < 0)$
- Segment 4 is defined as $\Delta \mathrm{RR}_{n+1} = 0 \quad (\Delta \mathrm{RR}_n < 0)$

According to this structure, the usual scenario for an ectopic beat is recovered: after a premature ventricular contraction (V), a compensatory pause is observed until the next sinus beat (R) such that RV+VR≈2RR [37]. Here RV designates the interval between the last sinus beat and the premature ventricular contraction and, VR the compensatory pause. We have thus RV=$T - \tau$, VR=$T + \tau$ and $\overline{\mathrm{RR}} = T$ where τ is the delay by which the ventricular contraction is prematured and T is the mean RR-interval. The value of delay τ cannot be predicted as clearly evidenced by the length of the first segment: many different delays can be obtained from a single point in

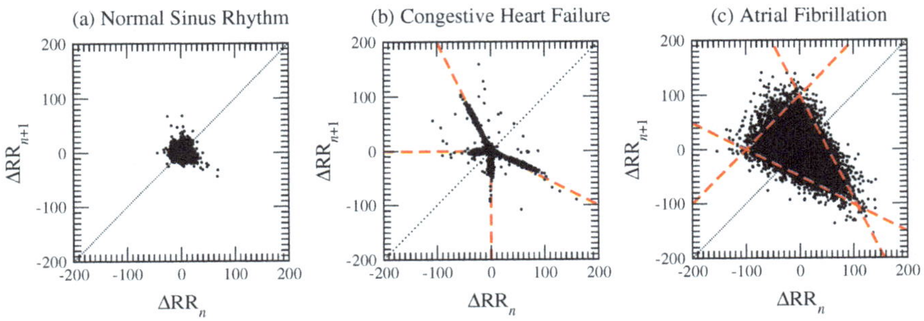

Fig. C.26 First-return map on the ΔRR for three different types of cardiodynamics

the central cloud. There is thus a deterministic structure combined with a stochastic process.

The third typical cardiodynamics is associated with atrial fibrillation which is induced by a high-frequency excitation of the atrium leading to both dyssynchronous atrial contraction and irregular ventricular excitation. This pathology results from a vulnerable atrial substrate [70]. For instance, atrial fibrillation is associated with increased risk of stroke and transient ischemic attack [83]. First-return map built from ΔRR_n observed in patients with atrial fibrillation (Fig. C.26c) reveals a triangular cloud of points bounded by the three segments as follows.

- Segment 1 is defined as $\Delta RR_{n+1} = -1.9\Delta RR_n + 100$
- Segment 2 is defined as $\Delta RR_{n+1} = -0.5\Delta RR_n - 50$
- Segment 3 is defined as $\Delta RR_{n+1} = \Delta RR_n + 100$

The fluctuations are quite large since they can reach very frequently 100 ms.

These three types of first-return maps were observed in adults, preterm newborns, infants with sudden-death risk, patients with chronic respiratory failures, etc. [54]. These pathologies can be automatically detected—for instance from an Holter—by using some markers as a Shannon entropy [18] or a complexity measure [54]. Cardiodynamics have not been proved to be chaotic but techniques developed for characterizing chaotic behaviors can be successfully applied to them.

C.8 Conclusion

Chaos is not only a pure theoretical concept. It corresponds to some behaviours encountered in various systems from the real world. Depending on the field, its characterization is more or less easy. There are many difficulties for observing chaos in experimental or observational systems. Chaos is expected between periodic motion and fully "turbulent" behaviors. Often, as encountered in fluid mechanics, the transition to turbulence is "non-smooth" and low-dimensional chaotic regimes are not observed. In certain cases, when the parameter values are sufficiently well controlled and when the dynamics is adequately contrained, chaos may be indeed present. Nevertheless, as encountered in ecology where there are many models producing chaotic regimes, it remains a tough challenge to evidence chaos in real ecosystems. These chaotic regimes could be very important in natural systems because they offer a flexibility (adaptability) as well as a regularity that is required for living systems. This is still an open field of research.

References

1. C.G. Abbot, A group of solar changes. Smithson. Misc. Collect. **80**(2), 2917.
2. L.A. Aguirre, C. Letellier, J. Maquet, Forecasting the time series of sunspot numbers. Sol. Phys. **241**, 103 (2008)
3. L.A. Aguirre, C. Letellier, Modeling nonlinear dynamics and chaos: a Review. Math. Probl. Eng. **2009**, 238960 (2009)
4. A.A. Andronov, A.A. Vitt, S. E. Chaikin, *Theory of Oscillators* (Pergamon Press, Oxford, 1966)
5. M.R. Bassett, J.L. Hudson, The dynamics of the electrodissolution of copper. Chem. Eng. Commun. **60**(1–6), 145–159 (1987)
6. M.R. Bassett, J.L. Hudson, Quasi-periodicity and chaos during an electrochemical reaction. J. Phys. Chem. **93**, 2731–2737 (1989)
7. S. Basu, The peculiar solar cycle 24 — where do we stand? J. Phys.: Conf. Ser. **440**, 012001 (2013)
8. E.R. Buley, F.W. Cummings, Dynamics of a system of N atoms interacting with a radiation field. Phys. Rev. A **134**(6), 1454–1460 (1964)
9. K.F. Bonhöffer, Activation of passive iron as a model for the excitation of nerve. J. Gen. Phys. **32**(1), 69–91 (1948)
10. G. Boulant, M. Lefranc, S. Bielawski, D. Derozier, Horseshoe templates with global torsion in a driven laser. Phys. Rev. E **55**(5), 5082–5091 (1997)
11. G. Boulant, M. Lefranc, S. Bielawski, D. Derozier, A non-Horseshoe template in a chaotic laser model. Int. J. Bifur. Chaos **8**(5), 965–976 (1998)
12. R.N. Bracewell, The sunspot number series. Nature **171**, 649–650 (1953)
13. M. Carbonell, R. Oliver, J.L. Ballester, A Search for chaotic behavior in solar activity. Astron. Astrophys. **290**, 983–994 (1994)
14. T. Dingyuan, Chaotische Amplituden- und Phasendynamik des optisch angeregten NH_3 Einmodenringlasers und Vergleich mit den reellen und komplexen Lorenz-Gleichangen. Ph.D. Thesis, Hannover University (1993)
15. C. Elton, M. Nicholson, The ten-year cycle in numbers of the lynx in Canada. J. Animal Ecol. **11**, 215–244 (1942)
16. R. FitzHugh, Impulses and physiological states in theoretical models of nerve membrane. Biophys. J. **1**, 445–466 (1961)
17. U. Freitas, E. Roulin, J.-F. Muir, C. Letellier, Identifying chaos from heart rate: the right task? Chaos **19**(2), 028505 (2009)
18. E. Fresnel, E. Yacoub, U. Freitas, A. Kerfourn, V. Messager, E. Mallet, J.-F. Muir, C. Letellier, An easy-to-use technique to characterize cardiodynamics from first-return maps on ΔRR-intervals. Chaos **25**(8), 083111 (2015)
19. G.J. Fromell, J. Root, R. Barish, The Wenckebach phenomenon. J. Emerg. Med. **4**(2), 115–118 (1986)
20. R. Gilmore, C. Letellier, *The Symmetry of Chaos* (Oxford University Press, Oxford, 2007)
21. L. Glass, Chaos and heart rate variability. J. Cardiovasc. Electrophysiol. **10**, 1358–1360 (1999)

22. L. Glass, Introduction to controversial topics in nonlinear science: is the normal heart rate chaotic? Chaos **19**(2), 028501 (2009)
23. A.Z. Grasiuk, A.N. Oraevsky, The Dynamics of quantum oscillators, in *Quantum Electronics and Coherent Light*, ed. by P.A. Miles (Academic Press, New York, 1964), pp. 192–197
24. A.Z. Grasiuk, A.N. Oraevsky, Transient processes in molecular oscillators. Radio Eng. Electron. Phys. **9**, 424- (1964)
25. K.R. Graziani, J.L. Hudson R.A. Schmitz, The Belousov—Zhabotinskii reaction in a continuous flow reactor. Chem. Eng. J. **12**(1), 9–21 (1976)
26. S. Guzzetti, M.G. Signorini, C. Cogliati, S. Mezzetti, A. Porta, S. Cerutti, A. Malliani, Nonlinear dynamics and chaotic indices in heart rate variability of normal subjects and heart-transplanted patients. Cardiovasc. Res. **31**, 441–446 (1996)
27. H. Haken, Analogy between higher instabilities in fluids and lasers. Phys. Lett. A **53**(1), 77–78 (1975)
28. H.B. Lin, *Elementary Symbolic Dynamics and Chaos in Dissipative Systems* (World Scientific Publishing, Singapore, 1989)
29. G.E. Hale, F. Ellerman, S.B. Nicholson, A.H. Joy, The magnetic polarity of sun-spots. Astrophys. J. **49**, 153–178 (1919)
30. D.H. Hathaway, R.M. Wilson, E.J. Reichmann, A synthesis of solar cycle prediction techniques. J. Geophys. Res. **104**(10), 22375–22388 (1999)
31. D. Hennequin, F. De Tomasi, F. De Tomasi, B. Zambon, E. Arimondo, Homoclinic orbits and cycles in the instabilities of a laser with a saturable absorber. Phys. Rev. A **37**(6), 2243–2246 (1988)
32. J.L. Hudson, J.C. Mankin, Chaos in the Belousov-Zhabotinsky reaction. J. Chem. Phys. **74**, 6171–6177 (1981)
33. K. Jinno, S. Xu, R. Berndtson, A. Kawaruma, M. Matsumoto, Prediction of sunspots, using reconstructed chaotic system equations. J. Geophys. Res. **100**(A8), 14773–14781 (1995)
34. J.K. Kanters, N.-H. Holstein-Rathlou, E. Agner, Lack of evidence for low-dimensional chaos in heart rate variability. J Cardiovasc. Electrophysiol. **5**, 591–601 (1994)
35. M. Kremliovsly, Can we understand time scales of solar activity? Sol. Phys. **151**, 351–370 (1994)
36. C. Lainscsek, C. Letellier, I. Gorodnitsky, Global modeling of the Rössler system from the z-variable. Phys. Lett. A **314**(5–6), 409 (2003)
37. R. Langendorf, Aberrant ventricular conduction. Amer. Heart J. **41**, 700–707 (1951)
38. L.J. Lanzerotti, Space weather effects on communications, in *Space Weather — Physics and Effects* (Springer, Berlin, 2007), pp. 247–268
39. M. Lefranc, D. Hennequin, D. Dangoisse, Homoclinic chaos in a laser containing a saturable absorber. J. Opt. Soc. Amer. B **8**(2), 239–249 (1991)
40. C. Letellier, P. Dutertre, B. Maheu, Unstable periodic orbits and templates of the Rössler system: toward a systematic topological characterization. Chaos **5**(1), 271–282 (1995)
41. C. Letellier, L. Le Sceller, P. Dutertre, G. Gouesbet, Z. Fei, J.L. Hudson, Topological characterization and global vector field reconstruction from an experimental electrochemical system. J. Phys. Chem. **99**, 7016–7027 (1995)
42. C. Letellier, J. Maquet, L. Le Sceller, G. Gouesbet, L.A. Aguirre, On the non-equivalence of observables in phase space reconstructions from recorded time series. J. Phys. A **31**, 7913–7927 (1998)
43. C. Letellier, O. Ménard, Th. Klinger, A. Piel, G. Bonhomme, Dynamical analysis and map modelling of a thermionic diode plasma experiment. Phys. D **156**, 169–178 (2001)
44. C. Letellier, R. Gilmore, Covering dynamical systems: Two-fold covers. Phys. Rev. E **63**, 16206 (2001)
45. C. Letellier, L.A. Aguirre, J. Maquet, Relation between observability and differential embeddings for nonlinear dynamics. Phys. Rev. E **71**, 066213 (2005)

46. C. Letellier, E. Roulin, O.E. Rössler, Inequivalent topologies of chaos in simple equations. Chaos Solitons Fractals **28**, 337–360 (2006)
47. C. Letellier, J. Maquet, L.A. Aguirre, R. Gilmore, Evidence for low dimensional chaos in the sunspot cycles. Astron. Astrophys. **449**, 379–387 (2006)
48. C. Letellier, J.-M. Ginoux, Development of the nonlinear dynamical systems theory from radio-engineering to electronics. Int. J. Bifur. Chaos **19**(7), 2131–2163 (2009)
49. C. Letellier, L.A. Aguirre, U.S. Freitas, Frequently asked questions about global modeling. Chaos **19**, 023103 (2009)
50. C. Letellier, R. Gilmore, Poincaré sections for a new three-dimensional toroidal attractor. J. Phys. A **42**, 015101 (2009)
51. C. Letellier, V. Messager, Influences on Otto E. Rössler's earliest paper on chaos. Int. J. Bifur. Chaos **20**(11), 3585–3616 (2010)
52. C. Letellier, L.A. Aguirre, Required criteria for recognizing new types of chaos: application to the "cord" attractor. Phys. Rev. E **85**, 036204 (2012)
53. C. Letellier, *Chaos in Nature*, 2nd edn. (World Scientific Publishing, Singapore, 2019)
54. C. Letellier, I. Leyva, I. Sendiña-Nadal, Dynamical complexity measure to distinguish organized from disorganized dynamics, Physical Review E, **101**, 022204, (2020)
55. J. Maquet, C. Letellier, Global models from the Canadian Lynx cycles as a first evidence for chaos in real ecosystems. J. Math. Biol. **55**(1), 21–39 (2007)
56. M.S. Weinstein, Hares, lynx, and trappers. Amer. Nat. **111**, 806–808 (1977)
57. T. Mausbach, T. Klinger, A. Piel, Chaos and chaos control in a strongly driven thermionic plasma diode. Phys. Plasmas **6**(10), 3817–3823 (1999)
58. P.J. Myrberg, Sur l'itération des polynomes réels quadratiques. Journal de Mathématiques Pures et Appliquées **41**, 339–351 (1962)
59. N.H. Packard, J.P. Crutchfield, J.D. Farmer, R.S. Shaw, Geometry from a time series. Phys. Rev. Lett. **45**(9), 712–716 (1980)
60. R. Rico-Martínez, K. Krischer, I.G. Kevrekidis, M.C. Kube, J.L. Hudson, Discrete- vs. continuous-time nonlinear signal proessing of Cu electrodissolution data. Chem. Eng. Commun. **118**(1), 25–48 (1992)
61. M. Rosalie, C. Letellier, Systematic template extraction from chaotic attractors. I-Genus-one attractors with an inversion symmetry. J. Phys. A **46**, 375101 (2013)
62. O.E. Rössler, D. Hoffmann, Repetitive hard bifurcation in a homogeneous reaction system, in *Analysis and Simulation of Biochemical Systems* (1972), pp. 91–102
63. O.E. Rössler, Chaotic behavior in simple reaction systems. Zeitschrift für Naturforschung A **31**, 239–264 (1976)
64. O.E. Rössler, K. Wegmann, Chaos in the Zhabotinskii reaction. Nature **271**, 89–90 (1978)
65. O.E. Rössler, R. Rössler, H.D. Landahl, Arrhythmia in a periodically forced excitable system, in *Proceedings of the Sixth International Biophysics Congress*, September 3–9, 1978, Kyoto, Japan (1978), p. 296
66. J.-C. Roux, A. Rossi, S. Bachelart, C. Vidal, Representation of a strange attractor from an experimental study of chemical turbulence. Phys. Lett. A **77**, 391–393 (1980)
67. R.A. Schmitz, K.R. Graziani, J.L. Hudson, Experimental evidence of chaotic states in the Belousov-Zhabotinsky reaction. J. Chem. Phys. **67**(7), 3040–3044 (1977)
68. S.H. Schwabe, Sonnenbeobachtungen im Jahre 1843. Astron. Nachrichten **21**(15), 233–236 (1843)
69. J.E. Skinner, C. Carpeggiani, C.E. Landisman, K.W. Fulton, Correlation dimension of heart beat intervals is reduced in conscious pigs by myocardial ischemia. Circ. Res. **68**, 966–976 (1991)
70. L. Staerk, J.A. Sherer, D. Ko, E.J. Benjamin, R.H. Helm, Atrial fibrillation. Circ. Res. **120**(9), 1501–1517 (2017)

71. F. Takens, Detecting strange attractors in turbulence. *Lecture Notes in Mathematics*, vol. 898 (1981), pp. 366–381
72. T.D. Tsankov, R. Gilmore, Strange attractors are classified by bounding tori. Phys. Rev. Lett. **91**(13), 134104 (2003)
73. T.D. Tsankov, R. Gilmore, Topological aspects of the structure of chaotic attractors in $\mathbb{R}^3$. Phys. Rev. E **69**, 056206 (2004)
74. N.B. Tufillaro, T. Abbott, J. Reilly, *An Experimental Approach to Nonlinear Dynamics and Chaos* (Addison-Wesley, New York, 1992)
75. B. van der Pol, J. van der Mark, Le battement du cœur considéré comme oscillation et un modèle électrique du cœur. L'onde électrique **7**, 365–392 (1928)
76. C. Vidal, J.-C. Roux, Pour la Science **39**, 50 (1981)
77. J. Weeks, *The Shape of Space* (Marcel Dekker, New York, 2002)
78. C.O. Weiss, J. Brock, Evidence for Lorenz-type chaos in a laser. Phys. Rev. Lett. **57**, 2804–2806 (1986)
79. K. Wegmann, O.E. Rössler, Different kinds of chaotic oscillations in the Belousov-Zhabotinsky reaction. Zeitchrift für Naturfoschung A **33**, 1179–1183 (1978)
80. K.F. Weinckbach, Zur Analyse des unregelmässigen Pulses, Zeitschrift für klinische Medizin **37**, 475–488 (1899)
81. N. Wessel, M. Riedl, J. Kurths, Is the normal heart rate "chaotic" due to respiration? Chaos **19**(2), 028508 (2009)
82. R. Wolf, Sur le Retour périodique de minimums de taches solaires; concordance entre ces périodes et les variations de déclinaisons magnétiques. Comptes-Rendus de l'Académie des Sciences (Paris) **35**(19), 704–705 (1852)
83. P.A. Wolf, R.D. Abbott, W.B. Kannel, Atrial fibrillation as an independent risk factor for stroke: the Framingham Study. Stroke 22, 983–988 (1991)

Index

O. E. Rössler and C. Letellier, *Chaos*, Understanding Complex Systems,
https://doi.org/10.1007/978-3-030-44305-4

The manufacturer's authorised representative in the EU is Springer Nature Customer Service Centre GmbH, Europaplatz 3, 69115 Heidelberg, Germany. If you have any concerns regarding our products, please contact ProductSafety@springernature.com

Printed and bound by CPI Group (UK) Ltd, Croydon, CR0 4YY
22/09/2026
02221057-0005